Methods in Molecular Biology

Series Editor
John M. Walker
School of Life and Medical Sciences
University of Hertfordshire
Hatfield, Hertfordshire, UK

For further volumes:
http://www.springer.com/series/7651

For over 35 years, biological scientists have come to rely on the research protocols and methodologies in the critically acclaimed *Methods in Molecular Biology* series. The series was the first to introduce the step-by-step protocols approach that has become the standard in all biomedical protocol publishing. Each protocol is provided in readily-reproducible step-by-step fashion, opening with an introductory overview, a list of the materials and reagents needed to complete the experiment, and followed by a detailed procedure that is supported with a helpful notes section offering tips and tricks of the trade as well as troubleshooting advice. These hallmark features were introduced by series editor Dr. John Walker and constitute the key ingredient in each and every volume of the *Methods in Molecular Biology* series. Tested and trusted, comprehensive and reliable, all protocols from the series are indexed in PubMed.

Plant-Virus Interactions

Edited by

Elizabeth P. B. Fontes

Departamento de Bioquimica e Biologia Molecular, Universidade Federal de Viçosa, Viçosa, Minas Gerais, Brazil

Kristiina Mäkinen

Department of Agricultural Sciences, University of Helsinki, Helsinki, Finland

Humana Press

Editors
Elizabeth P. B. Fontes
Departamento de Bioquimica e Biologia Molecular
Universidade Federal de Viçosa
Viçosa, Minas Gerais, Brazil

Kristiina Mäkinen
Department of Agricultural Sciences
University of Helsinki
Helsinki, Finland

ISSN 1064-3745 ISSN 1940-6029 (electronic)
Methods in Molecular Biology
ISBN 978-1-0716-3484-4 ISBN 978-1-0716-3485-1 (eBook)
https://doi.org/10.1007/978-1-0716-3485-1

Cover Illustration Caption: Begomovirus-infected tomato plant.

This Humana imprint is published by the registered company Springer Science+Business Media, LLC, part of Springer Nature.
The registered company address is: 1 New York Plaza, New York, NY 10004, U.S.A.

Paper in this product is recyclable.

Preface

Viruses are obligate intracellular parasites that rely on the host cellular machinery to replicate their genome, transcribe and translate viral genes, and package virions. In the case of plant viruses, they also need an active mechanism to move from the replication site to the plasmodesmata and infect neighboring uninfected cells. Additionally, viral molecular interactions must successfully evade the host's innate immune system. Any failure in these processes can jeopardize the virus's ability to infect the host. Therefore, understanding the protein–protein interaction network between the virus and the host is crucial for comprehending the virus's life cycle and developing strategies for broad-spectrum and long-lasting resistance against viral infections.

This book has been organized to provide practical guidance starting from targeting specific molecular interactions within the virus–host interactome and progressing to the identification of the complete virus–host protein–protein interaction network. The initial chapters primarily focus on acquiring the necessary molecular tools, including cloning and characterizing viral genomes, as well as techniques for inoculating DNA and RNA viruses. The book also covers biochemical and genetic approaches to confirming protein–protein interactions both in vivo and in vitro. To analyze the functional aspects of these interactions, the book presents procedures and protocols for assessing replication, translation, viral genome movement, and insect transmission. Furthermore, the book explores cutting-edge techniques for detecting multiple molecular interactions between the host and the virus, as well as monitoring immune hubs.

Viçosa, Brazil	*Elizabeth P. B. Fontes*
Helsinki, Finland	*Kristiina Mäkinen*

Contents

Contributors

MIGUEL A. ARANDA • *Department of Stress Biology and Plant Pathology, Centro de Edafología y Biología Aplicada del Segura (CEBAS)-CSIC, Murcia, Spain*
JOSÉ T. ASCENCIO-IBÁÑEZ • *Department of Molecular and Structural Biochemistry, North Carolina State University, Raleigh, NC, USA*
MAXIMILLER DAL-BIANCO • *Plant Genetics and Biochemistry Laboratory, BIOAGRO, Universidade Federal de Vicosa, Vicosa, Minas Gerais, Brazil; Department of Biochemistry and Molecular Biology, Universidade Federal de Vicosa, Vicosa, Minas Gerais, Brazil*
MARY M. DALLAS • *Department of Plant and Microbial Biology, North Carolina State University, Raleigh, NC, USA*
JOÃO PAULO HERRERA DA SILVA • *Dep. de Fitopatologia/BIOAGRO, Universidade Federal de Viçosa, Viçosa, MG, Brazil*
JUAN ANTONIO DIAZ-PENDÓN • *Institute for Mediterranean and Subtropical Horticulture La Mayora (IHSM), CSIC-UMA, Málaga, Spain*
SAVITHRAMMA P. DINESH-KUMAR • *Department of Plant Biology and The Genome Center, College of Biological Sciences, University of California, Davis, Davis, CA, USA*
LIVIA DONAIRE • *Abiopep S.L., Parque Científico de Murcia, Complejo de Espinardo, Murcia, Spain; Department of Stress Biology and Plant Pathology, Centro de Edafología y Biología Aplicada del Segura (CEBAS)-CSIC, Murcia, Spain*
PEDRO AUGUSTO BRAGA DOS REIS • *Department of Biochemistry and Molecular Biology/Bioagro, Universidade Federal de Viçosa, Viçosa, MG, Brazil; National Institute of Science and Technology in Plant-Pest Interactions, Viçosa, MG, Brazil*
CHRISTIANE ELIZA MOTTA DUARTE • *Glycobiology and Cell Signaling Laboratory, Universidade do Estado de Minas Gerais, Passos-MG, Brazil*
PINKY DUTTA • *Department of Agricultural Sciences, Viikki Plant Science Centre, University of Helsinki, Helsinki, Finland*
NIVEA COSTA EUCLYDES • *Plant Molecular Biology Laboratory/Bioagro, Universidade Federal de Viçosa, Viçosa – MG, Brazil*
YANCA F. M. FERREIRA • *Laboratory of Virology and Molecular Biology, Embrapa Vegetables, Brasília, DF, Brazil*
ELIZABETH P. B. FONTES • *National Institute of Science and Technology in Plant-Pest Interactions, Bioagro, Universidade Federal de Viçosa, Viçosa, Minas Gerais, Brazil; Departament of Biochemistry and Molecular Biology, Universidade Federal de Viçosa, Viçosa, Minas Gerais, Brazil*
MIRIAM GIL-VALLE • *Centro de Biotecnología y Genómica de Plantas UPM-INIA/CSIC and E.T.S. Ingeniería Agronómica, Alimentaria y de Biosistemas, Universidad Politécnica de Madrid, Madrid, Spain*
ANA GRANDE-PÉREZ • *Instituto de Hortofruticultura Subtropical y Mediterránea "La Mayora", Universidad de Málaga-Consejo Superior de Investigaciones Científicas (IHSM-UMA-CSIC), Málaga, Spain; Departamento de Biología Celular, Genética y Fisiología, Facultad de Ciencias, Universidad de Málaga, Málaga, Spain*
LINDA HANLEY-BOWDOIN • *Department of Plant and Microbial Biology, North Carolina State University, Raleigh, NC, USA*

MENGLING HUANG • *National Key Laboratory of Agricultural Microbiology, Hubei Key Lab of Plant Pathology, Hubei Hongshan Laboratory, College of Plant Science and Technology, Huazhong Agricultural University, Wuhan, Hubei, China*

ALICE K. INOUE-NAGATA • *Laboratory of Virology and Molecular Biology, Embrapa Vegetables, Brasília, DF, Brazil*

PULKIT KANODIA • *Plant Pathology, Entomology & Microbiology Department, Iowa State University, Ames, IA, USA; Interdepartmental Genetics & Genomics, Iowa State University, Ames, IA, USA; Santa Clara, CA, USA*

BO LI • *National Key Laboratory of Agricultural Microbiology, Hubei Key Lab of Plant Pathology, Hubei Hongshan Laboratory, College of Plant Science and Technology, Huazhong Agricultural University, Wuhan, Hubei, China*

YINZI LI • *London Research and Development Centre, Agriculture and Agri-Food Canada, London, ON, Canada*

YUANYUAN LI • *Department of Plant Biology and The Genome Center, College of Biological Sciences, University of California, Davis, Davis, CA, USA*

JUAN JOSÉ LÓPEZ-MOYA • *Centre for Research in Agricultural Genomics (CRAG, CSIC-IRTA-UAB-UB), Consejo Superior de Investigaciones Científicas (CSIC), Barcelona, Spain; Institute for Mediterranean and Subtropical Horticulture La Mayora (IHSM), CSIC-UMA, Málaga, Spain*

ROSA LOZANO-DURAN • *Department of Plant Biochemistry, Centre for Plant Molecular Biology (ZMBP), Eberhard Karls University Tübingen, Tübingen, Germany*

ROSA LOZANO-DURÁN • *Department of Plant Biochemistry, Centre for Plant Molecular Biology (ZMBP), Eberhard Karls University, Tübingen, Germany*

ZACHARY LOZIER • *Plant Pathology, Entomology & Microbiology Department, Iowa State University, Ames, IA, USA; Bioinformatics & Computational Biology, Iowa State University, Ames, IA, USA*

JOÃO PAULO BATISTA MACHADO • *National Institute of Science and Technology in Plant-Pest Interactions, Viçosa, MG, Brazil; Institute of Agricultural Sciences, Universidade Federal de Viçosa, Florestal, MG, Brazil*

KRISTIINA MÄKINEN • *Department of Agricultural Sciences, Viikki Plant Science Centre, University of Helsinki, Helsinki, Finland*

LAURA GONÇALVES COSTA MARTINS • *National Institute of Science and Technology in Plant-Pest Interactions, Bioagro, Universidade Federal de Viçosa, Viçosa, Minas Gerais, Brazil*

LAURA MEDINA-PUCHE • *Department of Plant Biochemistry, Centre for Plant Molecular Biology (ZMBP), Eberhard Karls University, Tübingen, Germany*

NATHAN MEIER • *Department of Plant Biology and The Genome Center, College of Biological Sciences, University of California, Davis, Davis, CA, USA*

W. ALLEN MILLER • *Plant Pathology, Entomology & Microbiology Department, Iowa State University, Ames, IA, USA; Molecular, Cellular & Developmental Biology, Iowa State University, Ames, IA, USA; Interdepartmental Genetics & Genomics, Iowa State University, Ames, IA, USA; Bioinformatics & Computational Biology, Iowa State University, Ames, IA, USA; Biochemistry, Biophysics & Molecular Biology Department, Iowa State University, Ames, IA, USA*

NURIA MONTES • *Plant Physiology, Pharmaceutical and Health Sciences Department, Faculty of Pharmacy, Universidad San Pablo-CEU, CEU-Universities, Boadilla del Monte, Madrid, Spain; Unidad de Metodología, Instituto de Investigación Sanitaria La Princesa (IIS-IP) and Servicio de Reumatología, Hospital Universitario de la Princesa, Instituto de Investigación Sanitaria (IIS-IP), Madrid, Spain*

Ivair J. Morais • *Department of Phytopathology, University of Brasília, Brasília, DF, Brazil*
Ugrappa Nagalakshmi • *Department of Plant Biology and The Genome Center, College of Biological Sciences, University of California, Davis, Davis, CA, USA*
Tatsuya Nagata • *Department of Molecular Biology, University of Brasília, Brasília, DF, Brazil*
Erich Y. T. Nakasu • *Laboratory of Virology and Molecular Biology, Embrapa Vegetables, Brasília, DF, Brazil*
Célio Cabral Oliveira • *Brazilian Center for Research in Energy and Materials, Brazilian Biorenewables National Laboratory, Campinas, SP, Brazil*
Irene Ontiveros • *Centre for Research in Agricultural Genomics (CRAG, CSIC-IRTA-UAB-UB), Consejo Superior de Investigaciones Científicas (CSIC), Barcelona, Spain; Institute for Mediterranean and Subtropical Horticulture La Mayora (IHSM), CSIC-UMA, Málaga, Spain*
Israel Pagán • *Centro de Biotecnología y Genómica de Plantas UPM-INIA/CSIC and E.T.S. Ingeniería Agronómica, Alimentaria y de Biosistemas, Universidad Politécnica de Madrid, Madrid, Spain*
Maija Pollari • *Faculty of Agriculture and Forestry, Department of Agricultural Sciences, University of Helsinki, Helsinki, Finland*
Edgar A. Rodríguez-Negrete • *Instituto Politécnico Nacional, CIIDIR Unidad Sinaloa, Departamento de Biotecnología Agrícola, Guasave, Sinaloa, Mexico*
Cristina Sáez • *Centro de Biotecnología y Genómica de Plantas UPM-INIA/CSIC and E.T.S. Ingeniería Agronómica, Alimentaria y de Biosistemas, Universidad Politécnica de Madrid, Madrid, Spain*
Fredy Davi Albuquerque Silva • *Department of Biochemistry and Molecular Biology/Bioagro, Universidade Federal de Viçosa, Viçosa, MG, Brazil; National Institute of Science and Technology in Plant-Pest Interactions, Viçosa, MG, Brazil*
Eduardo Bassi Simoni • *Department of Biology, University of North Carolina at Chapel Hill, Chapel Hill, NC, USA; Department of Biochemistry and Molecular Biology/Bioagro, Universidade Federal de Viçosa, Viçosa, MG, Brazil*
Jahanara Sonia • *Plant Pathology, Entomology & Microbiology Department, Iowa State University, Ames, IA, USA; Molecular, Cellular & Developmental Biology, Iowa State University, Ames, IA, USA*
Huang Tan • *Department of Plant Biochemistry, Centre for Plant Molecular Biology (ZMBP), Eberhard Karls University Tübingen, Tübingen, Germany*
Aiming Wang • *London Research and Development Centre, Agriculture and Agri-Food Canada, London, ON, Canada*
Junshi Yazaki • *RIKEN Center for Integrative Medical Sciences, Yokohama City, Kanagawa, Japan*
Xiao Yu • *National Key Laboratory of Agricultural Microbiology, Hubei Key Lab of Plant Pathology, Hubei Hongshan Laboratory, College of Plant Science and Technology, Huazhong Agricultural University, Wuhan, Hubei, China*
F. Murilo Zerbini • *Dep. de Fitopatologia/BIOAGRO, Universidade Federal de Viçosa, Viçosa, MG, Brazil*
Xi Zhang • *CAS Center for Excellence in Molecular Plant Sciences, Chinese Academy of Sciences, Shanghai, China; University of the Chinese Academy of Sciences, Beijing, China*
Yongliang Zhang • *State Key Laboratory of Plant Environmental Resilience and Ministry of Agriculture Key Laboratory of Soil Microbiology, College of Biological Sciences, China Agricultural University, Beijing, China*

Chapter 1

Computational Pipeline for the Detection of Plant RNA Viruses Using High-Throughput Sequencing

Livia Donaire and Miguel A. Aranda

Abstract

In this chapter, we describe a computational pipeline for the in silico detection of plant viruses by high-throughput sequencing (HTS) from total RNA samples. The pipeline is designed for the analysis of short reads generated using an Illumina platform and free-available software tools. First, we provide advice for high-quality total RNA purification, library preparation, and sequencing. The bioinformatics pipeline begins with the raw reads obtained from the sequencing machine and performs some curation steps to obtain long contigs. Contigs are blasted against a local database of reference nucleotide viral sequences to identify the viruses in the samples. Then, the search is refined by applying specific filters. We also provide the code to re-map the short reads against the viruses found to get information on sequencing depth and read coverage for each virus. No previous bioinformatics background is required, but basic knowledge of the Unix command line and R language is recommended.

Key words Plant viruses, Bioinformatics analysis, High-throughput sequencing, Short reads, Contig assembly, Virus detection

1 Introduction

High-throughput sequencing (HTS) is a powerful technique that, together with a well-designed bioinformatics pipeline, can identify both known and novel viruses infecting plants of virtually any species [1]. According to the purpose of the HTS project, some key aspects in the experimental design are critical and need to be decided prior to selecting the bioinformatics pipeline to follow. Broadly, these aspects include the selection of the nucleic acid type (e.g., total RNA, double-stranded RNA, virion-associated nucleic acids, and small RNAs) and isolation method, the sequencing platform (e.g., Illumina, Oxford Nanopore, and PacBio), and, according to these, the library preparation method and the sequencing strategy (e.g., length of reads, paired-end (PE) or single-end (SE) reads, and sequencing depth) [2]. There are no universal rules for correct bioinformatics analysis of HTS data for

Elizabeth P. B. Fontes and Kristiina Mäkinen (eds.), *Plant-Virus Interactions*, Methods in Molecular Biology, vol. 2724, https://doi.org/10.1007/978-1-0716-3485-1_1,

virus identification, nor even for any other application, but most of the published pipelines share the same basic steps. These include quality control (QC) of raw reads, which is essential to guarantee the success of subsequent analyses and can be performed using FastQC software [3]. Depending on the QC outcome, several software tools can be used to trim reads based on quality scores or remove adapter sequence contamination [4]. Either trimmed reads or de novo assembled contigs can be used to search for viruses against a database, being BLAST [5], one of the preferred algorithms [2]. In most cases, the results must be filtered and manually analyzed. Numerous pipelines are available for plant virus detection and discovery, with some able to detect new viruses, but most of them require skills in command-line interfaces; some examples of recently published pipelines are PVDP [6], VirHunter [7], and Viroscope [8].

Here, we described the bioinformatics pipeline we have successfully used to identify viruses from total RNA samples in different HTS projects [9, 10]. The pipeline is organized into four main steps: (i) raw data analysis and filtering; (ii) contig assembly and identification of plant viruses; (iii) calculation of sequencing depth and read coverage along the viral genomes; and (iv) visualization of alignments. The pipeline runs in Linux Operative System using either the command line or an R script; therefore, basic knowledge of the Linux command line and R language usage is highly recommended.

2 Materials

2.1 RNA Isolation, Library Construction, and Sequencing

1. A method for high-quality plant total RNA isolation.
2. 2100 Bioanalyzer Instrument (Agilent).
3. NanoDrop ND-1000 Spectrophotometer (NanoDrop Technologies).

2.2 In Silico Bioinformatics Pipeline

2.2.1 Hardware

1. Operating system: 64-bit Linux (*see* **Note 1**).
2. Available RAM: at least 16 GB (recommend 32 GB or higher).
3. Free disk space: around 30 GB per sample (recommended).

2.2.2 Software (*see* **Note 2**)

Generic Software

1. Java OpenJDK Runtime Environment (JDK 8 or higher). Instructions to download and install prebuilt OpenJDK packages can be found at https://openjdk.org/install/ (Accessed February 2023).
2. R v4.2.2 (or higher) and RStudio Desktop v2022.12 (or higher). Instruction for R installation of the most stable

Ubuntu releases can be found at https://cran.r (Accessed Feb 2023). RStudio can be downloaded at https://posit.co/download/rstudio-desktop/.

3. (Optional) GNU parallel. Instructions to install and use this tool can be found at https://www.gnu.org/software/parallel/parallel_tutorial.html (Accessed Feb 2023).

Specific Software

Install the same or a higher version of the following software tools:

1. FastQC v0.11.9. Installation instructions at https://www.bioinformatics.babraham.ac.uk/projects/download.html (Accessed Feb 2023).
2. Trimmomatic v0.40. Installation instructions at https://github.com/usadellab/Trimmomatic (Accessed Feb 2023).
3. (Optional) Bowtie 2 v2.4.4. Installation instructions can be found at https://bowtie-bio.sourceforge.net/bowtie2/manual.shtml (Accessed Feb 2023).
4. Trinity v2.13.2. Installation instructions and a complete tutorial can be found at https://github.com/trinityrnaseq/trinityrnaseq/wiki (Accessed Feb 2023).
5. BLAST Command Line Application v2.13.0. Installation instructions for any Operative System can be found at https://www.ncbi.nlm.nih.gov/books/NBK569861/ (Accessed Feb 2023). The BLAST+ package contains tools to manipulate local databases and core blast search programs like BLASTn or tBLASTx.
6. SeqKit, an ultrafast toolkit for FASTA/Q file manipulation v1.3-r106. Installation instructions can be found at https://bioinf.shenwei.me/seqkit/ (Accessed Feb 2023).
7. Burrow-Wheeler Aligner (BWA) v0.7.17. Complete installation instructions can be found at https://github.com/lh3/bwa (Accessed Feb 2023).
8. SAMtools v1.16.1. Instructions at http://www.htslib.org/download/ (Accessed Feb 2023).
9. (Optional) Tablet: short read alignment viewer. Instructions to download, install, and usage can be found at https://ics.hutton.ac.uk/tablet/ (Accessed Feb 2023).
10. Download the R script to run the complete pipeline in R language ("*HTSVdetection.R*") (optional) and the R script to make calculations on mapping reads and plots of aligned viral reads ("*HTSVcalculation.R*") from the GitHub page https://github.com/ldonaire/Rscripts (Accessed Feb 2023).

2.2.3 Local Databases

1. Obtain plant virus sequences in FASTA format. Virus reference genomes can be downloaded in FASTA format from the NCBI virus portal https://www.ncbi.nlm.nih.gov/labs/virus/vssi/ (Accessed Feb 2023). For this protocol, we created a plant virus nucleotide database applying the following filters: (i) sequence length: between 501 nts and 20,000 nts and (ii) host: Spermatophyta (seed plants), taxid: 58024. Download results in FASTA format (Nucleotide) with "*Accession*" as the unique identifier in the FASTA header (build custom option) to avoid future problems with long identifiers. Rename the file as "*pvDB.fasta*" and save it in a folder called "*virusDB*" (108,885 sequences on 02 Feb 2023).
2. Get additional information on the virus sequences. Information regarding the downloaded viral sequences can be obtained from the NCBI virus portal, using the same filters applied to obtain the sequences but selecting "Current table view result" in CSV format. Displayed columns are optional, except "*Accession*", which is mandatory. We usually select "*Isolate*", "*Species*", "*Genus*", "*Family*", "*Molecule type*", "*Length*", "*Segment*", "*Country*" and "*Host*". Importantly, select "*Accession with version (AY012345.1)*" to match the accession in the FASTA headers. Rename the file as "*pvDB.csv*" and save it in the folder "*virusDB*".
3. Build a blast database of the virus reference sequences to map contigs against the viral genomes typing in the Linux command line (*see* **Note 3**):

```
wd$ makeblastdb -in virusDB/pvDB.fasta -dbtype nucl -parse_-
seqids -out virusDB/pvDB

Options:
-in                  Input file/database name
-dbtype              Molecule type of input (nucl or prot)
-parse_seqids        Parse bar delimited sequence identifier in
                     FASTA input
-out                 Name of database to be created
```

The output is nine files with prefix "*pvDB.**" saved in the folder "*virusDB*".

4. (Optional) Create a local plant host database for subtracting plant reads. FASTA files for host plants are available for different species in several plant genomic databases. Here, we used the Tomato Heinz 1706 genome, release ITAG2.4 from Sol Genomics Network. The genomic FASTA file "ITAG2.4_genomic.fasta" was saved in a folder called "*plantDB*". To create the Bowtie 2 index, type:

```
wd$ bowtie2-build plantDB/ITAG2.4_genomic.fasta plantDB/
ITAG2.4_genomic
```

The output is six files with the "*.bt2" extension saved in the folder "*plantDB*".

2.2.4 Input Data

1. Raw FASTQ files. Use raw FASTQ files obtained directly from the sequencing company (*see* **Note 4**). To provide an example, we have used here raw FASTQ files from two tomato sample pools sequenced using Illumina NovaSeq 6000 platform (150 pb-PE reads) [10]. Files can be downloaded from the Short Read Archive (SRA) under the accessions SRR14066947 and SRR14066948 using the SRA Toolkit (*see* **Note 5**). We will refer to these two sets of data along the text as Tom2 and Tom1, respectively.
2. Sample names file. More than one sample can be analyzed at the same time using parallel processing. These simultaneous analyses can be done using several software tools like GNU parallel. To run GNU parallel, we recommend creating a file containing the name of each sample in different lines (without "_1" or "_2" in PE files or file extensions). In our example, we created the file "*sample.names.txt*", which looks like:

```
wd$ cat sample.names.txt
Tom1
Tom2
```

Here, the Linux command *cat* prints in the terminal the content of the file "*sample.names.txt*" located within the wd.

3 Methods

3.1 Total RNA Library Construction and Sequencing

In this section, we offer some tips for optimizing sequencing performance, but the reader needs to be aware that we are not providing a comprehensive protocol.

1. High-quality total RNA purification. Several methods are suitable for high-quality total RNA isolation, including commercial kits. Based on our experience, we suggest using guanidine isothiocyanate and acid-phenol-based methods such as TRI reagent (MRC), followed by one or more phenol:chloroform cleaning steps [10]. DNase treatment of the RNA samples is essential; we use DNase I (AMPD1, Sigma-Aldrich) following the manufacturer's instructions.
2. Total RNA QC. For QC of total RNA, we recommend measuring the RNA Integrity Number (RIN) with a Bioanalyzer

(we use the 2100 model from Agilent). Although a RIN value higher than seven is typically recommended, we have found that it may be difficult to reach this value for some plant species. According to our experience with melon and tomato samples, lower RIN values may be acceptable if discrete ribosomal RNA bands are visible in the gel-like image or after agarose gel electrophoresis. We quantified the amount of RNA using a NanoDrop ND-1000 Spectrophotometer (NanoDrop Technologies).

For the subsequent steps of the protocol, we typically rely on third-party companies to handle the library preparation, including library QC, multiplexing, and Illumina sequencing. Each sequencing company has specific RNA sample requirements in terms of concentration, purity, and volume.

3. cDNA library preparation and QC. For virus detection in total RNA samples, we choose the TrueSeq Stranded Total RNA with Ribo-Zero Plant Kit (Illumina), including the depletion of plant ribosomal RNAs. We do not recommend protocols for mRNA enrichment since many plant RNA viruses do not have a poly(A) tail. For multiplex sequencing, individual barcodes (or index adapters) are added to each DNA fragment during the library preparation following specific recommendations and kits provided by Illumina. To achieve high-quality sequencing data, an accurate quantitation of the cDNA in the libraries is required. This can be done by qPCR using the Illumina qPCR Quantification Protocol Guide (Illumina).

4. Illumina sequencing. For Illumina sequencing by synthesis technology, the library of cDNA fragments is loaded onto a flow cell, where the fragments are clonally amplified by bridge amplification to generate clusters of identical fragments. These clusters are the template for sequencing. Among the variety of sequencing options, we usually select (i) multiplex sequencing of maximum 50 cDNA libraries, (ii) 150 bp reads as minimum sequencing length, (iii) PE reads, and (iv) sequencing depth according to the number of multiplexed libraries. We found that a minimum sequencing depth of 1.5 million reads per sample for rRNA-depleted total RNA tomato samples was needed to detect low-abundance viruses [10]. Then, the selection of the specific Illumina platform is normally based on a combination of these requirements and the availability of the sequencing company. In our case, we normally use the NovaSeq 6000 platform. In our example, we pooled three libraries in a multiplex format and ordered 80 M of reads per sample, 150 bp-PE reads. In the following sections, we will describe our analysis of two of these samples, consisting of two pools of duplicated samples from the same 20 field tomato plants showing virus-like symptoms [10].

3.2 Raw Data Analysis and Filtering

The first steps of this pipeline are intended to analyze PE reads from raw FASTQ files (*see* **Notes 4** and **5**). FastQC [3] identifies potential issues, such as poor base quality or adapter contamination, and these bases or reads are filtered out in the next step using Trimmomatic [11]. If the plant host genome is available, we highly recommend subtracting plant reads to enrich the datasets in viral reads.

1. QC of raw reads using FastQC [3]. FastQC involves a comprehensive set of QC metrics of raw sequencing reads, including measures of sequence quality, GC content, read length distribution, adapter contaminations, N content, and overrepresented k-mers. To run FastQC using as input the FASTQ files, type:

```
wd$ fastqc -t 4 Original_files/*.fastq.gz -o FastQC_raw

Options:
-t      Specifies the number of files (4) which can be processed
        simultaneously
-o      Save all output files in a specific directory
```

   Note that it is not necessary to decompress the FASTQ files to run FastQC.

   The output is a series of interactive HTML reports saved in the folder "*FastQC_raw*" created previously in the wd. These files include summary statistics and several graphical visualizations that help with the interpretation of this analysis. In our example, we found that three QC metrics failed: per base sequence content, sequence duplication levels, and adapter content (Fig. 1a). Duplication of some sequences/reads is an expected outcome in RNA-Seq data since over-representation of some transcripts/viral sequences is expected in the samples. The other two issues will be solved in the next step.

2. Cleaning of raw reads using Trimmomatic [11]. After the FastQC analysis, we used Trimmomatic to remove the first 13 bases at the 5′ end of the reads and to remove the adapter contamination (*see* **Note 6**). Additionally, we applied other filters to remove low-quality bases (below a threshold of 30) and dropped reads shorter than 36 bases long. The code is as follows (*see* **Note 7**):

```
wd$ cat sample.names.txt | parallel 'java -jar
trimmomatic-0.39.jar PE -phred33 -threads 4 Original_files/{}
_1.fastq.gz Original_files/{}_2.fastq.gz trimming/{}_1.PEt-
rim.fastq.gz trimming/{}_1.UnPEtrim.fastq.gz trimming/{}_2.
PEtrim.fastq.gz trimming2/{}_2.UnPEtrim.fastq.gz ILLUMINA-
CLIP: full_path_to_adapters/TruSeq3-PE-2.fa:2:30:10 HEAD-
CROP:13 LEADING:30 TRAILING:30 SLIDINGWINDOW:4:30 MINLEN:36"
```

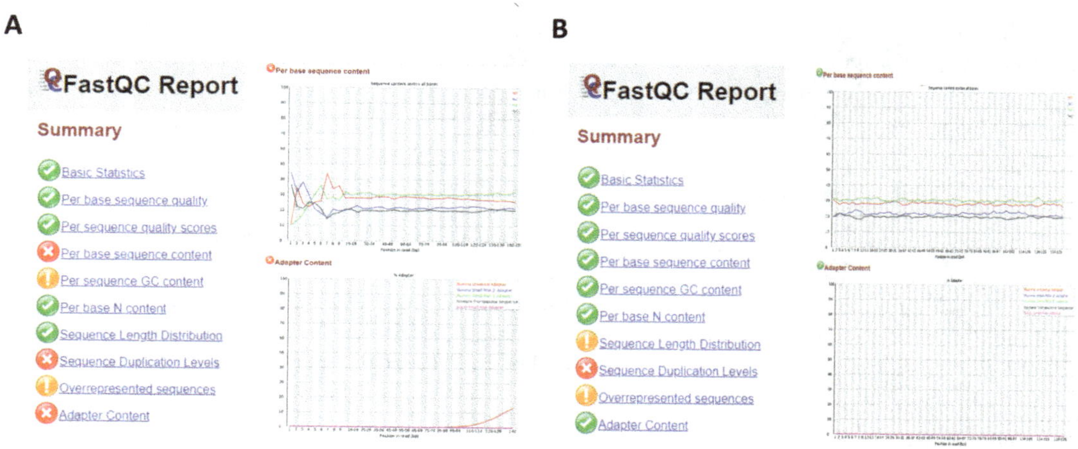

Fig. 1 Quality control (QC) of raw and filtered reads. Example of FastQC results of raw reads (**a**) and cleaned reads (**b**). Results from the file "*Tom1_1.fastq.gz*" are shown as an example. Metrics marked with green and orange signs passed the QC filters, while red signs indicate failure. After Trimmomatic, all metrics passed the QC, except sequence duplication levels, which are expected to be high in RNA-Seq experiments

```
Options:
ILLUMINACLIP    Remove adapters from the file "TruSeq3-PE-2.fa"
                file
HEADCROP        Cut 13 bases from the start of the reads
LEADING         Cut bases at the start of the read below a
                quality threshold of 30
TRAILING        Cut bases at the end of the read below a quality
                threshold of 30
SLIDINGWINDOW   Cut when the average quality threshold of a
                window of 4 bases is below 30
MINLEN          Drop the read shorter than 36 bases
```

Information about the Trimmomatic run is displayed in the standard output (*see* **Note 8**). The complete path to the "*adapters*" folder within the Trimmomatic directory should be added in the ILLUMINACLIP option. Option settings need to be adapted based on the dataset. The resulting FASTQ files are saved in a folder called "*trimming*" (created previously) and are referred to here as cleaned reads.

3. QC of cleaned reads using FastQC [3]. We rerun FastQC using the filtered reads to check Trimmomatic's results by typing:

```
wd$ fastqc -t 8 Original_files/*.fastq.gz -o FastQC_clean
```

```
Options:
-t    Specifies the number of files (8) which can be processed
      simultaneously
-o    Save all output files in a specific directory created
      previously
```

As shown in Fig. 1b, the QC metrics that failed for raw reads are acceptable after trimming.

4. (Optional) Filtering out plant reads using Bowtie 2 [12]. If the plant host genome is available, it is recommended to remove plant reads to enrich the datasets in virus reads and save computational time in the next steps. We kept reads that did not align with the plant genome using the Bowtie 2 program with default options:

```
plantDB$ cat ../sample.names.txt | parallel "bowtie2 --threads
8 -x ITAG2.4_genomic -1 ../trimming/{}_1.PEtrim.fastq.gz -2
../trimming/{}_2.PEtrim.fastq.gz -U ../trimming/{}_1.UnPE-
trim.fastq.gz,../trimming/{}_2.UnPEtrim.fastq.gz —un-gz ../
bowtie2_plant/{}_noPl.fastq.gz --un-conc-gz ../bowtie2_plant/
{}_noPlconc.fastq.gz -S ../{}_plant.sam"
```

```
--un         FASTQ  files containing unpaired reads that did
             not align with the plant genome
--un-conc    FASTQ files containing paired reads that did not
             align concordantly with the plant genome
```

Note that in this case, Bowtie 2 is executed within the "*plantDB*" directory instead of in the main wd (*see* **Note 3**). The double dot notation "../" is used to navigate one level up in the directory tree. Information about the Bowtie 2 run is displayed in the standard output, showing the number of unmapped reads that likely corresponds to viral reads. The resulting FASTQ files are saved in a folder called "*bowtie2_-plant*" (created previously) and are referred to here as filtered reads. We obtained files of paired ("**.1_noPlconc.fastq.gz*" or "**.2_noPlconc.fastq.gz*") and single ("**.noPl.fastq.gz*") unmapped reads.

3.3 Contig Assembly and Identification of Plant Viruses

The following steps are devoted to assembly short reads into long contigs using Trinity [13] and to identifying known plant viruses by BLAST [5] against the local virus database created previously (*see* Subheading 2.2.3, **item 3**). After that, we applied a series of filters to refine the search with the purpose to identify viruses with high confidence. Although the BLAST search results are in tab-delimited files that could be opened using a spreadsheet application like Excel, we use R language to filter the results.

1. Contig assembly using Trinity [13]. A fundamental step in the pipeline is the assembly of the short Illumina reads into long contigs using Trinity de novo assembler program. Trinity cannot work on paired and single reads at the same time, then we first join the files obtained in the previous step in a single file per each sample using the Linux command *zcat*:

```
wd$ cat sample.names.txt | parallel "zcat bowtie2_plant/{}
_noPlconc.1.fastq.gz bowtie2_plant/{}_noPlconc.2.fastq.gz
bowtie2_plant/{}_noPl.fastq.gz > bowtie2_plant/{}_allNoPl.
fastq"
```

The output files are not compressed since it is a prerequisite for Trinity input files. Also, Trinity cannot run with GNU parallel; then, we used the following bash loop to run the program for the two samples successively:

```
bowtie2_plant$ for FILE in *allNoPl.fastq; do Trinity --
seqType fq --single "$FILE" --run_as_paired --CPU 12 --max_-
memory 50G --output ../Trinity/${trinity_FILE%%.}; done

Options:
--seqType           Type of reads
--single            Single reads
--run_as_paired     Single reads contains pairs
--CPU               Number of CPUs to use
--max_memory        Maximum memory to use
--output            Name of the directory for output (must
                    include 'trinity' as a safety precaution)
```

The output files are saved in specific folders according to the input files used; in our example, the folders are called "*trinity_Tom1allNoPl.fastq*" and "*trinity_Tom2allNoPl.fastq*." Assembled contigs are saved in files with the suffix "** Trinity.fasta*" within the main "*Trinity*" directory.

2. Mapping contigs against the local virus nucleotide database using BLAST [5]. To identify plant viruses, we align the FASTA files of assembled contigs against the nucleotide database using BLASTn (BLAST for nucleotide sequences) as follows:

```
wd$ cat sample.names.txt | parallel "blastn -query Trinity/
trinity{}_allNoPl.fastq.Trinity.fasta -db virusDB/pvDB
-outfmt '6 qseqid qlen sseqid slen qstart qend length pident
evalue' -num_threads 12 -max_target_seqs 1 -word_size 7 >
blastn/{}_VirNt.txt"

Options:
-outfmt            Alignment view option as tabular
-num_threads       Number of threads to use in the blast search
-max_target_seqs   Number of aligned sequences to keep (only 1)
-word_size         Length of the exact initial match
```

The mapping results are saved in tab-delimited format files in the "*blastn*" folder (created previously) and contain the following customized columns: "*qseqid*" (query seq-id), "*qlen*" (query length), "*sseqid*" (subject seq-id or accession), "*slen*" (subject length), "*qstart*" (start of alignment in query), "*qend*" (end of alignment in query), "*length*" (alignment length), "*pident*" (percentage of identical matches), and "*evalue*" (expected value). We obtained 70,147 and 63,019 hits for Tom1 and Tom2, respectively, being the next filtering step a crucial one to identify viruses with enough confidence.

3. Filtering BLAST results using R. We applied the following filters to the BLAST output tables: (i) contig length between 500 and 14,000 nts; (ii) length of alignment higher than 300 nts, and (iii) E-value lower than 1e10^{-5}. For that, run the following lines in the R console (*see* **Notes 3** and **9**):

```
# List the path of the BLASTn output files into R
> myBlastList = list.files(path="blastn", pattern="*.txt",
full.names=TRUE)
# Load the BLASTn output table of the first sample analyzed
(in alphabetical order) as a data frame in R
> myBlastFile1 = read.delim(myBlastList[1], header=TRUE,
stringsAsFactors=FALSE, col.names=c("qseqid", "qlen", "sse-
qid", "slen", "qstart", "qend", "length", "pident", "evalue"))
# Save results of filtering BLASTn output in a new data frame
in R
> myBlastFile1 = subset(myBlastFile1, myBlastFile1$qlen >
500 & myBlastFile1$qlen < 14000 & myBlastFile1$length > 300 &
myBlastFile1$evalue < 0.00001)
# Export results of the first sample to a file to open in Excel
> write.csv(myBlastFile1, "blastn/Sample1_VirNt_filter.txt",
row.names=FALSE)
# Repeat these steps but change all occurrences of "1" by
successive numbers according to the number of samples ana-
lyzed.
```

Alternatively, this code can be run in RStudio using the R script "*HTSVdetection.R*" (*see* **Note 9**). The output are tab-delimited files for each sample (saved in the "*blastn*" folder) containing the same columns as the previous files but with filtered rows. The percentage of identity between the contig and the reference virus could help to distinguish between different strains or isolates of the same virus species or even different virus species according to the species' demarcation criteria for a particular species. After filtering, we obtained 77 and 80 hits for Tom1 and Tom2, respectively. Some accessions may appear more than once if various contigs for the same virus have been assembled. Then, the next steps of the pipeline are devoted to refining this search by calculating a series of parameters on viral reads.

3.4 Re-map the Reads against the Reference Sequences of the Identified Viruses

We next performed a number of steps to refine the actual virus identification by calculating (i) the number of mapping reads along each viral reference genome, (ii) the average sequencing depth for each viral reference genome, and (iii) the percentage of coverage along each viral reference genome. We also provide the R code to easily obtain summary information of the mapping reads and the code to add additional information about the virus species found.

1. Get the reference sequences for the identified viruses using SeqKit [14]. First, we need to copy the accession numbers of the viral reference sequences in the column "*sseqid*" of the filtered BLASTn output files to a file called "*myVirusIDs.txt.*" This file should contain only accession numbers of all samples analyzed, in different lines, without any additional characters, and without duplicate accession ids. Then, extract the sequences from the "*pvDB.fasta*" file (*see* Subheading 2.2.3, **item 1**) using SeqKit as follows:

```
wd$seqkit grep -f myVirusIDs.txt virusDB/pvDB.fasta > virusDB/
VirusFound.fasta
```

   This creates a FASTA file containing only the sequences of the virus identified by BLAST (43 sequences in this example). Additionally, using a similar code, it is possible to extract the sequences of the viral contigs from the Trinity contig FASTA files for future analyses (the R code is available in the R script "*HTSVdetection.R*" (*see* **Note 10**)).

2. Create an index to run BWA [15]. To run the BWA aligner, we need first to index the reference FASTA file as follows:

```
wd$ bwa index virusDB/VirusFound.fasta
```

   This command creates five additional files with "*Virus-Found.fasta*" as prefix and different file extensions in the folder "*virusDB*" (created before).

3. Align filtered reads against the reference viral sequences using BWA. For this, we used the MEM algorithm as follows:

```
wd$ cat sample.names.txt | parallel "bwa mem virusDB/Virus-
Found.fasta bowtie2_plant/{}_allNoPl.fastq > bwa/{}_bwaVirus-
Found.sam"
```

   The output is alignment files per each sample in Sequence Alignment Map (SAM) format saved in the folder "*bwa*" previously created. For the next steps, transform the output SAM files into BAM and sort and index them using SAMtools [16] using the following three command lines:

Table 1
Summary results of mapping reads of the examples

	Raw reads	Cleaned reads	Percentage	Filtered reads	Percentage	Viral reads	Percentage
Tom1	86,284,538	71,682,413	83.08	27,815,808	38.8	11,255,637	40.46
Tom2	84,739,174	70,769,593	83.51	27,935,399	39.47	12,034,529	43.08

Table 2
Viruses found in the Tom1 example dataset

Accession	Average depth	Length covered	Percentage covered	Counts	Isolate	Species	Genus	Family	Molecule type	Length	Segment	Country	Host
ON738336.1	7.73	2,665	92.85714286	172	VPS3	Duamitovirus boci3	Duamitovirus	Mitoviridae	ssRNA(+)	2,870		Russia	Vitis vinifera
ON738340.1	5.73	2,197	86.9410368	103	PeVMV3	Duamitovirus scsc3	Duamitovirus	Mitoviridae	ssRNA(+)	2,527		Russia	Vitis vinifera
OL472266.1	5.77	2,190	68.48030019	110	PAR19AT	Erysiphe necator associated ambiguivirus 2			RNA	3,198		Slovenia	Solanum lycopersicum
OL472300.1	5.93	2,394	91.13056719	128	KOP20AT	Erysiphe necator ourmia-like virus 82	Ourmiavirus	Botourmiaviridae	ssRNA(+)	2,627		Slovenia	Solanum lycopersicum
KC206082.1	18.99	121	18.81804044	29	Kar 1	Hibiscus leaf curl virus	Begomovirus	Geminiviridae	ssDNA(+/-	643		India	Hibiscus rosa-sinensis
OL472283.1	6.90	4,356	97.03720205	237	POR19ST	Leveillula taurica associated noda-like virus 1		Nodaviridae	ssRNA(+)	4,489		Slovenia	Solanum lycopersicum
KR068481.1	23.23	121	8.362128542	44	Theni MKU-2	Okra enation leaf curl virus	Begomovirus	Geminiviridae	ssDNA(+/-	1,447		India	Abelmoschus esculentus
MK376952.1	4.40	2,555	69.63750341	89	Anhui	Olive latent virus 1	Alphanecrovirus	Tombusviridae	ssRNA(+)	3,669		China	Nicotiana tabacum
OL472228.1	5.62	2,458	67.41634668	111	PAR19AT	Olive latent virus 1	Alphanecrovirus	Tombusviridae	ssRNA(+)	3,646		Slovenia	Solanum lycopersicum
MW594439.1	180.52	7,604	100	11,162	LRNV/Spanish	Ophiovirus lactucae	Ophiovirus	Aspiviridae	ssRNA(-)	7,604	RNA1	Spain	Solanum lycopersicum
MW594440.1	1006.42	1,826	100	15,027	LRNV/Spanish	Ophiovirus lactucae	Ophiovirus	Aspiviridae	ssRNA(-)	1,826	RNA2	Spain	Solanum lycopersicum
MW594442.1	1908.74	1,378	100	21,316	LRNV/Spanish	Ophiovirus lactucae	Ophiovirus	Aspiviridae	ssRNA(-)	1,378	RNA4	Spain	Solanum lycopersicum
MW594441.1	7756.77	1,503	99.86710963	94,459	LRNV/Spanish	Ophiovirus lactucae	Ophiovirus	Aspiviridae	ssRNA(-)	1,505	RNA3	Spain	Solanum lycopersicum
AY535019.1	704.30	1,414	99.78828511	8,363	Belg-2	Ophiovirus lactucae	Ophiovirus	Aspiviridae	ssRNA(-)	1,417	RNA4	Belgium	Lactuca sativa
AF484251.1	8945.15	6,450	100	498,384	Sp-13	Pepino mosaic virus	Potexvirus	Alphaflexiviridae	ssRNA(+)	6,450		Spain	Solanum lycopersicum
MK133092.1	192931.18	6,412	100	9,961,556	PepMV-P12-3G	Pepino mosaic virus	Potexvirus	Alphaflexiviridae	ssRNA(+)	6,412		Germany	Solanum lycopersicum
FJ820175.1	2738.16	2,411	99.66928483	54,288	5251	Pepino mosaic virus	Potexvirus	Alphaflexiviridae	ssRNA(+)	2,419		Spain	Solanum lycopersicum
JQ314459.1	2987.07	6,362	99.15835411	158,351		Pepino mosaic virus	Potexvirus	Alphaflexiviridae	ssRNA(+)	6,416		USA	Solanum lycopersicum
FJ940223.1	5767.05	6,329	98.58255452	294,509		Pepino mosaic virus	Potexvirus	Alphaflexiviridae	ssRNA(+)	6,420		Netherlands	Solanum lycopersicum
LN680393.2	362.66	391	19.87798678	1,175	Pit-MAIB	Pittosporum cryptic virus-1		Partitiviridae	dsRNA	1,967	RNA1		Pittosporum tobira
EF442780.1	47.09	3,416	99.38900204	1,332	Mexico-1	Southern tomato virus	Amalgavirus	Amalgaviridae	dsRNA	3,437		Mexico	Solanum lycopersicum
OL539725.1	9.32	1,101	47.31413838	101	IA-2016	Soybean ilarvirus I	Ilarvirus	Bromoviridae	ssRNA(+)	2,327	RNA3	USA	Glycine max
OL539724.1	2.76	535	18.10490694	16	IA-2016	Soybean ilarvirus I	Ilarvirus	Bromoviridae	ssRNA(+)	2,955	RNA2	USA	Glycine max
ON684329.1	11.99	7,344	94.63917526	824	Iraq-1	Tobacco vein clearing virus	Solendovirus	Caulimoviridae	dsDNA-RT	7,760		Iraq	Solanum lycopersicum
OL598362.1	2.12	754	13.4115973	17	TVCV_sa	Tobacco vein clearing virus	Solendovirus	Caulimoviridae	dsDNA-RT	5,622		South Africa	Nicotiana tabacum
MK517478.2	240.48	3,604	98.92945375	7,065	Fondi2018	Tomato blunervirus 1	Blunervirus	Kitaviridae	ssRNA(+)	3,643	RNA2	Italy	Solanum lycopersicum
KY810787.1	33.06	8,188	99.35687417	2,227		Tomato chlorosis virus	Crinivirus	Closteroviridae	ssRNA(+)	8,241	RNA2	United Kingdom	Solanum lycopersicum
OK058522.1	3.84	7,688	90.16066612	245	DSMZ PV-1023	Tomato chlorosis virus	Crinivirus	Closteroviridae	ssRNA(+)	8,527	RNA1	Spain	Solanum lycopersicum
KY810786.1	4.50	7,178	83.51367074	265		Tomato chlorosis virus	Crinivirus	Closteroviridae	ssRNA(+)	8,595	RNA1	United Kingdom	Solanum lycopersicum
KJ200304.1	3.72	7,053	82.07843594	222	2.5	Tomato chlorosis virus	Crinivirus	Closteroviridae	ssRNA(+)	8,593	RNA1	Spain	Solanum lycopersicum
MW594436.1	261.46	3,586	100	7,555	TFBV/Spanish	Tomato fruit blotch virus	Blunervirus	Kitaviridae	ssRNA(+)	3,586	RNA2	Spain	Solanum lycopersicum
MW594438.1	2480.06	1,926	100	38,786	TFBV/Spanish	Tomato fruit blotch virus	Blunervirus	Kitaviridae	ssRNA(+)	1,926	RNA4	Spain	Solanum lycopersicum
MZ401001.1	344.02	5,748	99.98260567	15,852	ToFBV_Tn92	Tomato fruit blotch virus	Blunervirus	Kitaviridae	ssRNA(+)	5,749	RNA1	Tunisia	Solanum tuberosum
MW594435.1	372.61	5,769	99.82695968	17,264	TFBV/Spanish	Tomato fruit blotch virus	Blunervirus	Kitaviridae	ssRNA(+)	5,779	RNA1	Spain	Solanum lycopersicum
MW594437.1	1682.36	2,858	99.61659115	43,121	TFBV/Spanish	Tomato fruit blotch virus	Blunervirus	Kitaviridae	ssRNA(+)	2,869	RNA3	Spain	Solanum lycopersicum
AJ489258.1	17.60	2,679	96.33225458	399	Almeria	Tomato yellow leaf curl virus	Begomovirus	Geminiviridae	ssDNA(+/-)	2,781		Spain	Capsicum annuum
MG715488.1	29.03	2,399	86.29496403	562	ES-Mlg-TY16-Tom-2017	Tomato yellow leaf curl virus	Begomovirus	Geminiviridae	ssDNA(+/-)	2,780		Spain	Solanum lycopersicum
KY971328.1	7.45	1,613	58.00071917	113	Florida_3.1	Tomato yellow leaf curl virus	Begomovirus	Geminiviridae	ssDNA(+/-)	2,781		USA	Solanum lycopersicum
FJ012359.1	5.17	1,105	49.9773858	58	Sinaloa	Tomato yellow leaf curl virus	Begomovirus	Geminiviridae	ssDNA(+/-)	2,211	DNA-A	Mexico	Physalis ixocarpa

The color scale indicates the percentage of coverage from low (red) to high (green).

```
wd$ cat sample.names.txt | parallel "samtools view -b -S -h
bwa/{}_bwaVirusFound.sam > bwa/{}_bwaVirusFound.bam"
wd$ cat sample.names.txt | parallel "samtools sort bwa/{}
_bwaVirusFound.bam -o bwa/{}_bwaVirusFound.sort.bam"
wd$ cat sample.names.txt | parallel "samtools index bwa/{}
_bwaVirusFound.sort.bam"
```

The output files with extensions "*.sort.bam*" and "*.sort.bam.bai*" are saved in the "*bwa*" folder. SAM files could be removed to save disk space.

4. Obtain summary information of mapped reads using the R script "*HTSVcalculation.R*". The first part of this R script is dedicated to counting reads using files obtained in different steps of this pipeline (raw reads, cleaned reads, filtered reads, and virus reads) to get a table summarizing the analysis results.

To run the script, we recommend the use of RStudio [17]. For these calculations, run the part of the script entitled "*Counting reads*" (*see* **Note 10**). The results are exported to a tab-delimited file ("*.csv" extension) that can be opened using Excel or any spreadsheet application and are shown in Table 1. Alternatively, the code is also written in the full pipeline R script "*HTSVdetection.R*" (*see* **Note 10**).

5. Calculate average sequencing depth and percentage of coverage using the R script "*HTSVcalculation.R*". The second part of this R script is devoted to calculating sequencing depth using SAMtools and, from this, calculating the percentage of coverage along the viral genomes using R commands. This script also provides the code to add information about the virus species found (*see* Subheading 2.2.3, **item 2**). For that, run the second part of the script entitled "Calculating basic parameters on viral reads and adding virus information" (*see* **Note 10**). The results are exported to one tab-delimited file ("*.csv" extension) per sample and can be opened using Excel or any spreadsheet application (Table 2). Alternatively, the code is also written in the full pipeline R script "*HTSVdetection.R*" (*see* **Note 10**).

3.5 Visualize Mapping Reads Along the Viral Genomes

Here, we offer two alternatives to visualize viral reads along viral genomes. One uses Tablet [18], a specific tool for graphical visualization of HTS data, including assemblies or read mappings. The other option is a code to graphically represent mapping reads along the viral genomes using the *ggplot2* R package.

1. Visualize mapping reads against viral genomes using Tablet [18]. In the main Tablet windows, click on "Open Assembly" and browse the BAM files in the "*bwa*" folder to open mapping reads files; as reference sequences, load the FASTA file "*VirusesFound.fasta*" created previously (located in "*bwa*" folder). The number of mapping reads across each viral reference sequence is displayed in the left panel, while the alignments can be visualized in the right panel (Fig. 2). Alignments can be displayed in different modes from the "Advanced tab": With the "Coverage" mode, it is possible to visualize the read coverage across the reference sequences; with the "Scaled" mode, it is possible to visualize individual reads and their alignments to the reference sequences. Customizing the display settings in "Home" and "Color Scheme" tabs is possible.
2. Create plots of read coverage along viral genomes. We also provide the R code to create plots of read coverage along the identified viral genomes using *ggplot2* R package. In the third part of the R script "*HTSVcalculation.R*" with the title: "Create plots of read coverage," there is the code to do so. Alternatively, the code is also written in the full pipeline R script "*HTSVdetection.R*" (*see* **Note 10**).

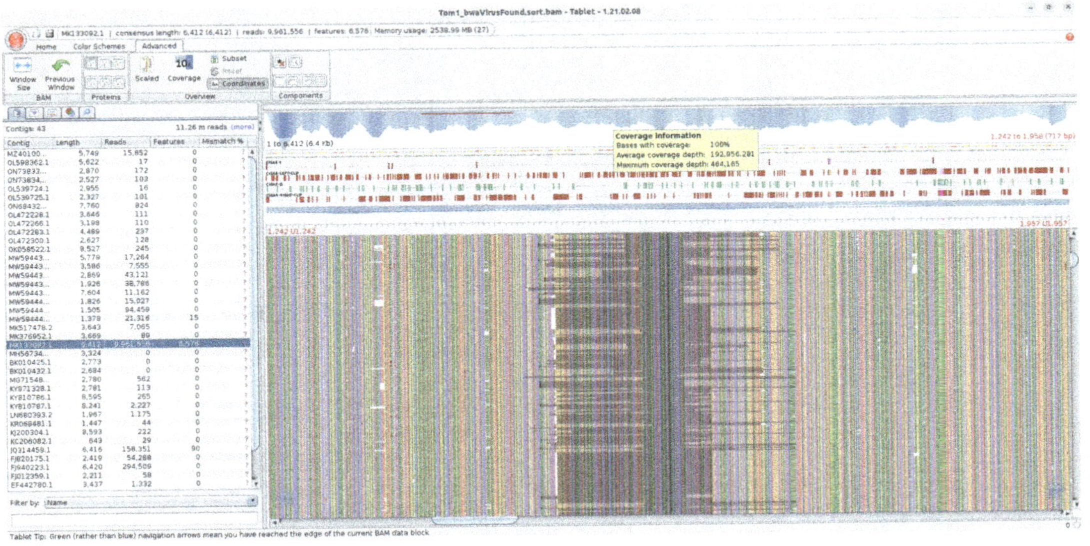

Fig. 2 Visualization of viral reads along the viral genomes using Tablet. Only one alignment can be visualized at a time. For example, here we show the alignment of reads along the pepino mosaic virus genome (Accession MK133092.1) in the Tom1 sample. The upper panel on the right shows the read coverage along the viral genome, which is 100%. The panel on the left shows the number of mapped reads per reference genome and the length of the genomes

Coverage plots for Tom1 are shown in Fig. 3, where one can easily notice that the coverage of reads along particular virus genomes is insufficient to be confident on the presence of that virus in the analyzed sample (e.g., for soybean ilarvirus 1); therefore, manual inspection and results curation are needed (Table 2).

4 Notes

1. The pipeline works on Linux Operative Systems with any Ubuntu distribution installed. We used an Intel Core i7–8700 CPU processor computer with 64 GB of RAM memory and Ubuntu 22.04.1 LTS.
2. Installing programs in Ubuntu is generally straightforward and can be done from the command line using the package management system Advanced Packing Tool (APT) or downloading from the Web and installing specific packages (generally with "*.deb*" extension) using the dpkg command. When possible, we recommend installing compiled binary programs instead of source code and installing the programs in PATH (list of directories where Ubuntu search for executable files), otherwise you would need to specify the full path to the executable files.

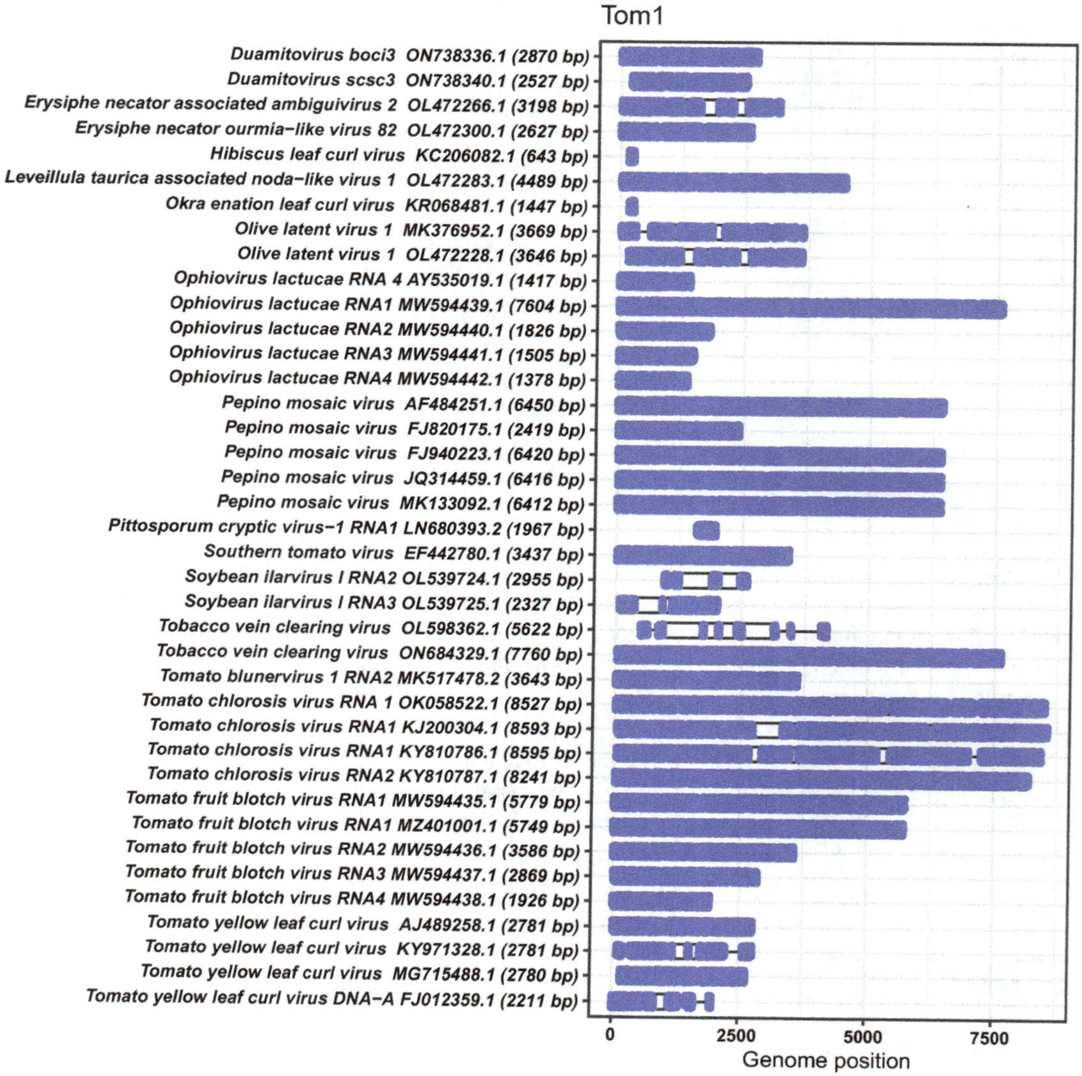

Fig. 3 Plots of read coverage of identified viruses. Plots show the position of the reads along the viral genomes (blue dots in x-axis) for all viruses identified in this analysis (represented in y-axis). The results of the Tom1 sample are shown as an example. The species, accession, and length (in pb) of the reference genomes are shown

3. The pipeline was developed to use text-based commands executed in the Linux command line (also referred to here as console or terminal). Optionally, the pipeline can be run using an R script. Then, basic knowledge of Linux command line and R language usage is required. To get basic skills in both, we recommend the HBC Training Program https://hbctraining.github.io/main/ Accessed Feb 2023).

Specific nomenclature used in this protocol is as follows:

- The symbol "$" is used to indicate that a command should be typed in the Linux terminal.
- "wd" is used here to refer to the current working directory, which is the complete path in your computer to the directory where input and output files are located and from which you should run the programs in the console (unless another path is specified). In our case, the wd in the Linux console looks like this:

```
bioinformatics@bioinformatics-cpu:/mnt/Linux_data/MiMB_proto-
col$
```

On the left of the colon, there are specified user and name of the computer we are using and, on the right, the full path to the wd where we are running the pipeline. From now on, we will use only "wd$."

- The symbol ">" is used to indicate an R command that can be executed in the R console. To start R, just type "R" in the Linux console and press enter. Alternatively, this pipeline can be run using a user-friendly interface for R; we recommend RStudio, and the provided R script "*HTSVdetection.R.*"

4. FASTQ files contain the sequence data from the base call of individual clusters generated using the Illumina technology that passes the filter criteria. In the case of multiplexed samples, a demultiplexing step is performed first. This process assigns clusters to specific samples based on the index sequences added to each library. For PE runs, two FASTQ files (R1 and R2) are generated for each sample.
5. To download data from SRA in FASTQ, use the SRA Toolkit following the instructions at the GitHub NCBI site (github.com/ncbi/sra-tools/wiki/02.-installing-SRA-Toolkit). Once installed and configured, use the tool fasterq-dump as follows:

```
wd$ fasterq-dump SRR14066947 SRR14066948 --outdir Original_-
files
```

The files are saved in a folder created previously called "Original_files" in the wd and are renamed to "*Tom1_1.fastq*" and "*Tom1_2.fastq*" (SRR14066948) or "*Tom2_1.fastq*" and "*Tom2_2.fastq*" (SRR14066947). Usually, FASTQ files obtained from the sequencing company are compressed ("*.*fastq.gz*"); to compress FASTQ files obtained from SRA, use the gzip program:

```
wd$ gzip -r Original_files
```

6. Illumina adapter sequence files are included with Trimmomatic distribution located in a folder called "*adapters*" within the Trimmomatic main directory. "*TruSeq3-PE-2.fa*" file contains adapter sequences used for PE reads obtained using different Illumina machines (like MiSeq or HiSeq). However, according to Trimmomatic's instructions, these sequences have not undergone extensive testing, and it is possible that other sequences may perform better for a given dataset. Then, Trimmomatic accepts a custom version of that FASTA file.
7. We used GNU parallel to run processes for Tom1 and Tom2 in parallel. The Linux command *cat* prints each line of the file "*sample.names.txt*" (*see* Subheading 2.2.4, **item 2**) that is subsequently piped to the replacement string of GNU parallel (appended by the symbol "{}") and used as input by different programs like Trimmomatic, Bowtie 2, etc.
8. In Unix, standard output (stdout) is the standardized data flow generated by command line programs, displayed by default on the terminal. The standard output provides information about the status of the running program or the result of a specific command line program. Redirecting the standard output to a file by including the option "> *file_name*" is possible.
9. In R and many programming languages, a line of code starting with the "#" symbol is considered a comment for human readers to explain the preceding or the following code and it is not executed.
10. To run an R code using RStudio, follow these steps:
    - In RStudio, click on File> New File > R Script or click on File> Open File to open an R script previously created.
    - Write or paste the R code to the new script window and save the R script on File> Save.
    - To run the R code, you can use different ways: (i) Click on "Run" at the top-right corner of the script window (will run the full script); (ii) use the keyboard shortcut "Ctrl+Enter" to run the current line of code; (iii) highlight a section of code and click on "Run"; or (iv) click on "Source" in the top-right corner of the script window (will run the full script). We recommend running the code line by line to avoid any bugs and to check that the variables are correctly created before continuing with the following line code. To visualize a variable in RStudio, simply type the variable name in the console and press enter or use the command "*print (variable)*."
    - The output of the R code will appear in the Console panel at the bottom of the RStudio window, viewed in the Plots panel, the Environment panel, or will be an exported file depending on the output command used.

The R script "*HTSVdetection.R*" performs the complete pipeline in R language by taking advantage of the "system()" R function that allows to execute a Linux command within the R code. The R script "*HTSVcalculation.R*" performs only some calculations from the following steps of the pipeline: in Subheading 3.4, **step 4**: Obtain summary information of mapped reads; **step 5**: Calculate average sequencing depth and percentage of coverage, where also there is the code to add virus information to the results and in Subheading 3.5; and **step 2**: Create plots of read coverage along the viral genomes.

Acknowledgments

We thank Ayoub Maachi, a former PhD student of Abiopep S.L., for his help with testing this bioinformatics pipeline. LD is a recipient of a fellowship of the Torres Quevedo Program (Programa de Contratación de Doctores y Tecnólogos, Ref. PTQ2021-011629) at Abiopep S.L. This research received funding from the European Union's Horizon 2020 research and innovation program under the Marie Skłodowska-Curie grant agreement no. 813542T.

References

1. Villamor DEV, Ho T, al Rwahnih M et al (2019) High throughput sequencing for plant virus detection and discovery. Phytopathology 109:716–725. https://doi.org/10.1094/PHYTO-07-18-0257-RVW
2. Kutnjak D, Tamisier L, Adams I et al (2021) A primer on the analysis of high-throughput sequencing data for detection of plant viruses. Microorganisms 9:841. https://doi.org/10.3390/microorganisms9040841
3. Andrews S (2010) FastQC: A quality control tool for high throughput sequence data. https://www.bioinformatics.babraham.ac.uk/projects/fastqc/. Accessed 23 Feb 2023
4. Blawid R, Silva JMF, Nagata T (2017) Discovering and sequencing new plant viral genomes by next-generation sequencing: description of a practical pipeline. Ann Appl Biol 170:301–314. https://doi.org/10.1111/aab.12345
5. Altschup SF, Gish W, Miller W et al (1990) Basic local alignment search tool. J Mol Biol 215:403–410. https://doi.org/10.1016/S0022-2836(05)80360-2
6. Gutiérrez P, Rivillas A, Tejada D et al (2021) PVDP: a portable open source pipeline for detection of plant viruses in RNAseq data. A case study on potato viruses in Antioquia (Colombia). Physiol Mol Plant Pathol 113. https://doi.org/10.1016/j.pmpp.2021.101604
7. Sukhorukov G, Khalili M, Gascuel O et al (2022) VirHunter: a deep learning-based method for detection of novel RNA viruses in plant sequencing data. Front Bioinform 2. https://doi.org/10.3389/fbinf.2022.867111
8. Valenzuela SL, Norambuena T, Morgante V et al (2022) Viroscope: plant viral diagnosis from high-throughput sequencing data using biologically-informed genome assembly coverage. Front Microbiol 13. https://doi.org/10.3389/fmicb.2022.967021
9. Maachi A, Donaire L, Hernando Y, Aranda MA (2022) Genetic differentiation and migration fluxes of viruses from melon crops and crop edge weeds. J Virol 96. https://doi.org/10.1128/jvi.00421-22
10. Maachi A, Torre C, Sempere RN et al (2021) Use of high-throughput sequencing and two RNA input methods to identify viruses infecting tomato crops. Microorganisms 9. https://doi.org/10.3390/microorganisms9051043
11. Bolger AM, Lohse M, Usadel B (2014) Trimmomatic: a flexible trimmer for Illumina sequence data. Bioinformatics 30:2114–2120.

https://doi.org/10.1093/bioinformatics/btu170

12. Langmead B, Trapnell C, Pop M, Salzberg SL (2009) Ultrafast and memory-efficient alignment of short DNA sequences to the human genome. Genome Biol 10. https://doi.org/10.1186/gb-2009-10-3-r25
13. Haas BJ, Papanicolaou A, Yassour M et al (2013) De novo transcript sequence reconstruction from RNA-seq using the trinity platform for reference generation and analysis. Nat Protoc 8:1494–1512. https://doi.org/10.1038/nprot.2013.084
14. Shen W, Le S, Li Y, Hu F (2016) SeqKit: a cross-platform and ultrafast toolkit for FASTA/Q file manipulation. PLoS One 11. https://doi.org/10.1371/journal.pone.0163962
15. Li H, Durbin R (2009) Fast and accurate short read alignment with burrows-wheeler transform. Bioinformatics 25:1754–1760. https://doi.org/10.1093/bioinformatics/btp324
16. Li H, Handsaker B, Wysoker A et al (2009) The sequence alignment/map format and SAMtools. Bioinformatics 25:2078–2079. https://doi.org/10.1093/bioinformatics/btp352
17. RStudio Team (2020) RStudio: integrated development for R. In: RStudio, PBC. http://www.rstudio.com/. Accessed 23 Feb 2023
18. Milne I, Stephen G, Bayer M et al (2013) Using tablet for visual exploration of second-generation sequencing data. Brief Bioinform 14:193–202. https://doi.org/10.1093/bib/bbs012

Chapter 2

Taxonomic Classification of Geminiviruses Based on Pairwise Sequence Comparisons

João Paulo Herrera da Silva and F. Murilo Zerbini

Abstract

Geminiviridae is the largest and one of the most diverse families of plant viruses, comprising 14 genera demarcated based on host range, type of insect vector, and phylogenetic relationships. The use of unbiased, whole-genome multiple displacement amplification techniques coupled with high-throughput sequencing has greatly expanded our knowledge of geminivirus diversity over the last two decades. As a result, a large number of new species have been described in recent years. Species demarcation criteria in the family are entirely based on sequence comparisons, but the specific cutoff values vary for each genus. The purpose of this chapter is to provide a step-by-step pipeline to classify new species in the family *Geminiviridae*.

Key words Multiple sequence alignment, Pairwise sequence comparisons, Phylogeny, SDT, MAFFT, Maximum likelihood, Species demarcation criteria, Taxonomy

1 Introduction

Virus taxonomy (the hierarchical classification and naming of virus taxa) is coordinated by the International Committee on Taxonomy of Viruses (ICTV). The ICTV is organized into seven Subcommittees, including the Plant Virus Subcommittee, and more than 120 Study Groups (SGs), including 22 plant virus SGs. Most SGs are organized at the family rank, such as the *Geminiviridae* and *Tolecusatellitidae* SG [1]. Thus, virus classification at the genus and species levels is usually established by SGs, which have considerable autonomy in establishing genus and species demarcation criteria within their respective families. For geminiviruses, the criteria for genera demarcation include host range (monocots or dicots), type of insect vector (leafhoppers, treehoppers, whiteflies, and aphids), genome organization, and phylogenetic relationships [2]. The *Geminiviridae* SG pioneered the demarcation of species based solely on sequence comparisons, establishing detailed guidelines on how pairwise identities must be calculated and determining

Elizabeth P. B. Fontes and Kristiina Mäkinen (eds.), *Plant-Virus Interactions*, Methods in Molecular Biology, vol. 2724, https://doi.org/10.1007/978-1-0716-3485-1_2,

the cutoff values for each genus [3–7]. Although not without criticism at its establishment [8], these demarcation criteria have stood the test of time and have been adopted by several other SGs [9, 10]. One aspect of this classification system is that it can be easily implemented in semi-automated (or even fully automated) forms. Here, we provide a pipeline, based on current species demarcation criteria, to classify species in the different genera of the *Geminiviridae*. It is worth remembering that taxonomy is a dynamic field, and the species demarcation criteria may change over time.

1.1 The Geminiviridae

The *Geminiviridae* family comprises viruses that infect plants and have a small genome composed of one or two circular, single-stranded DNA (ssDNA) molecules ranging from 2.5 to 5.2 kb in length [2, 7]. The family belongs to the order *Geplafuvirales*, class *Repensiviricetes*, phylum *Cressdnaviricota*, kingdom *Shotokuvirae*, and realm *Monodnaviria* [11]. Geminiviruses have unique, quasi-icosahedral twinned particles made up of 110 subunits of a single coat protein organized into 22 pentameric capsomeres [12–15].

The *Geminiviridae* family is the largest and one of the most diverse among all viruses. Its 14 genera (*Becurtovirus*, *Begomovirus*, *Capulavirus*, *Citlodavirus*, *Curtovirus*, *Eragrovirus*, *Grablovirus*, *Maldovirus*, *Mastrevirus*, *Mulcrilevirus*, *Opunvirus*, *Topocuvirus*, *Turncurtovirus*, and *Topilevirus*) are defined based on host range, type of insect vector, genome organization, and phylogenetic relationships [2]. The family contains 520 species as of May 2023 (ictv.global/taxonomy). The genus *Begomovirus*, with 445 species, is the largest in the entire virosphere. On the other extreme, there are six orphan geminivirus species, that is, which are not affiliated to a genus: *Common bean curly stunt virus*, *Limeum africanum-associated virus*, *Parsley yellow leaf curl virus*, *Polygala garcinii-associated virus*, *Opuntia virus 2*, and *Olea europaea geminivirus* [16–20]. The enormous diversity of geminiviruses results from high mutation rates and frequent recombination [21–28].

Geminiviruses have quite similar genomic organization in size and number of components. The exceptions are the bipartite representatives of the genus *Begomovirus* and Olea europaea geminivirus, which have two genomic components packaged separately [2, 20]. All geminivirus genomes have a long intergenic region (LIR) with a conserved stem-loop structure where the origin of replication is located. In the case of bipartite begomoviruses, there is a region of approximately 200 nucleotides in the LIR that is almost identical between cognate DNA-A and DNA-B components, known as the common region [2, 29]. Geminivirus genomes contain genes in both the viral and the complementary senses [30]. Gene composition varies slightly among the 14 genera. In general, geminiviruses encode at least six proteins, two or three in the viral sense (CP, V1/AV1, and V2/AV2) and three or four in the complementary sense (Rep, Trap, Ren, and C4/AC4).

Begomoviruses have intra-genus differences, such as the lack of AV2 in New World begomoviruses, the presence of C5/AC5 in some viruses, and the two genes (MP and NSP) encoded in the DNA-B component of bipartite begomoviruses [4]. Eragroviruses are the simplest geminiviruses regarding genomic organization, with only four genes (CP and MP in the viral sense and Rep and Trap in the complementary sense) [6]. Becurtoviruses, capulaviruses, citlodaviruses, grabloviruses, mastreviruses, mulcrileviruses, and topileviruses express Rep (C1) and Rep-A (C1:C2) from spliced transcripts [2, 31].

1.2 Species Demarcation Criteria

The main species demarcation criterion adopted for geminiviruses is based on pairwise comparisons of full-length nucleotide sequences. The cutoff values for species demarcation are determined from the distribution of pairwise identity percentages. This sequence-based classification is easy to implement and has proven to be robust so far. When available, information on the biological properties of a given geminivirus may help delimit species, especially in those cases where the percent identity between sequences is very close to the threshold [2, 4, 32].

The species demarcation cutoff is 77% for curtoviruses, 78% for capulaviruses, citlodaviruses, maldoviruses, mastreviruses, mulcrileviruses, opunviruses, and topileviruses, 80% for becurtoviruses, grabloviruses and turncurtoviruses, and 91% for begomoviruses [3–5, 7, 33]. There are no species demarcation criteria for *Eragrovirus* and *Topocuvirus*, since both genera contain a single species [2, 6].

1.3 Step-by-Step Procedure for the Classification of Species in the Family Geminiviridae

This chapter aims to provide a step-by-step guide to determining the taxonomic placement of geminiviruses based on their complete nucleotide sequences (Fig. 1). It is not the scope of this text to provide a detailed discussion on analysis assumptions, parameter configuration, or the accuracy of one method or another. It will be up to the user to decide on the software of their preference.

2 Materials

1. Datasets in fasta format. The datasets should be composed of the most closely related sequences (based on a Blast*n* search) to the sequence you want to classify.
2. Sequence Demarcation Tool Version 1.2 (SDT v1.2) software [34]. SDT v1.2 is available as a free download at http://web.cbio.uct.ac.za/SDT (*see* **Note 1**). This software was specifically developed to classify sequences based on identity values from pairwise alignments (as opposed to multiple sequence alignments). Furthermore, percent identities are calculated with

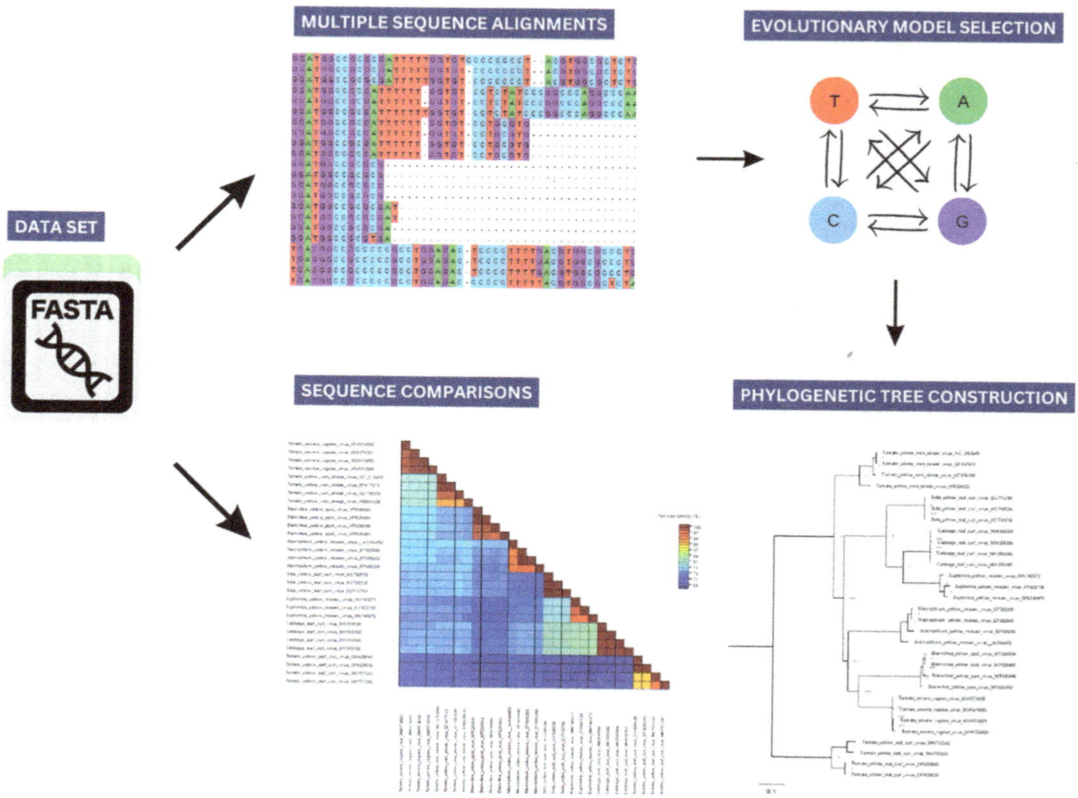

Fig. 1 Workflow for taxonomic classification of geminiviruses based on pairwise sequence comparisons. The dataset in FASTA format is comprised of the complete genomic sequence of the virus to be classified (Query) and sequences of related viruses, as determined in a BLAST*n* search. From the FASTA file, pairwise sequence comparisons are performed with Sequence Demarcation Tool (SDT). Species assignment is based on the percent identities between the Query sequence and the other sequences in the dataset. Phylogenetic analysis can be performed to corroborate the results of pairwise sequence comparisons. For this, a multiple sequence alignment is obtained with the appropriate program (such as MAFFT). Then, the best nucleotide substitution model is chosen with ModelTest-NG, and a phylogenetic tree based on the maximum-likelihood method is obtained with RaxML-NG. A distinct species will form a well-supported monophyletic clade with a considerable phylogenetic distance from other species

pairwise deletion of gaps (in other words, sites with gap characters are excluded from the pairwise identity calculations). The software is available for 32- and 64-bit Windows, Linux, and macOS operating systems (*see* **Note 2**).

3. Multiple DNA sequence aligner software. There are numerous computer programs developed to perform this task, such as MUSCLE, MAFFT, T-Coffee, and Clustal. They are available for different operating systems and implemented on different platforms, and many have online versions (e.g., at the CIPRES portal: https://www.phylo.org/). We will use MAFFT 7 as an example, but the user should feel free to choose the most

familiar software for them (*see* **Note 3**). MAFFT 7 versions for Linux, macOS, and Windows can be downloaded from https://mafft.cbrc.jp/alignment/software/. The program's manual is available at the link mentioned above and (as always) is recommended reading.

4. Software to test the nucleotide substitution model that best fits the data set. We will use ModelTest-NG, but the user should feel free to choose the software of their choice. ModelTest-NG versions for Linux, macOS, and Windows can be downloaded from https://github.com/ddarriba/modeltest. The software is also available on the CIPRES portal (https://www.phylo.org/). *See* **Note 3**.
5. Software for running phylogenetic analysis. As with sequence alignment tools, there are many software packages for phylogenetic analysis based on different methods (neighbor-joining, maximum parsimony, maximum-likelihood, Bayesian inference, among others). We will provide simple, step-by-step instructions to run a maximum-likelihood phylogenetic analysis in RaxML-NG. RaxML-NG is available as 64-bit Linux and 64-bit macOS versions at https://github.com/amkozlov/raxml-ng (*see* **Note 4**). The analysis can also be performed in the CIPRES portal (https://www.phylo.org/).

3 Methods

3.1 Building the Dataset

1. Submit the sequence(s) you wish to classify to a BLAST*n* search of the "non-redundant nucleotide" database. This analysis can be performed on the Web at https://blast.ncbi.nlm.nih.gov/Blast.cgi?PROGRAM=blastn&PAGE_TYPE=BlastSearch&LINK_LOC=blasthome. BLAST*n* will perform an alignment between the submitted sequence (Query) and the sequences from the National Center for Biotechnology Information (NCBI) database (Subject) and will return the result of the comparisons between the sequences and the respective statistical analysis (percent coverage, percent identity, E-value, and score).
2. The dataset should be composed of the sequences constituting the top *n* hits from the BLAST*n* search. *n* can vary from as few as 20 to as many as 100 sequences, depending on how similar these sequences are (*see* **Note 5**). The dataset must be saved in a file in FASTA format. It is imperative that the genomes are complete and that the added sequences are linearized in the conventional orientation; that is, all of them must start at the first nucleotide after the Rep cleavage site. This nucleotide (an adenine) is located within the nonanucleotide at the origin of replication (5′-TAATATT//AC-3′) (*see* **Note 6**).

3.2 Performing Pairwise Sequence Comparisons with SDT v1.2

1. Start SDT v1.2, go to the "Open" menu, and choose the dataset in FASTA format to be analyzed. The dataset can be loaded aligned or unaligned (*see* **Note 7**).
2. Once the dataset is loaded, the program will open a window with available analysis options. Choose the MAFFT alignment option and click on "Run." SDT v1.2 will perform the alignment of every unique pair of sequences, and then, the percent identity will be calculated from the alignment (*see* **Note 8**).
3. At the end of the analysis, STD v1.2 will generate a colored graphic matrix with pairwise identity values. The matrix can be saved as an image in .EMF (vector graphics) or .BMP (bitmap graphics) formats. There is also the option to save the identity values as .txt or .csv files. Another visualization option is to plot the distribution of percent identities on a graph.

3.3 Classification Based on Pairwise Nucleotide Sequence Identities

1. If the Query sequence has a pairwise identity value above the species demarcation threshold with a sequence from the dataset (corresponding to "virus X"), then the virus from which the Query sequence is derived is a member of the same species as virus X. On the other hand, if the Query sequence has percent identities below the species demarcation threshold with all other sequences in the dataset, then the corresponding virus should be considered a member of a new species. *See* **Note 9**.

3.4 Phylogenetic Analysis: Obtaining a Multiple Sequence Alignment

1. Execute MAFFT7 in the terminal or command prompt. There is also the option to carry out the analysis online at https://www.ebi.ac.uk/Tools/msa/mafft/ (*see* **Note 3**). From the terminal or command prompt, type: % mafft [arguments] *input_file_name* > *output_file_name* (*see* **Note 10**).
2. The input file must be in FASTA format. MAFFT automatically recognizes the type of dataset (nucleotide or amino acid sequences).
3. The user can choose the alignment method that is most suitable for the specific situation they are working on. MAFFT has different alignment methods designed to solve different problems, such as datasets of different sizes and complexities. These are explained in detail in the program manual. If, even after checking the manual, the user is unsure which method to use, choose: % mafft --auto *input_file_name* > *output_file_name* (*see* **Note 11**).
4. The output can be viewed using a sequence preview tool or as a text file. Detailed visualization of the result is crucial to evaluate the quality of the alignment. A poor alignment will result in a phylogeny with low credibility. *See* **Note 12**.

3.5 Phylogenetic Analysis: Running ModelTest-NG

1. Open ModelTest-NG in the terminal or command prompt. Run the following command: % modeltest-ng --input *alignment_file_name* --output *output_file_name* --processes 24 --datatype nt --schemes 11. *See* **Note 13**.
2. At the end of the analysis, ModelTest-NG will generate a table with details of the analysis. In the table, you will find various model selection criteria, such as Akaike (AIC), corrected Akaike (AICc), and Bayesian (BIC) criteria for each model. At the end of the page, you will see a summary with the models that best fit the data based on each criterion mentioned above. *See* **Note 14.**

3.6 Phylogenetic Analysis: Performing Maximum-Likelihood Analysis in RaxML-NG

1. Run raxml-ng in the terminal or command prompt: % raxml-ng --bootstrap --msa *alignment_file_name* --model GTR+G --seed 2 --threads 2 --bootstrap 1000 --bs-trees 100 --prefix *output_file_name*. *See* **Note 15**.
2. At the end of the analysis, the program will generate several files, among them a .tree file containing the best phylogenetic tree with the bootstrap supports values of the branches. The result of this analysis can be viewed using tools for viewing and editing phylogenetic trees, such as FigTree (http://tree.bio.ed.ac.uk/software/figtree/) or iTOL (https://itol.embl.de/). *See* **Notes 16** and **17**.

4 Notes

1. For details on installation and usage, consult the SDT v1.2 manual, available at the same download link. As a general rule, reading every program's manual is essential to choose the correct parameters and to be able to analyze the results critically.
2. The macOS version of SDT is a command line-based application written in Python 2 (which is no longer supported). It may be difficult to install and run successfully on ARM-based computers running macOS 11 or later due to multiple dependency issues. Until an updated version is released, we do not recommend using it.
3. Programs that are command line-based can be intimidating to users who do not have at least a basic knowledge of bioinformatics. These users may opt for online versions whenever they are available, as these versions are generally much easier to use. Nevertheless, we encourage users to use the command line versions, as they tend to be more powerful in many instances. For absolute beginners, the methods in this chapter provide the exact commands that need to be typed for each analysis (feel free to copy/paste them). As the user becomes more

comfortable with the command line, they will notice that the command line greatly facilitates repetitive tasks and gives the user much more control over the program.

4. There are separate macOS versions for Intel-based and ARM-based computers. In Windows computers, it is necessary to run the Linux binary via Windows Subsystem for Linux.
5. For example, if the 25th hit has high (>90%) coverage but low (<75%) identity with the query, using the top 25 hits will be satisfactory. However, if the 25th hit still has high coverage and very high identity with the query, sequences further down the list should be included in the dataset until a sequence with high coverage/low identity is included.
6. Sequences that have been submitted to GenBank after 2005 tend to begin at this nucleotide. Unfortunately, older sequences often do not. If the dataset contains a small number of those sequences, it is easy to adjust them manually. However, for a large number of sequences, it will be helpful to write a simple script to automate the process.
7. Since SDT performs the alignment itself, it is actually better to load the dataset unaligned.
8. The percent identity is given by the function 1-(M/N), where M is the number of mismatching nucleotides and N is the total number of sites of the alignment. The calculation is based on true alignments, that is, disregarding the gaps.
9. Some begomovirus species have a very large number of sequence isolates, and in these cases, it is not uncommon that the query sequence will have a percent identity above the species demarcation threshold with some isolates and below the threshold with other isolates. The Geminiviridae Study Group of the ICTV has provided a set of helpful "conflict resolution criteria" that address this and other issues. These are available in ref. [4]. Specifically for the case mentioned above, the conflict resolution criterion states that "The new isolate should be classified as belonging to the species with which it shares >91 % nt sequence identity with any one isolate from that species, even if it is <91 % identical to all other isolates from that species."
10. In this and in every subsequent command line input, the text in *italics* should be replaced with the user-given names. For example, if the user named the file that contains their dataset as New_begomovirus.fasta and would like the output file to be named New_begomo_alignment.txt, then the input text should be % mafft [arguments] New_begomovirus.fasta > New_begomo_alignment.txt.
11. These are the basic options for carrying out the analysis. Additional parameters can be configured to increase the alignment accuracy, such as gap opening penalty score and gap extension

penalty score: % mafft --auto --op <value> --ep <value> *input_file_name* > *output_file_name*. The default values for gap opening and gap extension penalty scores are 1.53 and 0.123, respectively. For more details, the user should check the program's manual.

12. The need to inspect the alignment cannot be overstated. With some experience, the user will be able to adjust the alignment manually, improving its accuracy. But with that said, the user should only do this if they are sure of what they are doing. Otherwise, manual adjustment can have the opposite effect, generating artifacts and reducing the reliability of subsequent analyses. Note that all the following analyses will be performed from the alignment, and poor alignments will increase the uncertainty of the inferences.

13. Meaning of the parameters:

    --input: This is the MAFFT output file (*i.e.*, the alignment file). You must input the dataset duly prepared, as described above, aligned in PHYLIP or FASTA format.

    --processes: Indicates the number of concurrent threads.

    --datatype: Indicates whether the dataset is comprised of nucleotide (nt) or amino acid (aa) sequences.

    --schemes: Indicates the number of predefined DNA substitution schemes to be evaluated.

14. For more details on parameter settings and analysis of results, we highly recommend reading the manual, available in the Releases section of the program's website (https://github.com/ddarriba/modeltest/releases), and/or checking the Wiki section (https://github.com/ddarriba/modeltest/wiki).

15. Meaning of the parameters:

    --msa: Indicates the input, which is the MAFFT output file (*i.e.*, the alignment file); in this example, the alignment is in FASTA format, but RaxML-NG recognizes FASTA, PHYLIP, and CATG file formats.

    --model: Indicates the evolutionary model determined based on ModelTest-NG, as described in the previous section.

    --seed: Specifies a random number seed for the parsimony inferences; this allows the user to reproduce their results.

    --threads: Specifies the number of CPU cores.

    --bootstrap: Indicates the number of bootstrap replicates to run.

    --bs-trees: Sets the number of trees to be saved for each bootstrap replicate.

    --prefix: Specifies the prefix to be used for the output files generated by RaxML; this parameter is important to avoid overwriting previously generated output files.

16. A distinct species will form a well-supported monophyletic clade with a considerable phylogenetic distance from other species and will corroborate the result from the SDT analysis.
17. For more details, the user is referred to the RaxML-NG manual (https://github.com/amkozlov/raxml-ng/wiki), where they will find more information about analysis settings and advanced parameters.

References

1. Siddell SG, Smith DB, Adriaenssens E, Alfenas-Zerbini P, Dutilh BE, Garcia ML, Junglen S, Krupovic M, Kuhn JH, Lambert AJ, Lefkowitz EJ, Łobocka M, Mushegian AR, Oksanen HM, Robertson DL, Rubino L, Sabanadzovic S, Simmonds P, Suzuki N, Van Doorslaer K, Vandamme A-M, Varsani A, Zerbini FM (2023) Virus taxonomy and the role of the International Committee on Taxonomy of Viruses (ICTV). J Gen Virol 104:001840
2. Fiallo-Olive E, Lett JM, Martin DP, Roumagnac P, Varsani A, Zerbini FM, Navas-Castillo J (2021) ICTV virus taxonomy profile: *Geminiviridae* 2021. J Gen Virol 102:001696
3. Muhire B, Martin DP, Brown JK, Navas-Castillo J, Moriones E, Zerbini FM, Rivera-Bustamante R, Malathi VG, Briddon RW, Varsani A (2013) A genome-wide pairwise-identity-based proposal for the classification of viruses in the genus *Mastrevirus* (family *Geminiviridae*). Arch Virol 158:1411–1424
4. Brown JK, Zerbini FM, Navas-Castillo J, Moriones E, Ramos-Sobrinho R, Silva JC, Fiallo-Olive E, Briddon RW, Hernandez-Zepeda C, Idris A, Malathi VG, Martin DP, Rivera-Bustamante R, Ueda S, Varsani A (2015) Revision of *Begomovirus* taxonomy based on pairwise sequence comparisons. Arch Virol 160:1593–1619
5. Varsani A, Martin DP, Navas-Castillo J, Moriones E, Hernandez-Zepeda C, Idris A, Zerbini FM, Brown JK (2014) Revisiting the classification of curtoviruses based on genome-wide pairwise identity. Arch Virol 159:1873–1882
6. Varsani A, Navas-Castillo J, Moriones E, Hernández-Zepeda C, Idris A, Brown JK, Zerbini FM, Martin DP (2014) Establishment of three new genera in the family *Geminiviridae*: *Becurtovirus*, *Eragrovirus* and *Turncurtovirus*. Arch Virol 159:2193–2203
7. Roumagnac P, Lett J-M, Fiallo-Olivé E, Navas-Castillo J, Zerbini FM, Martin DP, Varsani A (2022) Establishment of five new genera in the family *Geminiviridae*: *Citlodavirus*, *Maldovirus*, *Mulcrilevirus*, *Opunvirus*, and *Topilevirus*. Arch Virol 167:695–710
8. Van Regenmortel MHV (2018) The species problem in virology. Adv Virus Res 100:1–18
9. Varsani A, Krupovic M (2017) Sequence-based taxonomic framework for the classification of uncultured single-stranded DNA viruses of the family *Genomoviridae*. Virus Evol 3: vew037
10. Roux S, Adriaenssens EM, Dutilh BE, Koonin EV, Kropinski AM, Krupovic M, Kuhn JH, Lavigne R, Brister JR, Varsani A, Amid C, Aziz RK, Bordenstein SR, Bork P, Breitbart M, Cochrane GR, Daly RA, Desnues C, Duhaime MB, Emerson JB, Enault F, Fuhrman JA, Hingamp P, Hugenholtz P, Hurwitz BL, Ivanova NN, Labonte JM, Lee KB, Malmstrom RR, Martinez-Garcia M, Mizrachi IK, Ogata H, Paez-Espino D, Petit MA, Putonti C, Rattei T, Reyes A, Rodriguez-Valera F, Rosario K, Schriml L, Schulz F, Steward GF, Sullivan MB, Sunagawa S, Suttle CA, Temperton B, Tringe SG, Thurber RV, Webster NS, Whiteson KL, Wilhelm SW, Wommack KE, Woyke T, Wrighton KC, Yilmaz P, Yoshida T, Young MJ, Yutin N, Allen LZ, Kyrpides NC, Eloe-Fadrosh EA (2019) Minimum Information about an Uncultivated Virus Genome (MIUViG). Nat Biotechnol 37:29–37
11. Krupovic M, Varsani A, Kazlauskas D, Breitbart M, Delwart E, Rosario K, Yutin N, Wolf YI, Harrach B, Zerbini FMJJOV (2020) *Cressdnaviricota*: a virus phylum unifying seven families of rep-encoding viruses with single-stranded, circular DNA genomes. J Virol 94: e00582–e00520
12. Zhang W, Olson NH, Baker TS, Faulkner L, Agbandje-McKenna M, Boulton MI, Davies JW, McKenna R (2001) Structure of the maize streak virus geminate particle. Virology 279:471–477
13. Bottcher B, Unseld S, Ceulemans H, Russell RB, Jeske H (2004) Geminate structures of

African cassava mosaic virus. J Virol 78:6758–6765

14. Hesketh EL, Saunders K, Fisher C, Potze J, Stanley J, Lomonossoff GP, Ranson NA (2018) The 3.3 Å structure of a plant geminivirus using cryo-EM. Nat Commun 9:2369
15. Hipp K, Grimm C, Jeske H, Böttcher B (2017) Near-atomic resolution structure of a plant geminivirus determined by electron cryomicroscopy. Structure 25:1303–1309
16. Claverie S, Bernardo P, Kraberger S, Hartnady P, Lefeuvre P, Lett JM, Galzi S, Filloux D, Harkins GW, Varsani A, Martin DP, Roumagnac P (2018) From spatial metagenomics to molecular characterization of plant viruses: a geminivirus case study. Adv Virus Res 101:55–83
17. Hasanvand V, Heydanejad J, Massumi H, Kleinow T, Jeske H, Varsani A (2020) Isolation and characterization of a novel geminivirus from parsley. Virus Res 286:198056
18. Zhang R, Wu X, Jiang X, Wu X, Luan X, Cheng X (2020) Molecular characterization of common bean curly stunt virus: a novel recombinant geminivirus in China. Arch Virol 165: 257–260
19. Fontenele RS, Salywon AM, Majure LC, Cobb IN, Bhaskara A, Avalos-Calleros JA, Arguello-Astorga GR, Schmidlin K, Khalifeh A, Smith K, Schreck J, Lund MC, Kohler M, Wojciechowski MF, Hodgson WC, Puente-Martinez-R, Van Doorslaer K, Kumari S, Oyeniran KA, Verniere C, Filloux D, Roumagnac P, Lefeuvre P, Ribeiro SG, Kraberger SP, Martin DP, Varsani A (2021) New world cactaceae plants harbor diverse geminiviruses. Viruses 13:694
20. Chiumenti M, Greco C, De Stradis A, Loconsole G, Cavalieri V, Altamura G, Zicca S, Saldarelli P, Saponari M (2021) Olea europaea geminivirus: a novel bipartite geminivirid infecting olive trees. Viruses 13:481
21. Lima ATM, Silva JCF, Silva FN, Castillo-Urquiza GP, Silva FF, Seah YM, Mizubuti ESG, Duffy S, Zerbini FM (2017) The diversification of begomovirus populations is predominantly driven by mutational dynamics. Virus Evol 3:vex005
22. Crespo-Bellido A, Hoyer JS, Dubey D, Jeannot RB, Duffy S (2021) Interspecies recombination has driven the macroevolution of cassava mosaic begomoviruses. J Virol 95:e00541–e00521
23. Duffy S, Holmes EC (2009) Validation of high rates of nucleotide substitution in geminiviruses: phylogenetic evidence from East African cassava mosaic viruses. J Gen Virol 90: 1539–1547
24. Duffy S, Holmes EC (2008) Phylogenetic evidence for rapid rates of molecular evolution in the single-stranded DNA begomovirus tomato yellow leaf curl virus. J Virol 82:957–965
25. Duffy S, Shackelton LA, Holmes EC (2008) Rates of evolutionary change in viruses: patterns and determinants. Nat Rev Genet 9: 267–276
26. Lefeuvre P, Moriones E (2015) Recombination as a motor of host switches and virus emergence: geminiviruses as case studies. Curr Opin Virol 10:14–19
27. Pita JS, Fondong VN, Sangare A, Otim-Nape GW, Ogwal S, Fauquet CM (2001) Recombination, pseudorecombination and synergism of geminiviruses are determinant keys to the epidemic of severe cassava mosaic disease in Uganda. J Gen Virol 82:655–665
28. Rojas MR, Hagen C, Lucas WJ, Gilbertson RL (2005) Exploiting chinks in the plant's armor: evolution and emergence of geminiviruses. Annu Rev Phytopathol 43:361–394
29. Hanley-Bowdoin L, Bejarano ER, Robertson D, Mansoor S (2013) Geminiviruses: masters at redirecting and reprogramming plant processes. Nat Rev Microbiol 11: 777–788
30. Revington GN, Sunter G, Bisaro DM (1989) DNA sequences essential for replication of the B genome component of tomato golden mosaic virus. Plant Cell 1:985–992
31. Yazdi HRB, Heydarnejad J, Massumi H (2008) Genome characterization and genetic diversity of beet curly top Iran virus: a geminivirus with a novel nonanucleotide. Virus Genes 36:539–545
32. Zerbini FM, Briddon RW, Idris A, Martin DP, Moriones E, Navas-Castillo J, Rivera-Bustamante R, Varsani A, ICTV Consortium (2017) ICTV virus taxonomy profile: *Geminiviridae*. J Gen Virol 98:131–133
33. Varsani A, Roumagnac P, Fuchs M, Navas-Castillo J, Moriones E, Idris A, Briddon RW, Rivera-Bustamante RF, Zerbini FM, Martin DP (2017) *Capulavirus* and *Grablovirus*: two new genera in the family *Geminiviridae*. Arch Virol 162:1819–1831
34. Muhire BM, Varsani A, Martin DP (2014) SDT: a virus classification tool based on pairwise sequence alignment and identity calculation. PLoS One 9:e108277

Chapter 3

Initiation of Plant Virus Infections by Agroinfiltration of Infectious Viral cDNAs (icDNAs)

Maija Pollari

Abstract

Agroinfiltration uses *Agrobacterium* to deliver T-DNA-based gene expression constructs into plants. This chapter focuses on the standard method, specifically from the perspective of plant virus research, and describes a protocol for the initiation of virus infections in plants via infiltration of *Agrobacterium* strains carrying infectious viral cDNAs (icDNAs). The method outlines the culture and preparation of *Agrobacterium* for infiltration, the infiltration procedure, optimization of the optical density of the *Agrobacterium* suspension, and sampling of infected plants post-agroinfiltration. The advantages of the agroinfiltration method compared to traditional mechanical inoculation using sap from infected plants are discussed. The protocol is applicable for different pathosystems, although case-specific optimization of infiltration parameters and sampling is recommended.

Key words Agroinfiltration, *Agrobacterium tumefaciens*, Transient gene expression, Infectious cDNA, Plant–virus infection

1 Introduction

The genus *Agrobacterium* encompasses a widespread group of gram-negative bacteria best known for their ability to transfer part of their DNA into plant hosts. This characteristic has been explored and utilized by researchers since the 1980s, and it has firmly established the genus as the workhorse of the genetic transformation of plants.

Agrobacterium tumefaciens (hereafter *A. tumefaciens*), initially described as a bacterium inducing crown gall tumors in daisies, is the type species of the genus *Agrobacterium* [1–3]. The genus currently encompasses 16 species [4, 5], many of which are soil-borne or root-associated phytopathogens. Additionally, the diverse genus also includes non-pathogenic species from a variety of habitats and different lifestyles. Indeed, before the availability of molecular techniques and genome sequencing, the taxonomic

Elizabeth P. B. Fontes and Kristiina Mäkinen (eds.), *Plant-Virus Interactions*, Methods in Molecular Biology, vol. 2724, https://doi.org/10.1007/978-1-0716-3485-1_3,

organization of the clade was often disputed, which led to the transfer of species into the genus *Rhizobium* and vice versa [4]. A curious example illustrating the diversity of the genus is a recent addition, *Agrobacterium tomkonis*, which was isolated from the surface of the observation dome on the International Space Station [5, 6].

The ability of *Agrobacterium* strains to transfer and integrate part of their DNA into the host genome defines their unique relationship with plants. The tumor- and root-inducing (Ti and Ri, respectively) plasmids of *Agrobacteria* contain T-DNA, a region flanked by conserved 25 bp regions that define the area transferred into the plant cell and facilitate its integration into the host genome. Ti plasmids also encode genes for oncogenesis and virulence factors that facilitate the delivery of the T-DNA into plants. In nature, phytopathogenic *Agrobacteria* are chemotactically attracted by phenolic compounds secreted from plant lesions and wound sites. In addition to attracting the bacteria, plant-derived phenolics simultaneously activate the transcription of the *vir* genes in the Ti plasmid. In the agroinfiltration procedure described here, acetosyringone, a phenolic metabolite produced in plants as part of pattern-triggered immune responses, plays the role of the inducer. Acetosyringone is not only a potent activator of the *vir* genes but also enhances Ti plasmid copy number, which can promote transformation efficiency [7]. Interestingly, when combined with peroxidases, acetosyringone was found to inhibit the growth of some pathogenic bacteria [8]. In *Agrobacteria*, the newly synthesized *vir* gene products excise the T-DNA from the Ti plasmid. The T-DNA is then transported to the plant cell and targeted to the nucleus for mRNA synthesis and eventual integration into the genome. An enormous variety of gene expression constructs, including full-length infectious viral cDNAs (icDNAs), can be inserted between the T-DNA's left and right borders for delivery into and expression in plant cells (Fig. 1).

The molecular details of the T-DNA transfer and integration processes have been recently reviewed in detail by Gelvin [9] and Hwang et al. [10]. Currently, strains commonly used for agroinfiltration are non-oncogenic. In these strains, the tumorigenic nopaline synthesis and *vir* genes have been removed from the Ti plasmids. Instead, the T-DNA borders and *vir* genes are present on separate plasmids [11–13]. Because they assist in the T-DNA transfer process, the plasmids hosting the *vir* genes are called "helper plasmids" (Fig. 1). The initiation of virus infections by agroinfiltration depends not only on the successful expression of the viral icDNA in the host cell but also on the coordinated replication, translation, encapsidation, and movement of the virus, a process outlined in Fig. 2 using potato virus A (PVA) as an example. PVA is a potyvirus with a circa 10 kb single-stranded positive-sense RNA genome [14]. Replicated potyviral genomes have a

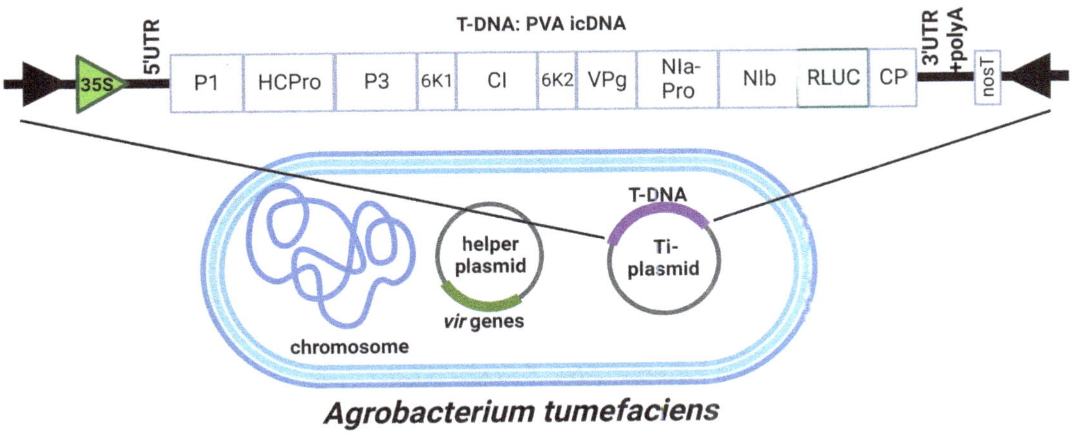

Fig. 1 *A. tumefaciens* carrying a Ti plasmid with the potato virus A (PVA) full-length icDNA construct. In this construct, the PVA icDNA is tagged with the *Renilla* luciferase (RLUC) reporter gene. Black arrows on the construct denote the T-DNA left and right borders. In addition, the bacterium harbors a helper plasmid encoding the *vir* genes required for the expression and transfer of the T-DNA into the host plant. (Figure created in BioRender.com)

characteristic viral protein, VPg (viral protein genome-linked), covalently attached to the 5′ end of the genome and a poly-A tail.

Agroinfiltration-mediated viral infections were first accomplished with the DNA virus cauliflower mosaic virus in turnips in 1986 [15] and a few years later with the RNA virus tobacco mosaic virus in the model plant *Nicotiana benthamiana* (*N. benthamiana*) [16]. Traditionally, virus infections were initiated by mechanical inoculation with material from infected plants. The mechanical inoculation approach is robust and effective but has drawbacks and limitations. First, it requires sufficient intact and infectious virus particles in the source material. This requirement excludes the possibility of studying virus mutants that do not produce stable particles. Second, it is difficult to quantitate the virus load introduced during the inoculation, which is performed by rubbing the leaf with the inoculum mixed with an abrasive such as carborundum. Third, in mechanical inoculation, the applied pressure can damage the inoculated leaf, so it is impossible to obtain samples from where the infection was introduced.

Because it is independent of the availability of virions, initiating infections via icDNA in an *Agrobacterium*-hosted binary vector enables the study of non-replicating, movement-deficient, or otherwise compromised virus mutants that cannot be mechanically inoculated. Moreover, it does not require specialized equipment such as electroporators or particle bombardment instruments for biolistic transformation. It is currently the most efficient method for conducting controlled plant–virus infections [17]. Agroinfiltration allows the convenient control of the quantity of bacteria introduced to the plant by adjusting the infiltration parameters,

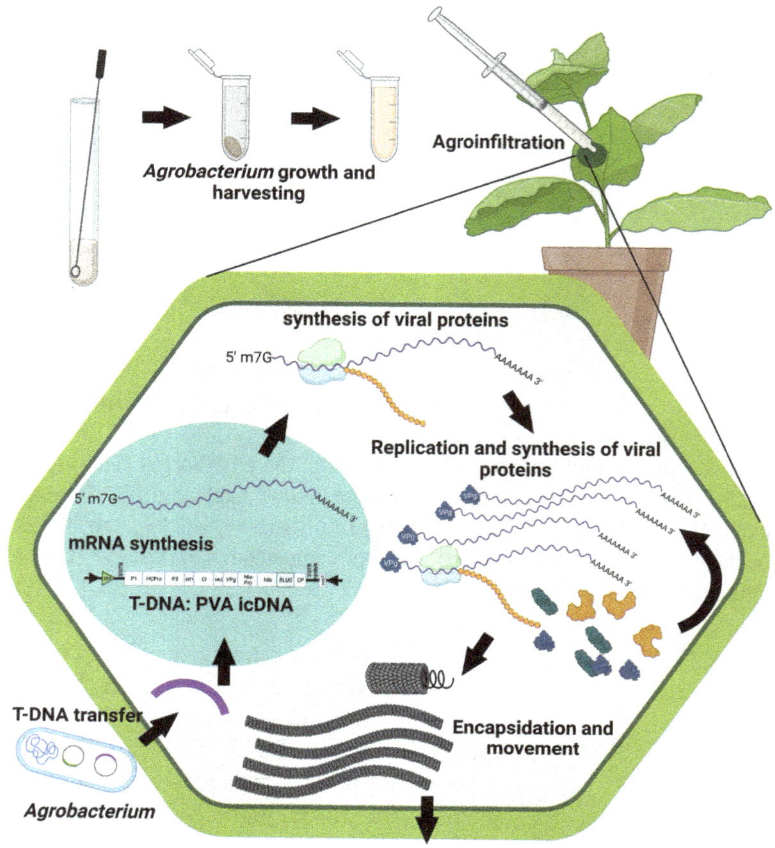

Fig. 2 Overview of the initiation of plant virus infections using agroinfiltration. *Agrobacterium* carrying the viral icDNA in the T-DNA region of the Ti plasmid is cultivated, harvested, and resuspended in induction buffer containing acetosyringone as the activator of *vir* gene expression. The *Agrobacterium* suspension is infiltrated into host plant leaves using a needleless syringe. Infiltrated *Agrobacterium* cells attach to the host cell surface and initiate the excision and transfer of the T-DNA segment. In the nucleus, the T-DNA including the viral icDNA expression construct, is transcribed into mRNA. Capped transcripts are exported into the cytoplasm for protein synthesis. Viral proteins, together with auxiliary host factors, engage in the replication of the viral genome. New genomes provide templates for further translation of virus proteins, engage in cell-to-cell movement, and at later stages are encapsidated into particles that carry the infection to upper leaves via the phloem. (Figure created in BioRender.com)

including the optical density of the suspension (optical density at 600 nm, OD_{600}). For experimental purposes, initiating infections at different saturation levels is often desirable. For example, a low percentage of initially infected cells is advantageous when investigating the short-distance movement of the virus to neighboring cells. On the other hand, the detection of non-replicating mutant viruses may require that a high ratio of the cells initially receives the viral icDNA.

The method consists of three parts: [1] the culture and preparation of *Agrobacteria*, [2] the infiltration procedure, and [3] recommendations for the optimization of agroinfiltration-mediated infection assays and sampling. In addition, it includes suggestions for the safe management of infected plant material and agrobacterium waste. First, *Agrobacterium* carrying the viral icDNA is inoculated and cultured in selective media to increase cell density. The cells are then harvested, washed, and resuspended at the desired OD_{600} in induction buffer containing a high acetosyringone concentration to activate the transcription of *vir* genes. After a short recovery period, the *Agrobacterium* suspension is infiltrated with a small syringe into the abaxial side of *N. benthamiana* leaves. The infiltrated plants are maintained in standard conditions until sampling at the chosen time, days post-infiltration (dpi), and location.

2 Materials

2.1 Agrobacterium Cultures and the Preparation of Infiltration Suspensions

1. *A. tumefaciens* strain carrying the viral icDNA of interest (*see* **Note 1**).
2. Selective Luria–Bertani (LB) plates supplemented with appropriate antibiotics (store at 4 °C for up to 2 weeks).
3. 0.5 M 2-(N-morpholino)ethanesulfonic acid (MES, CAS no. 4432-31-9) pH 6.5. Filter-streilize and store at 4 °C.
4. 150 mM acetosyringone (1-(4-hydroxy-3,5-dimethoxyphenyl) ethan-1, (CAS no. 2478-38-8). Dissolve in DMSO or ethanol, aliquot, and store at −20 °C (*see* **Note 2**).
5. Relevant antibiotic stock solutions, store at −20 °C. For the PVA construct used as an example in this chapter, prepare kanamycin (10 mg/mL) and carbenicillin (10 mg/mL).
6. MES-buffered LB medium supplemented with acetosyringone and antibiotics: 10 mM MES pH 6.5, 20 μM acetosyringone. Supplement with antibiotics as appropriate and store at 4 °C for up to 2 weeks. Warm to room temperature before use (*see* **Note 3**).
7. Tape or parafilm for sealing agar plates with *A. tumefaciens* cultures.
8. Induction buffer: 10 mM MES, pH 6.5, 10 mM $MgCl_2$, 150 μM acetosyringone. Prepare fresh (*see* **Note 4**).
9. 1 M $MgCl_2$, sterilize by autoclaving and store at room temperature.

10. Spectrophotometer or tabletop photometer capable of measuring absorbance at 600 nm and matching cuvettes.

2.2 Syringe Infiltration of N. benthamiana Leaves

1. Plants: *N. benthamiana* at the 4–6 leaf development stage (*see* **Note 5**).
2. Controlled and contained greenhouse or growth chamber environment for growing and maintaining experimental *N. benthamiana* plants (*see* **Note 6**).
3. Freshly prepared induction buffer: 10 mM MES, pH 6.5, 10 mM $MgCl_2$, and 150 μM acetosyringone (*see* **Note 7**).
4. *Agrobacterium* infiltration suspensions.
5. 1-mL disposable syringes.

2.3 Waste Management

1. Chlorine-based disinfectant, 10% working solution.

3 Methods

Perform all steps of the agroinfiltration protocol at room temperature. Use appropriate personal protective equipment: for example, when infiltrating plants, wear a laboratory coat, gloves, and safety goggles. When preparing plates and media, work in a biological safety cabinet to avoid contamination. Prior to starting, obtain relevant permissions for working with genetically modified (GM) viruses, *Agrobacterium* strains, and plants infected with modified viruses. Similarly, ensure that laboratories, greenhouses, and other working facilities' safety and containment levels comply with national and international legislation for GM organisms.

3.1 Agrobacterium Culture and Preparation of Agrobacterium Suspensions for Infiltration

1. Streak the *A. tumefaciens* strain carrying the viral icDNA construct and/or other constructs of interest on LB agar plates supplemented with appropriate antibiotics or refresh the culture by streaking from a glycerol stock stored at −70 °C.
2. Incubate the plates for 16–48 h at 28 °C (see **Note 8**).
3. Prepare primary cultures by inoculating a small amount of bacteria from the LB agar plate into 3–5 mL MES-buffered LB medium supplemented with acetosyringone and appropriate antibiotics. Incubate overnight at 28 °C with shaking at 200–220 rpm.
4. The next day, prepare secondary cultures by diluting the primary culture 1:10 in fresh MES-buffered LB medium with appropriate antibiotics. For example, add 500 μL of the overnight primary culture to 4.5 mL of fresh medium. The volumes

can be scaled down or up according to the final target volume and OD_{600} (*see* **Note 9**).

5. Incubate at 28 °C with shaking (200–220 rpm) for at least 3 h (see **Note 10**).
6. Harvest the bacteria by centrifugation at room temperature for 5 min at 5000×g. Discard the supernatant.
7. To remove traces of culture medium and antibiotics, wash the cells twice with distilled water. To wash, resuspend the pellet by gentle pipetting in 1 mL of distilled water, centrifuge for 5 min at 5000×g, and discard the supernatant.
8. After the final washing step, resuspend the cells into 1 mL of freshly prepared induction buffer.
9. Measure the OD_{600} of the suspension with a spectrophotometer or a tabletop photometer.
10. Dilute the suspension to the desired optical density (OD_{600}) and volume with the induction buffer (*see* **Note 11**).
11. Let the cells recover at room temperature for 0.5–2 h before infiltration (*see* **Note 12**).

3.2 Agroinfiltration of N. benthamiana Leaves

1. Perform the infiltration into one or two fully developed *N. benthamiana* leaves with a 1-mL syringe. Mix the bacterial suspensions by gently inverting the tube and fill the syringe with the suspension. Remove any bubbles that remain in the syringe.
2. Press the syringe on the abaxial side of the leaf and simultaneously exert gentle counter-pressure with your index finger on the other side. Steadily inject the desired volume of bacterial suspension into the leaf.
3. Successful infiltration can be observed by the spreading of the infiltration mixture in the leaf. If a larger agroinfiltrated area is required, repeat the infiltration at a new location until the target leaf area is soaked with the bacterial suspension.
4. Infiltrate at least three plants per treatment or experimental condition.
5. In each experiment, include an appropriate mock control treatment by, for example, infiltrating selected plants with the induction buffer only. Take care to avoid contamination of the healthy control plants with *Agrobacteria* carrying the viral icDNA construct.
6. After agroinfiltration, water and maintain the treated plants in standard growth conditions (*see* **Notes 6** and **13**).

3.3 Optimization of Agroinfiltration-Based Virus Infection Assays

3.3.1 Optimization of the Concentration of the Agrobacterium Suspension

1. Perform trials with different OD_{600} values of the *Agrobacterium* suspension to optimize the concentration according to the needs of your experiment and the viral icDNA in use.
2. Begin the optimization with agroinfiltration using a tenfold dilution series ranging from OD_{600} 1 to 0.01 (*see* **Notes 11** and **14**).
3. Use replication-deficient or movement-deficient viruses as reference constructs to assess the effects of new mutations.

3.3.2 Optimization of Sample Collection in Plants Infected by the Agroinfiltration Method

1. To optimize the sampling time for locally infected leaves, obtain samples on consecutive days from agroinfiltrated leaves. A time-course analysis will reveal the progress and peak of the infection in the local leaves. For a shorter range, you may consider obtaining samples already at consecutive hours post-infiltration.
2. To study the local infection in the agroinfiltrated leaves, sample the agroinfiltrated area around the infiltration point at, for example, 1–7 dpi.
3. Avoid the syringe scar and, to reduce variance, obtain samples from an equal distance from the application site (*see* **Note 15**).
4. To study the spread of the infection in the plant and to find the optimal sampling time for the systemic infection, obtain samples from upper leaves on consecutive days (*see* **Note 16**).
5. Obtain samples from near the base of young upper leaves at, for example, 4–10 dpi (*see* **Note 17**).

3.3.3 GM Waste Management

1. Even contained work with GM organisms requires permission. Work in compliance with the appropriate safety regulations for the use of GMs.
2. Sterilize agroinfiltrated and subsequent systemically infected plant material by autoclaving.
3. Sterilize reusable plastics and glassware by autoclaving or disinfecting with a chlorine-based solution by submerging and soaking the equipment for 24 h in a 10% solution (*see* **Note 18**).

4 Notes

1. The protocol has been optimized for the PVA-*N. benthamiana* pathosystem, but it can be readily applied and adapted with case-specific modifications or preferences for use in other positive-sense RNA virus–host combinations. A range of *A. tumefaciens* strains with modified characteristics, for example, improved transformation efficiency in recalcitrant hosts,

reduced recombination rate, etc., are also available [18]. Here, *A. tumefaciens* strain C58C1 ((pGV2260 (pTiB6S3ΔT-DNA)) is used as the host for the binary, disarmed Ti plasmid pBIN19 carrying the full-length PVA icDNA (Gene Bank accession no. AJ296311) tagged with a *Renilla* luciferase reporter (Fig. 3) [19, 20]. pBIN19 confers kanamycin resistance (concentration for selection 50 μg/mL), and the host strain carries a helper plasmid, which requires carbenicillin (100 μg/mL) for selection [21].

2. Precipitate may form in the acetosyringone stock during storage at −20 °C. Vigorous vortexing combined with warming to 30 °C may help dissolve the precipitates. Ensure that all precipitates have dissolved prior to using the stock solution. DMSO is harmful; wear appropriate protective equipment when handling the chemical.
3. Prepare LB medium (tryptone 10 g/L, NaCl 5 g/L, yeast extract 5 g/L, adjust pH to 7.0) and sterilize by autoclaving. Cool the medium to room temperature. To prepare 100 mL of buffered LB medium for *Agrobacterium* cultivation, add 2 mL of 0.5 M MES buffer pH 6.5, 13 μL of 150 mM AS, 500 μL of kanamycin (stock 10 mg/mL), and 1 mL of carbenicillin (stock 10 mg/mL) to 90 mL of LB medium. Fill to 100 mL with LB and use immediately or store at 4 °C.
4. To prepare 100 mL of the induction buffer, mix 2 mL 0.5 M MES buffer pH 6.5, 100 μL 150 mM acetosyringone, and 1 mL of 1 M $MgCl_2$ and fill with distilled water up to 100 mL. Use immediately, and do not store.
5. The 4–6 leaf development stage is recommended for the infiltration of virus constructs. In standard growth conditions, the seedlings reach this stage approximately 4–5 weeks after sowing. The expression of both a PVA icDNA construct and a standard protein expression construct, 35S-firefly luciferase, was less efficient when larger leaves in older plants were infiltrated [20]. For a single experiment, use plants of the same age and select plants as uniform in size as possible. Do not let the seedlings dry at any point during growth, as drying may render the leaves difficult to infiltrate. Wilted plants should not be infiltrated at all.
6. Standard growth conditions for *N. benthamiana* in a controlled greenhouse environment: 16-h light/ 8-h dark photoperiod at 22/20 °C respectively, 60% relative humidity.
7. The pH of the MES buffer varies between laboratories using agroinfiltration techniques, and it seems that a slightly acidic range works well. Preferably choose a pH between 5.6 and 6.5. Agroinfiltration tests in the laboratory of Teemu Teeri (Department of Agricultural Sciences, University of Helsinki) showed

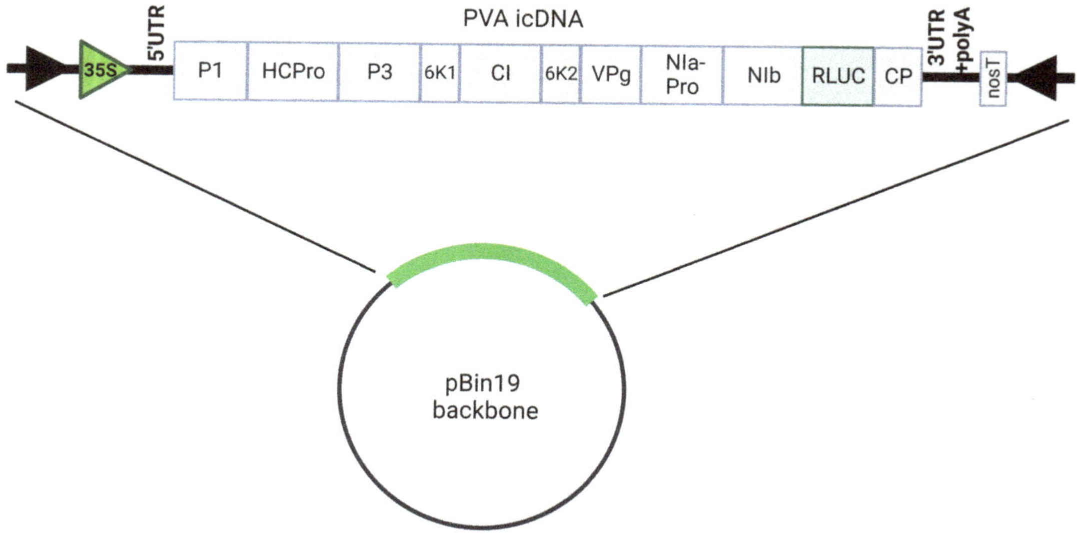

Fig. 3 Schematic presentation of the PVA icDNA construct used as the example in this chapter. The full-length PVA icDNA is in the pBin19 binary vector under the control of the 35S promoter and the *nos* terminator. The icDNA includes the *Renilla* luciferase reporter gene (RLUC) between the RNA-dependent RNA polymerase (NIb) and coat protein (CP) cistrons. Black arrows on the construct denote the T-DNA left and right borders. (Figure created in BioRender.com)

there was no difference in the transient expression of a *Petunia hybrida* dihydroflavonol 4-reductase using a pH 5.6 or 6.0 induction buffer (T. Teeri, University of Helsinki Department of Agricultural Sciences, personal communication, February 14, 2022). This gives some indication of the flexible range of the pH of the induction buffer.

8. The plates may be sealed with tape or parafilm and stored at 4 °C. *A. tumefaciens* stays fresh on appropriately stored plates for circa 1 week, after which new plates should be streaked.
9. The next day, check the growth of the primary culture by visually assessing the cloudiness of the culture as compared to the non-inoculated culture medium. Optional: measure the OD_{600} of the primary overnight culture to verify sufficient cell density.
10. In LB medium at 28 °C, *A. tumefaciens* cells divide at approximately 90 min intervals. Grow the *Agrobacteria* for at least two cell cycles in the secondary culture but do not allow the cells to reach the stationary phase. The best results are obtained using *Agrobacterium* in the logarithmic growth phase, which in the strain C58C1 corresponds to OD_{600} 0.2–1.4 [20].
11. The optimal OD_{600} varies between constructs, viruses, and hosts. Thus, infiltration parameters should be experimentally verified for each virus construct. For example, a wild-type PVA

icDNA construct with the *Renilla* luciferase reporter is often infiltrated in *N. benthamiana* at a low OD_{600} (0.05), while the replication-deficient mutant $PVA^{\Delta GDD}$ requires a fivefold higher OD_{600} (0.2) [20]. All gene expression from the non-replicating virus is derived from the primary mRNA transcript generated from the T-DNA in the nucleus, and to achieve a detectable level of the virus it is, therefore, necessary to saturate the host tissue with the icDNA construct. To illustrate the effect of OD_{600} on the density of initially infected cells, *N. benthamiana* leaves were infiltrated at decreasing OD_{600} 0.1, 0.01, and 0.001 with *Agrobacterium* carrying a binary Ti plasmid with non-replicating PVA, which does not move from cell to cell. Visualization of virus-derived GFP at 3 dpi showed that with the highest OD_{600} (0.1), almost all cells expressed GFP and therefore contained the viral icDNA (Fig. 4). Agroinfiltration with a tenfold dilution, 0.01, resulted in a marked reduction in the ratio of infected vs. non-infected cells and with the most dilute suspension (0.001), only scattered individual cells were observed to express virus-derived GFP.

12. Although the recovery period is flexible, shorter times may be beneficial and result in more robust expression levels. Compared to a 3-h recovery, a 0.5-h recovery resulted in improved expression of a *Petunia hybrida* dihydroflavonol 4-reductase (T. Teeri, University of Helsinki Department of Agricultural Sciences, personal communication, February 14, 2022). The effect of the recovery period on the initiation of virus infections has not been tested, but we recommend using a consistent recovery time between experiments.

13. Optionally, the infiltrated plants may be covered overnight with a light cloth or plastic to reduce transpiration. In the case of mild or asymptomatic viruses, the agroinfiltrated areas will recover quickly, although a small scar will form at the site of the syringe application. Depending on the virus, strain, and robustness of the infection, over time, more severe symptoms and even necrosis may occur in the agroinfiltrated area.

14. The timing and location of sampling should be optimized independently for each virus construct and host plant combination. In addition, consider the sensitivity of detection methods and the possible differences in the turnover rate of reporters and viral proteins.

15. Due to the dilution of the *Agrobacterium* solution during spread into the leaf tissue, expression levels decrease with increasing distance from the infiltration site [20]. Therefore, it is recommended to obtain samples at a uniform distance and from as close to the infiltration site as possible. In addition, use

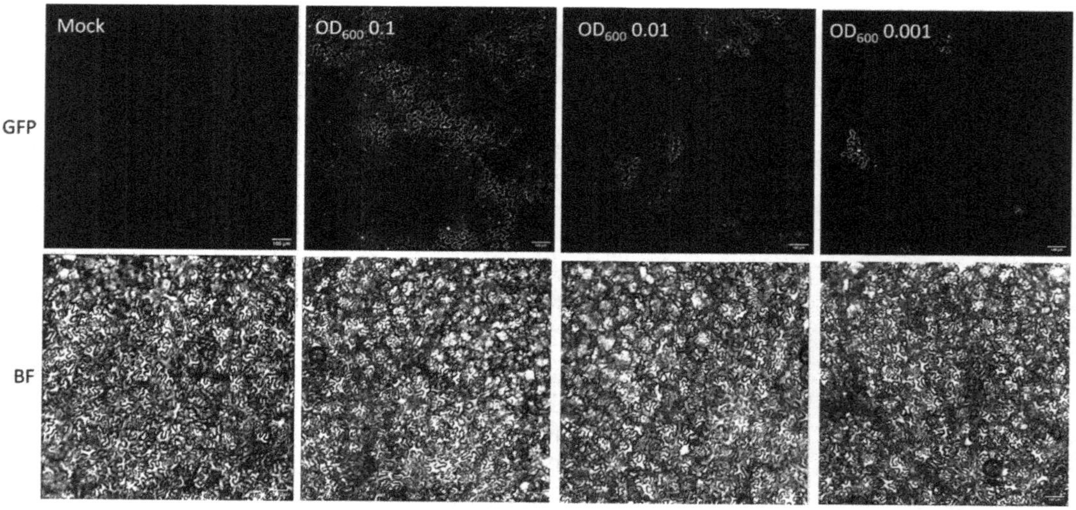

Fig. 4 Effect of the OD_{600} of the *Agrobacterium* suspension on the density of initially infected epidermal cells in *N. benthamiana* at 3 dpi. Leaves were infiltrated with *Agrobacterium* hosting non-replicating potato virus A tagged with a GFP reporter at OD_{600} 0.1, 0.01, and 0.001. GFP fluorescence (top row) was visualized from leaf disks at 3 dpi with epifluorescence microscopy using a FITC filter, and brightfield images (BF, bottom row) were obtained with a VIS filter. Mock control plants were infiltrated with buffer. The scale bars represent 100 μm

suitable analysis methods to distinguish between active viral gene expression and virus titers (i.e., the accumulation of virus particles). For example, the *Renilla* luciferase reporter signal, indicating viral gene expression, is detectable at 24 h after infiltration, but an ELISA test detecting the accumulation of PVA coat protein was first positive at 52 h post-infiltration [20]. For quantitative real-time PCR analysis of viral transcripts in the agroinfiltrated leaves, be aware that there may still be agrobacteria present that carry the Ti plasmid copies of the viral icDNA. Any such icDNA copies will affect the outcome of the analysis and may do so significantly, especially at early sampling times.

16. The timing of the systemic spread may vary from less than a week to several weeks, depending on the virus, *Agrobacterium* strain, and host. It should therefore be experimentally verified on a case-by-case basis.

17. Long-distance movement of plant viruses occurs primarily via the phloem, where virus particles are passively transported among the flow of photoassimilates. The direction of movement is from source to sink tissues. The systemic infection is often initiated near the base of the upper leaves, possibly because that is where viruses first exit the phloem.

18. Ensure the disinfectant has been tested and approved to eliminate bacteria and viruses. For example, within the EU select a virucide in compliance with the European standard EN 14476.

References

1. Smith EF, Townsend CO (1907) A plant-tumor of bacterial origin. Science 25(643): 671–673. https://doi.org/10.1126/science.25.643.671
2. Conn HJ (1942) Validity of the genus Alcaligenes. J Bacteriol 44(3):353–360
3. Skerman VBD, McGowan V, Sneath PHA (eds) (1989) Approved lists of bacterial names (amended). ASM Press, Washington DC
4. Flores-Félix JD, Menéndez E, Peix A, García-Fraile P, Velázquez E (2020) History and current taxonomic status of genus agrobacterium. Syst Appl Microbiol 43(1):126046. https://doi.org/10.1016/j.syapm.2019.126046
5. Mousavi SA, Young JPW (2022) International Committee on Systematics of Prokaryotes, Subcommittee on the taxonomy of rhizobia and agrobacteria, minutes of the annual meeting by videoconference, 5 July 2021, followed by online discussion until 31 December 2021. Int J Syst Evol Microbiol 72(6). https://doi.org/10.1099/ijsem.0.005453
6. Singh NK, Lavire C, Nesme J, Vial L, Nesme X, Mason CE, Lassalle F, Venkateswaran K (2021) Comparative genomics of novel agrobacterium G3 strains isolated from the international space station and description of agrobacterium Tomkonis Sp. Nov. Front Microbiol 12:765943. https://doi.org/10.3389/fmicb.2021.765943
7. Cho H, Winans SC (2005) VirA and VirG activate the Ti plasmid RepABC operon, elevating plasmid copy number in response to wound-released chemical signals. Proc Natl Acad Sci 102(41):14843–14848. https://doi.org/10.1073/pnas.0503458102
8. Szatmári Á, Móricz ÁM, Schwarczinger I, Kolozsváriné Nagy J, Alberti Á, Pogány M, Bozsó Z (2021) A pattern-triggered immunity-related phenolic, acetosyringone, boosts rapid inhibition of a diverse set of plant pathogenic bacteria. BMC Plant Biol 21:153. https://doi.org/10.1186/s12870-021-02928-4
9. Gelvin SB (2017) Integration of agrobacterium T-DNA into the plant genome. Annu Rev Genet 51:195–217. https://doi.org/10.1146/annurev-genet-120215-035320
10. Hwang H-H, Yu M, Lai E-M (2017) Agrobacterium-mediated plant transformation: biology and applications. Arab Book 2017(15). https://doi.org/10.1199/tab.0186
11. Shaw CH, Leemans J, Shaw CH, van Montagu M, Schell J (1983) A general method for the transfer of cloned genes to plant cells. Gene 23(3):315–330. https://doi.org/10.1016/0378-1119(83)90021-5
12. Pasternak JJ, Gruber MY, Thompson JE, Glick BR (1983) Development of DNA-mediated transformation systems for plants. Biotechnol Adv 1(1):1–15. https://doi.org/10.1016/0734-9750(83)90297-5
13. Hoekema A, van Haaren MJ, Fellinger AJ, Hooykaas PJ, Schilperoort RA (1985) Non-oncogenic plant vectors for use in the agrobacterium binary system. Plant Mol Biol 5(2):85–89. https://doi.org/10.1007/BF00020090
14. Inoue-Nagata AK, Jordan R, Kreuze J, Li F, López-Moya JJ, Mäkinen K, Ohshima K, Wylie SJ, ICTV Report ConsortiumYR (2022) ICTV virus taxonomy profile: potyviridae 2022. J Gen Virol 103(5):001738. https://doi.org/10.1099/jgv.0.001738
15. Grimsley N, Hohn B, Hohn T, Walden R (1986) "Agroinfection," an alternative route for viral infection of plants by using the Ti plasmid. Proc Natl Acad Sci U S A 83(10): 3282–3286. https://doi.org/10.1073/pnas.83.10.3282
16. Yamaya J, Yoshioka M, Meshi T, Okada Y, Ohno T (1988) Expression of tobacco mosaic virus RNA in transgenic plants. Mol Gen Genet MGG 211(3):520–525. https://doi.org/10.1007/BF00425710
17. Peyret H, Lomonossoff GP (2015) When plant virology met agrobacterium: the rise of the deconstructed clones. Plant Biotechnol J 13(8):1121–1135. https://doi.org/10.1111/pbi.12412
18. De Saeger J, Park J, Chung HS, Hernalsteens J-P, Van Lijsebettens M, Inzé D, Van Montagu M, Depuydt S (2021) Agrobacterium strains and strain improvement: present and outlook. Biotechnol Adv 53:107677. https://doi.org/10.1016/j.biotechadv.2020.107677
19. Puurand Ü, Valkonen JPT, Mäkinen K, Rabenstein F, Saarma M (1996) Infectious

in vitro transcripts from cloned CDNA of the potato a potyvirus. Virus Res 40(2):135–140. https://doi.org/10.1016/0168-1702(95)01263-X

20. Eskelin K, Suntio T, Hyvärinen S, Hafren A, Mäkinen K (2010) Renilla luciferase-based quantitation of potato virus A infection initiated with agrobacterium infiltration of *N. Benthamiana* leaves. J Virol Methods 164(1–2):101–110. https://doi.org/10.1016/j.jviromet.2009.12.006

21. Deblaere R, Bytebier B, De Greve H, Deboeck F, Schell J, Van Montagu M, Leemans J (1985) Efficient octopine Ti plasmid-derived vectors for agrobacterium-mediated gene transfer to plants. Nucleic Acids Res 13(13):4777–4788. https://doi.org/10.1093/nar/13.13.4777

Chapter 4

Construction of Geminivirus Infectious Clones for Agroinoculation into Plants

Ivair J. Morais, Alice K. Inoue-Nagata, and Erich Y. T. Nakasu

Abstract

The production of geminiviral infectious clones provides a standardized inoculum for use in several host–virus studies. Geminiviruses present either one (monopartite) or two (bipartite) circular single-stranded DNA components, which commonly range from 2.6 to 2.8 kb. Cloning of a monomeric genome is useful for obtaining its precise sequence. For infectious clones, however, it is essential that more than one copy of the genome, more specifically of the origin of replication, is present in order to guarantee the production of full-length progeny DNA. Here, the complete process of preparing infectious geminiviral clones is described starting from the DNA extraction and selection of restriction endonucleases followed by two protocols for constructing dimeric clones: restriction endonuclease digestion and ligation (1) and Gibson Assembly (2).

Key words Rolling circle amplification, Geminivirus, Infectious clone, Circular ssDNA virus, Gibson assembly

1 Introduction

Geminiviruses comprise a significant group of viruses that infect plants. Their genomes consist of either one or two single-stranded circular DNA molecules, each ranging from 2.6 to 2.8 kilobases in size. Most geminivirus genomes have a 9-nucleotide motif (TAA-TATT/AC), known as the origin of replication, in which the viral protein Rep cleaves the DNA during replication. The production of geminivirus infectious clones involves the dimeric cloning of each genomic component (implying two copies of the origin of replication) and insertion into a binary vector for agroinoculation into plants using *Agrobacterium tumefaciens*.

Two protocols are presented here for the construction of infectious clones of geminiviruses. The first (*see* Subheading 3.6) consists of ligating dimeric genomic DNA to a binary vector, as described in Ferreira et al. (2008) [1]. These dimeric molecules

Elizabeth P. B. Fontes and Kristiina Mäkinen (eds.), *Plant-Virus Interactions*, Methods in Molecular Biology, vol. 2724, https://doi.org/10.1007/978-1-0716-3485-1_4,

are produced by partial digestion of the rolling circle amplification (RCA) products with a single-cutting restriction endonuclease (RE) and isolation of the dimer DNA fragments from agarose gels. Prior to this step, a pilot test is needed to define the appropriate RE and adjust the digestion protocol to enrich the preparation with partially digested dimeric units.

The second protocol (*see* Subheading 3.7) is based on Gibson Assembly [2], which was optimized for generating geminiviral infectious clones [3, 4]. A sequence of the genome is used to design two sets of primers. These primers are used in high-fidelity PCR protocols to generate two full-size genome amplicons (termed U1 and U2) with complementary ends. The 5′ and 3′ ends of the U1 amplicon will overlap at least 15 nucleotides between the vector and the 5′ end of U2, respectively, and the U2 copy will overlap the 5′ part of the vector at its 3′ end. The three fragments (U1, U2, and vector) are ligated with each other by a sequence of steps involving reactions of exonuclease, polymerase, and ligase before insertion into bacterial cells.

The first step for both protocols consists of total DNA extraction from an infected plant. We present a protocol based on the different solubility of polysaccharides and DNA developed by Doyle and Doyle (1987) [5]. Then, RCA is employed to generate several multimers of circular DNAs in the sample—in this case, geminiviral genomes—even from very low amounts of template [6]. Due to its strand displacement nature during polymerization [7], single-stranded DNAs are readily available as templates.

Once amplified, a preliminary analysis of the viral genome is carried out using restriction fragment length polymorphism (RFLP) to determine whether the sample contains a monopartite, bipartite, or even more than one geminivirus infecting the plant. Then, a single-cutter RE is identified for cloning if based on digestion and ligation protocol; for bipartite geminiviruses, a Southern blot is used to determine the RE specific for each component. For the construction of geminivirus infectious clones using Gibson assembly, however, the sequence of at least part of the genome is required for primer design. For this purpose, two sets of universal begomovirus primers (DNA-A and DNA-B), designed by Rojas et al. (1993) [8], are convenient for PCR amplification and sequencing.

2 Materials

Prepare all solutions using ultrapure water (e.g., Milli-Q purified water) and analytical-grade reagents. Use DNAse-free consumables. Store all reagents at −20 °C (unless otherwise indicated). Before use, thaw all reagents on ice completely. Diligently follow all applicable waste disposal regulations when disposing of waste materials.

2.1 Total DNA Extraction from Plant Tissue

1. 100 mg fresh or frozen geminivirus-infected plant tissue.
2. CTAB isolation buffer: 100 mM Tris–HCl (pH 8.0), 1.4 M NaCl, 20 mM EDTA, 2% CTAB, and 0.2% 2-mercaptoethanol (*see* **Note 1**).
3. Chloroform: isoamyl alcohol 24:1.
4. Pestle and mortar (*see* **Note 2**).
5. Isopropanol.
6. 70% ethanol in water.
7. 10 mM Tris–HCl, pH 8 adjusted with HCl.

2.2 Rolling Circle Amplification (RCA)

1. 10× reaction buffer (330 mM Tris-acetate (pH 7.9 at 37 °C), 100 mM Mg-acetate, 660 mM K-acetate, 1% (v/v) Tween-20, 10 mM DTT).
2. dNTPs (10 mM each).
3. n-hexamer primers (10 μM) (*see* **Note 3**).
4. phi29 DNA polymerase (10 units/μL).
5. Purified total DNA (50–100 ng/μL).

2.3 Detection of Geminiviral DNA and Partial Genome Sequencing

1. 10× RE buffer.
2. Frequently cutter RE (e.g., MspI).
3. RCA product.
4. 5× TBE buffer: 0.45 M Tris base, 0.45 M boric acid, 0.01 M EDTA (pH 8.0) (*see* **Note 4**).
5. 0.8% Agarose gel (Molecular Biology grade) in 0.5× TBE buffer.
6. Primers PAR1c496 (AATACTGCAGGGCTTYCTRTACATRGG) and PAL1v1978 (GCATCTGCAGGCCCACATYGTCTTYCC NGT) for DNA A and primers PCRc2 (CTAGCTGCAGC ATATTTACRARWATGCC) and PBLv2040 (GCCTCTGCA GCARTGRTCKATCTTCATACA) for DNA- B [8].

2.4 Digestion with RE for Identification of Single-Cutting RE and Southern Blotting for Bipartite Begomoviruses

1. 10× RE buffer.
2. RE (*see* **Note 5**).
3. 1% agarose gel (Molecular Biology grade) in 0.5× TBE buffer.
4. 5× TBE buffer.
5. Filter paper.
6. Denaturation solution: 0.5 M NaOH, 1.5 M NaCl.
7. Neutralization solution: 0.5 M Tris–HCl, 1.5 M NaCl, pH 7.2 adjusted with HCl.
8. 20× SSC: 3 M NaCl, 0.3 M sodium citrate, pH 7.0 adjusted with NaOH.

9. Nitrocellulose or nylon membrane.
10. PCR Master Mix (10 μL 10× PCR buffer, 8 μL 50 mM $MgCl_2$, 4 μL 2.5 mM dNTP, 1 μL 10 μM of each primer, and 1 μL *Taq* DNA polymerase (5 U/ μL), complete with water to 100 μL).
11. Hybridization solution (e.g., ULTRAhyb Ultrasensitive Hybridization Buffer from ThermoFisher Scientific).
12. Probe: known geminivirus DNA template, PCR primers, PCR reagents (10× PCR buffer, 25 mM $MgCl_2$, *Taq* DNA polymerase, water, Digoxigenin-11-dUTP, alkali-labile (Roche)).
13. Washing solution 1: 2× SSC, 0.5% SDS.
14. Washing solution 2: 0.5× SSC, 0.5% SDS.
15. 10× Tris-buffered saline—TBS: 0.2 M Tris–HCl, 1.5 M NaCl, pH 7.6 adjusted with HCl.
16. TBS-T: TBS with 0.05% Tween-20.
17. Blocking solution: 0.5× TBS, 2% low-fat skimmed milk.
18. Anti-digoxigenin-AP antibody, Fab fragments (Roche).
19. Developing buffer: 100 mM Tris–HCl, 100 mM NaCl, 5 mM $MgCl_2$, 0.05% Tween 20, pH 9.5.
20. 5-Bromo-4-chloro-3-indolyl phosphate (BCIP) solution: 50 mg/mL in 100% dimethylformamide.
21. Nitro blue tetrazolium (NBT) solution: 50 mg/mL in 70% dimethylformamide.

2.5 Cloning a Dimeric (Infectious) Clone in a Binary Vector

2.5.1 Preparation of Dimeric Viral-Derived DNA

1. RE buffer.
2. Single-cutting RE (enzyme selected in Subheading 3.5).
3. RCA product of a geminivirus.
4. 3 M sodium acetate, pH 5.2 adjusted with acetic acid.
5. Ethanol.
6. 10 mM Tris–HCl, pH 8.0 adjusted with HCl.
7. 1% agarose gel (Molecular Biology grade) in 0.5× TBE buffer.
8. 5× TBE buffer.
9. DNA purification kit from agarose gels.

2.5.2 Vector Preparation

1. RE buffer.
2. Binary vector (e.g., pCAMBIA 0380) (*see* **Note 6**).
3. RE (the same as in 2.5.1).
4. Calf intestinal alkaline phosphatase (CIAP or CIP) or shrimp intestinal alkaline phosphatase.
5. Phenol/chloroform/isoamyl alcohol (25:24:1).
6. 3 M sodium acetate buffer, pH 5.2 adjusted with acetic acid.

7. Ethanol.
8. 10 mM Tris–HCl, pH 8.0 adjusted with HCl.
9. DNA purification kit from agarose gels.

2.5.3 Ligation of Dimer and Plasmid

1. Dimeric genomic DNA fragments (*see* Subheading 2.5.1).
2. Linearized plasmid vector (*see* Subheading 2.5.2).
3. 10× T4 DNA ligase buffer.
4. T4 DNA ligase.

2.5.4 Transformation and Selection of Bacteria with Recombinant Plasmids

1. Ligation reaction.
2. Filter membrane, pore size 0.22 μm for dialysis (e.g., nitrocellulose Millipore GSWP29325).
3. Electrocompetent *Escherichia coli* cells (*see* **Note 7**).
4. SOC medium (*see* **Note 8**).
5. Luria–Bertani (LB) medium: 10 g tryptone, 10 g NaCl, 5 g yeast extract, pH 7.0 adjusted with NaOH, and volume for 1 L; sterilize by autoclaving.
6. Petri dish with LB medium and kanamycin (50 μg/mL).
7. Alkaline lysis solution I: 50 mM glucose, 25 mM Tris–HCl (pH 8), 10 mM EDTA.
8. Alkaline lysis solution II: 0.2 N NaOH, 1% SDS (freshly prepared).
9. Alkaline lysis solution III: 60 mL 5 M potassium acetate and 11.5 mL glacial acetic acid and make up the final volume of 100 mL with water. Autoclave and store it at 4 °C.
10. Isopropanol.
11. 70% ethanol in water.
12. 10 mM Tris–HCl, pH 8.0 adjusted with HCl.
13. Selected RE buffers.
14. Selected REs.
15. Plasmid DNA purification kit.
16. Phenol–chloroform–isoamyl alcohol (25:24:1) solution.

2.6 Cloning a Dimeric Genome by Gibson Assembly

1. Binary vector (e.g., pJL 89 vector).
2. Total DNA containing the target geminivirus genome (50–100 ng/μL).
3. 2 pairs of abutting primers (U1-F/U1-R and U2-F/U2-R) for the geminivirus insert (*see* **Note 9**).
4. 1 pair of abutting primers (pJL 89-F and pJL 89-R) for the vector (*see* **Note 10**).

5. High-fidelity DNA polymerase.
6. dNTPs (10 mM each).
7. Staining dye (e.g., crystal violet) (*see* **Note 11**).
8. Agarose gel DNA extraction kit.
9. DpnI buffer and enzyme.
10. Gibson Assembly Master Mix.
11. Electrocompetent *E. coli* strain DH10β.

3 Methods

Carry out all procedures at room temperature unless otherwise specified.

3.1 Extraction of Total DNA from an Infected Plant

Steps 2 to 9 must be carried out in a fume cupboard.

1. Place ~100 mg of plant tissue into an extraction tube.
2. Add 750 μL CTAB isolation buffer (*see* **Note 12**).
3. Macerate the sample with a pestle or in an automated extractor.
4. Incubate samples at 65 °C for 15 min up to 1 h.
5. Add 750 μL chloroform:isoamyl alcohol 24:1 and vortex for ~1 min.
6. Centrifuge at 10,000 rpm in a microcentrifuge for 10 min.
7. Collect the aqueous phase (~600 μL) and transfer it to a new tube.
8. Add 0.6 to 1 volume of isopropanol and incubate for 5–10 min at RT.
9. Centrifuge at 13,000 rpm in a microcentrifuge for 10 min and discard the supernatant (a pellet can be seen at the bottom of the tube) (*see* **Note 13**).
10. Add 1 mL of 70% cold ethanol, invert the tube, and centrifuge at 13,000 rpm for 5 min.
11. Discard the supernatant, remove any trace of ethanol with a micropipette, and bench-dry the pellet (*see* **Note 14**).
12. Resuspend the pellet using water or 10 mM Tris–HCl, pH 8 (usually 200 to 300 μL).
13. Store at −20 °C in aliquots (*see* **Note 15**).

3.2 Amplification of Circular DNA by RCA

1. Add reagents in the following order in a 200-μL tube: 2.35 μL water, 1 μL 10× reaction buffer, 1 μL dNTPs (10 mM each), 5 μL n-hexamer primers (10 μM), 0.15 μL phi29 DNA polymerase (10 U/μL), and 0.5 μL total DNA (50–100 ng/μL).

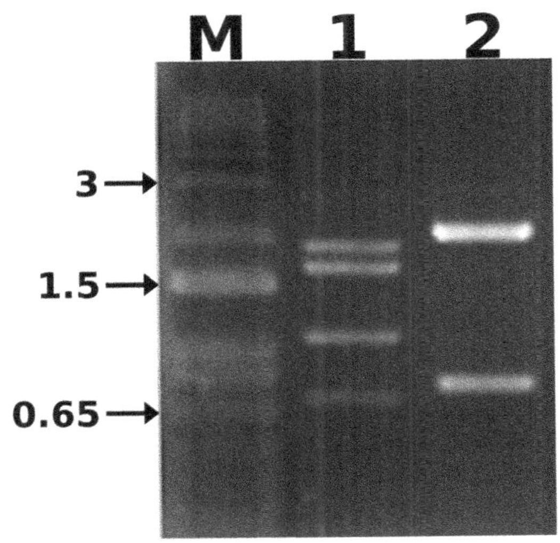

Fig. 1 Agarose gel electrophoresis of RCA products digested with MspI restriction endonuclease. A bipartite begomovirus genome was amplified by RCA, resulting in the digestion of DNA-A and DNA-B fragments totaling 5.2 kb (1.8 kb + 1.6 kb + 1.0 kb + 0.7 kb + 0.1 kb) in sample 1; a monopartite begomovirus in sample 2 with DNA fragments of 2.0 kb + 0.7 kb (total of 2.7 kb). M. 1 kb plus DNA size marker (ThermoFisher Scientific). Numbers on the left side represent fragment DNA sizes in kb

2. Incubate reaction at 30 °C for 4–48 h (*see* **Note 16**).
3. Inactivate the enzyme at 65 °C for 15 min.
4. Store at 4 °C for short-term storage or < −20 °C for a longer period.

3.3 Analysis of the DNA Profile After Digestion with a Frequent-Cutter RE

1. RCA products are visualized in agarose gels by loading 1 μL in a 0.8% agarose gel in TBE buffer (*see* **Note 17**).
2. On the same gel, run digested RCA products: Digest 1 μL of RCA product in a reaction with 1 μL 10× reaction buffer, 1–5 U of MspI RE, and adjust the volume to 10 μL. Incubate for >1 h. Load the reaction in the agarose gel.
3. Visualize the gel under UV light using the preferred DNA staining dye (*see* **Note 18**). An example is given in Fig. 1.

3.4 Partial Sequencing Using Universal Begomovirus Primers

1. Dilute the RCA products to 100 ng/μL (usually 1–4 μL RCA/ 10 μL) and Sanger sequence using primers PAR1c496 and PAL1v1978 for DNA-A and primers PCRc2 and PBLv2040 for DNA-B (*see* **Note 19**).
2. Analyze by BLAST (https://blast.ncbi.nlm.nih.gov/Blast.cgi) [9].
3. If appropriate, use the partial sequence for designing primers for Gibson assembly cloning (*see* Subheading 3.7).

3.5 Identification of Single-Cutting RE

3.5.1 RCA Digestion and Agarose Gel Electrophoresis Separation

1. Select the binary vector for cloning (e.g., pCAMBIA 0380) and several REs in the vector's multiple cloning site (MCS).
2. Digest 1 μL of RCA product with each RE separately.
3. Load the digestion reaction on a 0.8–1.0% agarose gel.
4. Stain the DNA with the preferred DNA stain.
5. Identify those REs that produced genome size DNA fragments of 2.6–2.8 kb. (*see* **Note 20**).
6. Select the preferred RE for cloning using the method of RE digestion and ligation (*see* Subheading 3.6).
7. If the geminivirus is a bipartite begomovirus, proceed to Subheadings 3.5.2, 3.5.3, and 3.5.4 for Southern blotting.

3.5.2 DNA Transfer and Membrane Fixation for Bipartite Begomoviruses

1. After choosing the REs to identify those present in DNA-A or DNA-B, run the digested RCA with the REs in a 0.9–1% agarose gel. Prepare two identical gels, one for the DNA-A probe and another for the DNA-B probe.
2. Stain the DNA and observe if the digestions were complete for each enzyme.
3. Remove parts of the gel that are not necessary. Soak the gel in denaturation solution, enough to cover the entire gel in a tray, and incubate at RT under slow agitation on an orbital shaker for 20 min.
4. Remove the denaturation solution and repeat the process once.
5. Remove the denaturation solution and cover the gel with the neutralization solution. Keep the tray at RT for 15 min on an orbital shaker under slow agitation. Repeat the process three times.
6. For the transference of the DNA from the gel to the membrane, half-fill a tray with 20× SSC. Make a platform in the solution using a glass or plastic plate/tray. Soak one layer of filter paper in 20× SSC and lay it over the plate, dipping the edges into the SSC solution. Remove any air bubbles between the filter and the plate.
7. Stack 3 gel-sized pieces of wet filter paper soaked in 20× SSC over the first long filter paper. Lay the gel over the filter papers. Remove any air bubbles that remain between the gel and the filter papers. Place PVC plastic pieces around the gel (this will prevent the gel from moving and eventually falling out).
8. Cut a piece of nylon or nitrocellulose membrane precisely the size of the gel, soak it in 2× SSC, and place it on top of the gel, removing any air bubbles that may have remained. Do not move the membrane once it is placed over the gel.
9. Place two sheets of wet (2× SSC) filter papers of the gel size on top of the membrane. Remove air bubbles.

10. Place a wad of paper towel (~10 cm tall) on top of the filter paper sheets, and make sure they do not move (*see* **Note 21**).
11. Let the transfer take place overnight.
12. Carefully remove the paper towel (they are expected to be wet) and then the membrane with a pair of tweezers.
13. Turn the membrane and make a small cut in the upper left corner to identify the correct side.
14. Quickly and carefully wash the membrane with 20× SSC to remove any agarose pieces that may have remained attached to the membrane.
15. Air-dry the membrane and store it wrapped in filter paper inside a plastic bag.
16. Proceed to the hybridization step.

3.5.3 Pre-Hybridization and Hybridization

1. Place the membranes separately in a hybridization vessel. The DNA side should face the tube inward. Add 5–10 mL of pre-warmed (68 °C) ULTRAhyb buffer (*see* **Note 22**).
2. Pre-hybridize at 42 °C (hybridization oven) for >30 min. At this stage, it is essential to keep the vessel rotating so that the liquid comes into contact with the entire membrane.
3. Prepare the probes: In a PCR tube, add 10 ng template DNA to a PCR Master Mix, and adjust the volume to 100 μL. Mix well. Separate 10 μL to a new tube, and add 0.5 μL dig-dUTP (e.g., Roche) to the remaining solution. Perform PCR as usual for each primer pair. After PCR, load 5 μL of each PCR product in an agarose gel. The probe moves slightly slower than the non-labeled PCR product (*see* **Note 23**).
4. Boil the probes for 5 min at 95 °C in a water bath, and immediately place them on ice for 5 min (*see* **Note 24**).
5. Add 50 μL of denatured probe (the volume can vary depending on probe concentration) in the hybridization buffer. Incubate overnight (14–24 h) at 42 °C.
6. Remove all ULTRAhyb mixed with the probe. Store the probe at −20 °C for future use.
7. Wash the membrane twice for 5 min each with washing solution 1 at 42 °C.
8. Wash the membrane twice for 15 min with washing solution 2 at 42 °C.

3.5.4 Detection

1. Discard the wash buffer and add the blocking solution (variable volume, but enough to cover the entire membrane). Incubate for 1 h at RT.
2. Add 0.5 μL of anti-digoxigenin-AP per 5 mL of blocking solution (dilution 1:10,000). Incubate for 1 h at RT.

3. Discard the blocking solution and wash the membrane three times (10 min each) with TBS-T (variable volume). Each wash should last 10 min at RT.
4. Add the substrate of alkaline phosphatase enzyme (60 μL NBT and 30 μL BCIP per 10 mL) in the developing buffer and incubate at RT for 10–60 min. Expect purple DNA bands revealing specific reactions (*see* **Note 25**).
5. Store the dried membranes protected from light at RT.
6. Choose the RE that produces one genome size DNA fragment that hybridizes with the DNA-A probe and another with the DNA-B probe (*see* **Note 26**).

3.6 Production of an Infectious Clone in a Binary Vector After Partial Digestion of RCA Product with a RE

3.6.1 Preparation of Dimeric Viral Genome

1. Determine the best reaction condition for producing dimeric molecules: Prepare 10 tubes (1.5 mL) and identify them from 1 to 10. In one tube, make a master mix containing 70 μL water, 10 μL 10× RE buffer, and 10 μL RCA product (*see* Subheading 3.2). Mix well, and distribute 9 μL in each tube (10 tubes with 9 μL). Add 1 μL (4 U/μL) of the selected single-cutting RE (see Subheading 3.5) in tube 10; 1 μL (2 U/μL) in tube 9; 1 μL (1 U/μL) in tube 8; 1 μL (0.5 U/μL) in tube 7; 1 μL (0.25 U/μL) in tube 6; 1 μL (0.12 U/μL) in tube 5; 1 μL (0.06 U/μL) in tube 4; 1 μL (0.03 U/μL) in tube 3; 1 μL (0.01 U/μL) in tube 2; and 1 μL water in tube 1 (*see* **Note 27**). Incubate at the recommended incubation temperature for 30 min. Heat-inactivate the reaction (if the RE is heat-inactivated) or keep it on ice.
2. Separate the digested fragments in an agarose gel (1%) (*see* **Note 28**). Select the RE concentration that provides the highest amount of the dimeric molecule (~5.2 kb Fig. 2).
3. Producing the dimers: Scale up the reaction 10- to 20-fold (*see* **Note 29**) and keep the same selected incubation time and inactivation method. Precipitate the DNA by adding 10% sodium acetate and two volumes of ethanol, and incubate at −80 °C for >1 h or − 20 °C overnight. Centrifuge at 10,000 rpm for 10 min, discard the supernatant, and wash the pellet once by adding 1 mL of ice-cold 70% ethanol and centrifuging at 10,000 rpm for 5 min. Dry the pellet and resuspend in 20 μL 10 mM Tris-HCl buffer, pH 8, and separate the DNA in an agarose gel. Excise the dimer DNA band from the gel with a sharp razor blade and elute the DNA using a DNA gel extraction kit. Quantify the DNA by spectrophotometry or by agarose gel electrophoresis. Repeat the process until a final solution of >20 ng/μL is obtained. If needed, reduce the volume using a Speed-Vac.

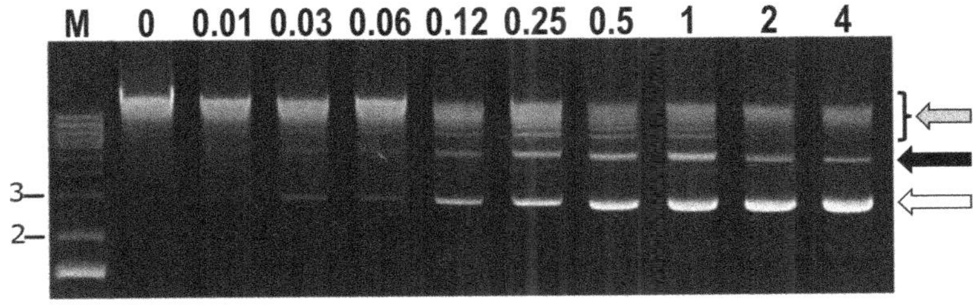

Fig. 2 RCA-amplified begomovirus DNA digested for 30 min with varying amounts of RE BglII (in enzyme units, above the gel). The reaction was performed with 7 μL water, 1 μL 10× buffer, 1 μL RCA product, and 1 μL diluted RE. The digestion produced monomeric (white arrow), dimeric (black arrow), and other partially or undigested (gray arrow) DNA molecules. In this case, the use of 1 unit per reaction produced the highest amount of dimeric molecules. M. 1 kb plus DNA size marker (ThermoFisher Scientific). Numbers on the left side represent fragment DNA sizes in kb. (Photograph by Dorian Y. M. Silva)

3.6.2 Preparation of Linearized Plasmid

1. Plasmid linearization: In a tube, add 20 μL RE buffer, ~10 μg pCAMBIA 0380 (*see* **Note 30**), 15–30 U selected RE, and water to 200 μL. Incubate at the optimal temperature for 3 h, add more 10 U RE, and incubate overnight. Run 5 μL of the digestion solution in an agarose gel to check if digestion is complete. If still partial, add another aliquot of RE and incubate >3 h until complete digestion. High quality restriction enzymes with the appropriate buffer (e.g., New England Biolabs) are preferred to avoid star activity.
2. Plasmid dephosphorylation: After digestion is completed, add 1 U CIAP and incubate for 5 min at 50 °C for 5′ recessed and blunt-ended DNA and for 5 min at 37 °C for DNA with 5′ overhang.
3. Inactivation of CIAP: Add 200 μL phenol:chloroform:isoamyl alcohol, vigorously mix for 5 min, and centrifuge >10,000 rpm for 10 min. Collect the aqueous phase and precipitate the DNA by adding 10% sodium acetate and two volumes of ethanol. Incubate at −80 °C for >1 h or − 20 °C overnight, and centrifuge 14,000 g for 5 min. Resuspend in 20 μL 10 mM Tris-HCl, pH 8 (*see* **Note 31**).
4. Isolation of the plasmid: Separate the linearized and dephosphorylated plasmid DNA in an agarose gel. Excise the linearized plasmid DNA band from the gel with a sharp razor blade and elute the DNA using a DNA gel extraction kit. Quantify the DNA by spectrophotometry or by agarose gel electrophoresis. The desired concentration is 50 ng/μL.

3.6.3 Ligation, Transformation, and Selection of Recombinant Plasmids

1. Ligation and transformation: In a tube, mix 1 μL 10× T4 DNA ligase buffer, 50 ng linearized and dephosphorylated pCAMBIA 0380 (*see* Subheading 3.6.2), 150 ng dimer (*see* Subheading 3.6.1), and 1 U T4 DNA ligase and adjust with water to 10 μL. For sticky ends, incubate at 22 °C for 10 min. For blunt ends, add 1 μL 50% PEG 4000 to the mixture and 5 U of T4 DNA ligase and incubate at 22 °C for 1 h.
2. Dialysis of the ligation reaction: Float a piece of a filter membrane (0.22 μm) on water and place the ligation reaction on top. Incubate for 15 min. Then, remove the ligation reaction and carefully transfer it to a tube. Use 1–2 μL for electrocompetent cells or up to 5 μL to transform 50 μL of chemically competent cells. Plate 50–200 μL in LB/agar plates containing kanamycin or a suitable antibiotic and incubate for 1–2 days at 28 °C or RT (*see* **Note 32**).
3. Selection of recombinant plasmids: Select six colonies and purify the plasmids. Grow selected bacteria in 5 mL liquid LB medium with the appropriate antibiotics at 28 °C overnight. Pellet the bacterial cells (10,000 rpm, 5 min) and resuspend in 100 μL alkaline lysis solution I. Add 200 μL alkaline lysis solution II, invert the tube, and incubate on ice for 5 min. Add 300 μL alkaline lysis solution III and invert the tube. Centrifuge at 10,000 rpm for 5 min, collect the supernatant (500–600 μL), and add 1 volume of isopropanol. Invert the tube. Centrifuge at 12,000 rpm for 10 min, eliminate supernatant, and rinse the pellet with 70% cold ethanol. Air-dry the pellet for 5 min and resuspend in 50 μL 10 mM Tris-HCl, pH 8.
4. Digest with 2 REs separately. The first is the RE used to prepare the dimer and plasmid, resulting in 1 vector size (~6.8 kb) fragment and a monomer size (~2.6 kb) fragment. Select a second RE present only once in the vector and absent in the virus genome. This will produce a single fragment of ~12.0 kb. Purify at least two clones using a plasmid purification kit, such as GenElute Plasmid Miniprep Kit (Sigma-Aldrich), and sequence to check if they were correctly cloned (*see* **Note 33**).

3.7 Production of an Infectious Clone in a Binary Vector by Gibson Assembly

1. Perform PCR for each of the three components using Q5 DNA polymerase. Use 98 °C for denaturing and allow ~30 s/kb during extension. Amplify a whole-genome copy of the geminivirus using primers U1-F and U1-R, another using U2-F and U2-R, and the vector using pJL89-F and pJL89-R (see **Note 10**).
2. Run PCR in a 1% agarose gel and stain with crystal violet dye (*see* **Note 34**).
3. Proceed to DNA purification from excised gel bands using an appropriate gel elution kit. Digest gel-purified pJL 89 with DpnI for 1 h at 37 °C to eliminate methylated DNA (i.e., undigested circular vectors).

4. Quantify each purified PCR product using spectrophotometry at 260 nm and calculate the amount needed for each, ranging from 0.02 to 0.5 pmols (*see* **Note 35**). In this case, use at least 60 ng of PCR-amplified pJL 89 and a two- to threefold molar excess of each insert. Keep the tubes on ice.
5. Add inserts and the vector in a 0.2-mL tube and water to a final volume of 10 μL if needed. Add 10 μL of 2× Gibson Assembly Master Mix and immediately incubate at 50 °C for 15 to 60 min. The reaction can then be kept on ice or at −20 °C prior to transformation.
6. Transform electrocompetent *E. coli* DH10β with 2 μL of the reaction (as described in Subheading 3.5.3).
7. To confirm that the dimer was correctly cloned, select up to 10 colonies and extract plasmids, as described in Subheading 3.6.3. Use a single-cut enzyme specific to pJL 89 (*see* **Note 36**) and check whether DNA size is >9 kb; monomers should yield a ~ 7.5 kb product.

4 Notes

1. Store the buffer without 2-mercaptoethanol at RT. Add 2-mercaptoethanol before use; this compound is toxic and must be handled carefully.
2. The cell lysis step can be performed manually using a pestle and mortar or using a cell homogenizer for fresh or frozen tissue powdered in liquid nitrogen.
3. A complete set of reagents is provided by some companies (e.g., Illustra TempliPhi Amplification Kit, Cytiva). If n-hexamer primers are not provided, purchase random hexamers that are thiol-modified in their two last nucleotides at the 3′ end.
4. Dilute to 0.5× before use. Use it for all electrophoresis separation steps described in this chapter.
5. Select the binary vector (e.g., pCambia 0380), and use unique REs that are present in the vector.
6. Select a plasmid vector with a high copy number and containing origins of replication for both *Escherichia coli* and *Agrobacterium tumefaciens* (syn. *Rhizobium radiobacter*), i.e., a binary vector. pCAMBIA 0380 (6813 bp) is a plant expression binary vector, and its presence induces kanamycin resistance to transformed bacteria.
7. We routinely use *E. coli* strains DH5α or DH10β for this purpose.

8. Preparation of SOC medium. To 950 mL of deionized H_2O, add 20 g tryptone, 5 g yeast extract, 0.5 g NaCl, 10 mL 250 mM KCl, pH 7.0 adjusted with NaOH, for 1 L. Sterilize by autoclaving. Just before use, add 5 mL of a sterile solution of 2 M $MgCl_2$. This medium is known as SOB medium. For the preparation of the SOC medium, add 20 mL of a sterile 1 M solution of glucose (sterilized by passing it through a 0.22-μm filter) per liter of SOB.
9. Designing of abutting primers for the geminivirus genome (*see* Fig. 3): one pair of primers for each monomer. The nonanucleotide region must be avoided when designing primers, as it

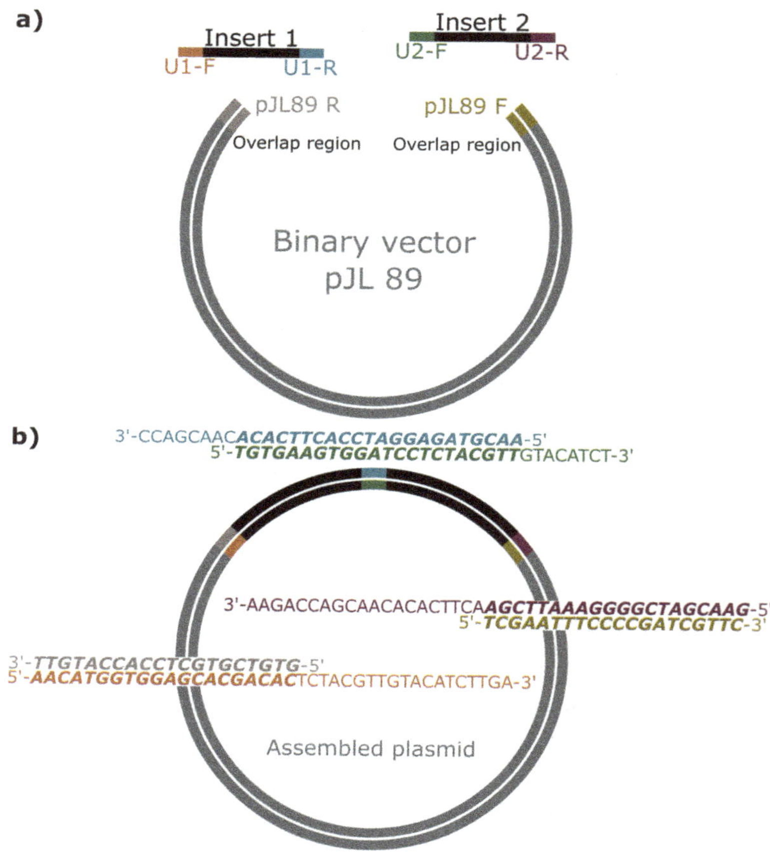

Fig. 3 Dimeric agroinfectious clone construction scheme using Gibson Assembly strategy. (**a**) Gibson reaction elements after DnpI vector digestion, just before assembly. Inserts 1 and 2 (black) correspond to the first and second viral units. U1-F (orange), U1-R (blue), U2-F (green), and U2-R (wine) terminal ends represent the virus primers containing overlap regions, and pJL89 F (dark yellow) and pJL89 R (brown), the vector primers. (**b**) Successfully assembled dimeric agroinfectious clone following Gibson assembly and primer sequences as described in Ferro et al. (2019) [4] with overlap regions in bold italic

must be present at least twice for correct processing. Aim for annealing temperatures above 60 °C (considering PCR with Q5 DNA polymerase). The U1-F primer must include a complementary region to the vector pJL89 at its 5′ (e.g., AACATGGTGGAGCACGACAC) end, followed by the target sequence in the geminivirus genome at its 3′ end. The reverse primer (U1-R) should be 30–40-nt long and must be designed immediately upstream (abutting) the forward primer, containing a ~ 15 nucleotide overlap sequence with AU2-F primer. The second primer pair must comprise a forward primer (AU2-F) complementary to part (~15 nt) of U1-R and an abutting reverse primer complementary to both pJL89 at its 5′ end (e.g., GAACGATCGGGGAAATTCGA) and target begomovirus at the 3′ end. There is no need to produce a dimeric genome for infectious clones. It is important to introduce two copies of the origin of replication, resulting in more than one copy of the genome.

10. It is advisable to amplify pJL89 vector by PCR using primers F (TCGAATTTCCCCGATCGTTC) and R (GTGTCGTGCTCCACCATGTT), as described by Ferro et al. (2019) [4].
11. Exposure to UV light may damage DNA molecules. We recommend staining with crystal violet stain (0.01 mg/mL) and exposure to white light for DNA band excision. Note that crystal violet is less sensitive than other DNA staining dyes.
12. Samples can be kept at RT before incubation at 65 °C.
13. The production of a whitish translucent pellet is expected. If a dark pellet is formed, issues in downstream procedures may occur.
14. A speed-vacuum equipment can be used to optimize this step, but do not over-dry, as DNA might not resuspend later.
15. Avoid repeated freeze-thawing procedures as they damage the DNA.
16. Alternatively, the tubes may also be kept at RT. Avoid incubation for more than 48 h.
17. A high molecular weight DNA band is observed near the wells. An amount of >1 μg is expected per μL of reaction.
18. MspI digests DNA at the site CCGG, but any other 4-base cutter RE can be used. After staining, ≥1 DNA band is expected. Sum up the digested fragment sizes. Non-digested DNA fragments of high molecular weight may be present as well. If the result is ~2.6 kb, there is a monopartite geminivirus in the sample. If the sum is >5.2, there is a bipartite begomovirus or a mixed infection of ≥2 geminiviruses (*see* Fig. 1).

19. Universal primers for begomoviruses: PAR1c496 (AATACTG CAGGGCTTYCTRTACATRGG), PAL1v1978 (GCATCTG CAGGCCCACATYGTCTTYCCNGT), PCRc2 (CTAGCTG CAGCATATTTACRARWATGCC) and PBL1v2040 (GCCT CTGCAGCARTGRTCKATCTTCATACA), as described in Rojas et al., (1993) [7]. If it is not a begomovirus, search for a universal primer pair for the genus.
20. If a RE site is present in only one site in the viral genome, there will be 1 DNA band of genome size in the gel. Carefully check if small DNA bands are not present in the bottom of the gel. These REs are potentially helpful for the construction of infectious clones. If a bipartite begomovirus is the target for cloning, go to Subheading 3.5.2 to identify REs specific to DNA-A and DNA-B.
21. If necessary, isolate the filter paper stack above the gel from the filter under the gel. This method will avoid movement of the SSC solution from the tray to the paper towel sheets.
22. Add enough buffer to cover the membrane entirely.
23. Prepare a probe for DNA-A and another for DNA-B. In general, probes produced using a geminivirus as a template will hybridize efficiently with other geminiviruses of the same genus but wash less stringently when dealing with an unknown geminivirus.
24. This denaturation step is only done when the probe has never been used. The probes can be used up to four times.
25. The substrate is photosensitive and should be protected from light. This process can also be done using other substrates, such as chemiluminescence CDP-Star (Sigma-Aldrich).
26. If the same RE site is present in both DNA-A and DNA-B, it can be conveniently used to clone both components.
27. Dilute RE to each final concentration in water immediately before use.
28. For these reactions, a decreasing amount of RE is added to obtain the desired partial digestion of the RCA products. For the best RE concentration, it is expected to observe a ladder, consisting of DNA fragments of genome size (~2.6 kb), dimer (~5.2 kb), trimer (~7.4 kb), and so on (*see* Fig. 2). Usually, the most abundant DNA size is that of a monomeric genome size.
29. If the digestion pattern changes when scaled up, adjust the RE concentration or digestion time to achieve the highest amounts of dimeric molecules.
30. Plasmid can be conveniently amplified by RCA.
31. This step is unnecessary if the alkaline phosphatase is 100% heat-inactivated (e.g., Sigma's recombinant shrimp alkaline phosphatase GEE70092X).

32. Use SOC medium for incubation of transformed cells at 28 °C for 1 h before plating the cells in LB agar medium with kanamycin.
33. If no colonies are obtained, check the amount of the DNA and plasmid used for ligation, the ligation efficiency, and cells competence. If none of the digests results in the expected patterns, repeat the process with other colonies until all colonies are screened.
34. Only highly concentrated DNA can be visualized using this dye under white light. Run a small aliquot in another gel and stain, as usual, to check amplification before staining with crystal violet. The primary purpose of using this method over ethidium bromide (or other dyes) is to avoid DNA damage caused by both the chemical and the UV light.
35. Agarose gels can also be used to estimate DNA concentration. To calculate pmols, use the formula: pmols = (ng of DNA) × 1000 / (bp × 650 da). Alternatively, online calculators for molecular biology can be used (e.g., https://nebiocalculator.neb.com/).
36. pJL 89 sequence and restriction sites can be selected on the webpage https://www.addgene.org/98558/. Avoid enzymes that also cut the viral genome, as a cloned dimer might be indistinguishable from a monomer.

Acknowledgments

We thank all the people in the laboratory, from the past to the present, who engaged in the task of producing infectious clones of geminiviruses. We also thank Dorian Yest Melo Silva for the gel picture in Fig. 2. This work was funded by FAP-DF grants 00193-00000229/2021-21, 0193.000872/2021-54 and 0193.001023/2021-18, and all other grants related to projects on geminivirus diseases. The first author has a scholarship from Coordenação de Aperfeiçoamento de Pessoal de Nível Superior (CAPES). AKI-N is a CNPq fellow.

References

1. Ferreira PTO, Lemos TO, Nagata T, Inoue-Nagata AK (2008) One-step cloning approach for construction of agroinfectious begomovirus clones. J Virol Methods 147:351–354. https://doi.org/10.1016/j.jviromet.2007.10.001
2. Gibson DG, Young L, Chuang RY et al (2009) Enzymatic assembly of DNA molecules up to several hundred kilobases. Nat Methods 6:343–345. https://doi.org/10.1038/nmeth.1318
3. Blawid R, Nagata T (2015) Construction of an infectious clone of a plant RNA virus in a binary vector using one-step Gibson assembly. J Virol Methods 222:11–15. https://doi.org/10.1016/J.JVIROMET.2015.05.003
4. Ferro MMM, Ramos-Sobrinho R, Xavier CAD et al (2019) New approach for the construction of infectious clones of a circular DNA plant virus using Gibson assembly. J Virol Methods 263:

20–23. https://doi.org/10.1016/J.JVIROMET.2018.10.017

5. Doyle JJ, Doyle JL (1987) A rapid DNA isolation procedure for small quantities of fresh leaf tissue. Phytochem Bull 19:11–15
6. Inoue-Nagata AK, Albuquerque LC, Rocha WB et al (2004) A simple method for cloning the complete begomovirus genome using the bacteriophage φ29 DNA polymerase. J Virol Meth 116:209–211. https://doi.org/10.1016/j.jviromet.2003.11.015
7. Kamtekar S, Berman AJ, Wang J et al (2004) Insights into strand displacement and processivity from the crystal structure of the protein-primed DNA polymerase of bacteriophage φ29. Mol Cell 16:609–618. https://doi.org/10.1016/j.molcel.2004.10.019
8. Rojas MR, Gilbertson RL, Russel DR et al (1993) Use of degenerate primers in the polymerase chain reaction to detect whitefly-transmitted geminiviruses. Plant Dis 77:340–347. https://doi.org/10.1094/PD-77-0340
9. Johnson M, Zaretskaya I, Raytselis Y et al (2008) NCBI BLAST: a better web interface. Nucleic Acids Res 36:W5–W9. https://doi.org/10.1093/NAR/GKN201

Chapter 5

Agroinoculation of Geminiviral Infectious Clones into Plants

Ivair J. Morais, Yanca F. M. Ferreira, Erich Y. T. Nakasu, Tatsuya Nagata, and Alice K. Inoue-Nagata

Abstract

Most geminiviruses are not transmitted by mechanical inoculation. Therefore, pathogenicity and plant–pathogen interaction studies rely on agroinoculation using infectious clones, which involves cloning the geminiviral genome in a binary vector (see previous chapter for details). A suspension containing the infectious clone inserted into *Agrobacterium tumefaciens* cells is then inoculated into plants, i.e., agroinoculated. Below is a simple protocol for agroinoculation of an infectious geminivirus clone into plants.

Key words *Agrobacterium tumefaciens*, *Agrobacterium radiobacter*, Infectious clone inoculation, Agroinjection, Geminivirus

1 Introduction

Agroinoculation has been used since the mid-1980s [1, 2] as a reliable method for completing Koch's postulates and studying plant host/virus interactions using geminiviral infectious clones. An effective geminiviral infectious clone usually has at least two geminiviral species-specific origins of replication and is cloned in a binary vector for agroinoculation. The vector is first inserted into a suitable *Escherichia coli* strain and then transferred to an *Agrobacterium tumefaciens* (syn. *Agrobacterium radiobacter*) strain for agroinoculation into plants.

2 Materials

Prepare all solutions using ultrapure water (e.g., Milli-Q purified water) and analytical-grade reagents. Use DNase-free consumables. Store all reagents at −20 °C and competent cells at −80 °C. Diligently follow all applicable waste disposal regulations when disposing of waste materials.

Elizabeth P. B. Fontes and Kristiina Mäkinen (eds.), *Plant-Virus Interactions*, Methods in Molecular Biology, vol. 2724, https://doi.org/10.1007/978-1-0716-3485-1_5,

2.1 Preparation of Chemically Competent A. tumefaciens Cells

1. LB broth (10 g/L tryptone, 5 g/L yeast extract, 0.5 g/L NaCl).
2. Rifampicin (20 mg/mL).
3. YEP liquid medium (5 g/L beef extract, 1 g/L yeast extract, 5 g/L peptone, 5 g/L sucrose, 0.5 g/L $MgCl_2$) supplemented with 0.5% glucose.
4. 150 mM NaCl.
5. 20 mM $CaCl_2$.
6. An appropriate *A. tumefaciens* strain (e.g., GV3101 and LBA4404) (*see* **Note 1**).

2.2 Transformation of Chemically Competent A. tumefaciens Cells

1. 1 tube with chemically competent *A. tumefaciens* cells (100 μL).
2. 1 μg plasmid (concentration of >100 ng/μL of infectious clone in nuclease-free water).
3. SOC liquid medium (*see* **Note 2**).
4. 90 mm petri dishes with 20 mL LB agar medium (10 g/L tryptone, 5 g/L yeast extract, 0.5 g/L NaCl, and 15 g/L agar) containing the proper antibiotics (*see* **Note 3**).

2.3 Agroinoculation

1. LB liquid medium with antibiotics (50–100 μg/mL) (*see* **Note 3**).
2. Acetosyringone stock solution (100 mM in ethanol or DMSO).
3. LB liquid medium with appropriate antibiotics and 20 μM acetosyringone, pH 7.5 (*see* **Note 4**).
4. Resuspension solution: 10 mM $MgCl_2$, 10 mM MES-K (pH 5.6 adjusted with KOH), and 150 μM acetosyringone (*see* **Note 5**).
5. 1-mL syringe.

3 Methods

Carry out all procedures at room temperature unless otherwise specified. All procedures are carried out with sterilized materials on a clean bench. The schematic illustration of the procedures is shown in Fig. 1.

3.1 Preparation of Chemically Competent A. tumefaciens Cells

1. Inoculate a single colony of an *A. tumefaciens* strain in 5 mL LB liquid medium containing 20 μg/mL rifampicin and incubate overnight at 28 °C with vigorous shaking (180–220 rpm).
2. Transfer 2.5 mL of the overnight culture to 250 mL YEP medium with 0.5% glucose and grow the bacteria at 28 °C

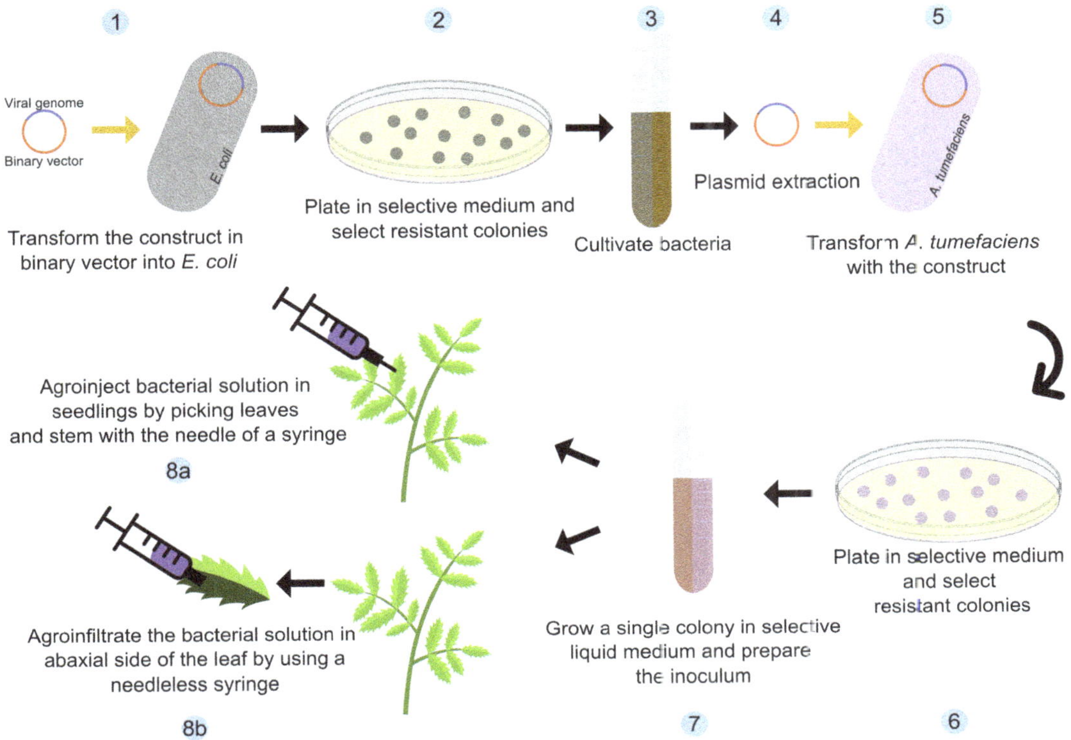

Fig. 1 Schematic illustration of the procedures for agroinoculation of an infectious viral construct in a binary vector

with vigorous shaking until the culture reaches the log phase at OD_{600} between 0.5 and 0.6 (*see* **Note 6**). Allow it to cool on ice for 15 min.

3. Harvest the bacteria by centrifugation at 4000 × *g* for 5 min at 4 °C. Discard the supernatant, resuspend each pellet with 25 mL of ice-cold 150 mM NaCl, and transfer it to a 50-mL tube.
4. Centrifuge the suspension at 4000 × *g* for 5 min at 4 °C. Discard the supernatant and resuspend the sediment in 5 mL of ice-cold 20 mM $CaCl_2$. Store the cells on ice.
5. Aliquot 100 μL of the bacterial suspension into 1.5 mL tubes on ice and quick-freeze in liquid nitrogen (optional). Store the competent cells at −80 °C.

3.2 Transformation of Chemically Competent A. tumefaciens

1. Take one tube of competent *A. tumefaciens* kept at −80 °C, thaw on ice and add approximately 1 μg of plasmid. Mix gently by tapping the tube. Incubate the mixture for 20–30 min on ice.
2. Heat-shock cells by placing the tube in a water bath at 37 °C for 5 min.

3. Mix the contents by gently flicking the tube and add 1 mL SOC medium. Shake the tube at 28 °C for 2 h.
4. Plate 100–200 μL per LB agar plate containing the appropriate antibiotics.
5. Incubate the plates upside down at 28 °C for 2–3 days.
6. Using sterile toothpicks, transfer individual colonies to new plates with LB agar containing the appropriate antibiotics.
7. Incubate the plates upside down at 28 °C for 2–3 days (*see* **Note 7**).

3.3 Agroinfiltration

1. Prepare a pre-culture by inoculating a single colony of *A. tumefaciens* transformed with your target clone in 5 mL LB medium with antibiotics. Incubate overnight with agitation at 28 °C.
2. Use 1 mL of the culture to inoculate 24 mL LB medium with antibiotics and 20 μM acetosyringone. Incubate overnight (*see* **Note 8**).
3. Pellet the bacteria by centrifugation at 5000 × *g* for 15 min and resuspend the pellet in resuspension solution. Measure the OD_{600} (*see* **Note 9**).
4. Dilute suspension to $OD_{600} = 1.0$ for agroinoculation (*see* **Note 10**). Concentration may be adjusted according to the objectives and target plants to increase infection rates.
5. Infiltrate true leaves using a 1-mL syringe without the needle. Gently press the syringe on the abaxial side of the leaf and apply counter-pressure with a finger on the other side. Successful infiltration is indicated by the appearance of a wet area on the leaf (*see* **Note 11**).
6. Alternatively, the infectious clone can be agroinoculated by pinning true leaves and stems of plantlets using a syringe needle while injecting the bacterial solution.

4 Notes

1. Verify if the *A. tumefaciens* strain is resistant to rifampicin. If not, do not add it to the medium.
2. Prepare SOC medium with 20 g/L tryptone, 5 g/L yeast extract, 0.5 g/L NaCl, and 10 mL 250 mM KCl (pH 7.0). Autoclave. To the cold medium, add 5 mL 2 M $MgCl_2$ (autoclaved). Make small aliquots and store them frozen (−20 °C) until use.
3. The specific antibiotics will depend on the bacterial strain and the plasmid used for transformation. Usually, two antibiotics

are used, one to maintain *A. tumefaciens* helper (Ti) vector, for example, 100 μg/mL gentamicin, and another specific to the binary vector, for example, 50 μg/mL spectinomycin or kanamycin. Popular *A. tumefaciens* strains, such as GV3101 and LBA4404, are resistant to rifampicin. Antibiotics should be added to a ready-to-use medium.

4. Add antibiotics and acetosyringone after autoclaving and use the medium immediately.
5. The resuspension solution must be autoclaved before adding acetosyringone. Tween-80 or Tween-20 (up to 0.05% final concentration) can be used to facilitate agroinfiltration into leaves.
6. This takes approximately 4–5 h. You may calculate the doubling time as 2 h in the described condition.
7. These individually transformed *A. tumefaciens* clones will be used for leaf inoculation.
8. The overnight growth step can be skipped if sufficient bacteria are available.
9. The OD_{600} should be within the range of 0.6–1.0 for optimal infiltration. Dilute the bacterial solution if $OD_{600} > 1.0$. Higher bacterial concentration may induce undesired "pattern-triggered immunity" in plants. Resuspension of the pellet in sterile water may also be used [3].
10. For a bipartite begomovirus inoculation, mix the bacterial solution of each component (DNA-A clone and DNA-B clone) in the same concentration.
11. Avoid infiltrating the cotyledons of the leaves. Use at least 0.2 mL for each agroinfiltration. For cucurbits, cotyledons are also target for agroinoculation in many cases.

Acknowledgments

We thank all the people in the laboratory, from the past to the present, who engaged in the task of producing infectious clones of geminiviruses. This work was funded by FAP-DF grants 0193.000872/2021-54 and 0193.001023/2021-18 and all other grants related to projects on geminivirus diseases. The first author has a scholarship from Coordenação de Aperfeiçoamento de Pessoal de Nível Superior (CAPES). AKI-N is a CNPq fellow.

References

1. Grimsley N, Hohn B, Hohn T, Walden R (1986) "Agroinfection", an alternative route for viral infection of plants by using the Ti plasmid. Proc Natl Acad Sci U S A 83:3282–3286
2. Grimsley N, Hohn T, Davies JW, Hohn B (1987) Agrobacterium-mediated delivery of infectious maize streak virus into maize plants. Nature 325:177–179
3. Navot N, Pichersky E, Zeidan M, Zamir D, Czosnek H (1991) Tomato yellow leaf curl virus: a whitefly-transmitted geminivirus with a single genomic component. Virology 185:151–161

Chapter 6

Begomovirus Inoculation in Arabidopsis and Cassava

José T. Ascencio-Ibáñez, Mary M. Dallas, and Linda Hanley-Bowdoin

Abstract

The use of infectious clones to inoculate plant viruses allows for controlled studies that lead to a better understanding of plant–virus interactions. The main methods used for laboratory inoculation of geminiviruses are agroinoculation and biolistics. We describe how to successfully inoculate geminiviruses, focusing on Arabidopsis as a model plant and cassava as a crop.

Key words Geminivirus, Agroinoculation, Biolistics, Infectious clones, Arabidopsis, Cassava

1 Introduction

Geminiviruses belong to an expanding family of plant viruses that create many agricultural problems in tropical and subtropical areas of the world. They are non-enveloped viruses with small, circular single-stranded DNA genomes that replicate through rolling circle replication and recombination-dependent replication in the nuclei of infected plant cells [1, 2]. These viruses are transmitted by insects, have single or two-component genomes (mono- and bipartite), and are sometimes associated with satellites with different characteristics [3]. These viruses can occur in mixed infections, which in some cases lead to recombination and increased viral diversity [4]. Geminiviruses are divided into multiple genera, of which *Begomovirus* and *Mastrevirus* are the most diverse within the family [5]. The *Geminiviridae* also includes *Becurtovirus, Capulavirus, Citlodavirus, Curtovirus, Eragrovirus, Grablovirus, Maldovirus, Mulcrilevirus, Opunvirus, Topilevirus, Topocuvirus, and Turncurtovirus* [5].

One of the primary needs for virological studies is to be able to infect plants under controlled conditions so that the viruses can be

Supplementary Information The online version contains supplementary material available at https://doi.org/10.1007/978-1-0716-3485-1_6.

Elizabeth P. B. Fontes and Kristiina Mäkinen (eds.), *Plant-Virus Interactions*, Methods in Molecular Biology, vol. 2724, https://doi.org/10.1007/978-1-0716-3485-1_6,

studied independently from other pathogens or stresses. Lately, the use of infectious clones has been more prevalent, but it is not the only way to inoculate viruses onto plants [6]. The most common method is mechanical inoculation, but very few geminiviruses can be mechanically transmitted. One can also inoculate plants by grafting, which is very efficient for transmission but labor-intensive for implementation [7]. Another way is using the insect vector, which is also very efficient. However, you need access to the insect, which requires special facilities and expertise for rearing and containment—both expensive and time-consuming investments. Viruses can also be transmitted through soil, nematodes, fungi, etc. [8], but these avenues have not been explored for geminiviruses. With advances in molecular biology, the use of cloned geminivirus genomes to generate infectious clones to infect plants has become popular due to its simplicity and efficiency. Infectious clones for geminiviruses are designed to include two copies of the viral origin of replication, which is recognized by the viral Rep protein (aka AL1 or C1) that catalyzes nicking and ligation to release the geminivirus circular genome.

Two major inoculation methods are commonly used in geminivirus studies—agroinoculation and biolistic inoculation. For agroinoculation, a hemi-dimer or a dimer of the viral genome is cloned into an *Agrobacterium* binary vector. Plants are inoculated with an *Agrobacterium* culture with an infectious clone, which is transferred into the plant cells. The viral genome in the T-DNA is transcribed to produce Rep, which then catalyzes the release of the viral genome to initiate viral infection [9]. Biolistic inoculation directly shoots cloned viral genomic DNA into the plant [10]. In this case, particle acceleration allows the cloned viral DNA precipitated onto gold or tungsten particles to enter plant cells, where the viral genome is released, and the infection process starts. The plasmids used for biolistic inoculation have a similar design as agroinoculation clones, except that the vector can be propagated at high levels in *E. coli*. It is important to note that not all geminiviruses can be inoculated by biolistics, but most can be inoculated via agroinoculation. This may reflect that phloem-limited geminivirus is difficult to inoculate via particle acceleration.

Another derivation of these molecular biology advances is the use of naked DNA to inoculate geminiviruses directly, but this method is much lower in efficiency. Moreover, not all geminiviruses can be inoculated using naked DNA [11]. Rolling circle amplified DNA has also been used to inoculate plants [12]. In some instances, total DNA from geminivirus-infected plants can also be used for inoculations. This approach can result in the loss of some viruses in a mixed infection; e.g., tomato yellow leaf curl virus cannot be inoculated in this way.

This protocol focuses on the inoculation of bipartite begomoviruses into Arabidopsis (Agrobacterium and biolistics) and cassava (biolistics). Inoculation of begomoviruses with bipartite genomes requires two infectious clones, one for the A component and one for the B component, to achieve full infection. Inoculation of the A component clone alone can be used to examine viral replication separate from viral movement functions. The agrobacterium protocol can also be used to inoculate Arabidopsis with *beet curly top virus* or *tomato yellow leaf curl virus*, which have monopartite genomes [13]. Agroinoculation has not yet worked in our hands for cassava begomoviruses.

2 Materials

2.1 Agroinoculation in Arabidopsis thaliana Col-0

1. Infectious clones in agrobacterium-compatible vectors transformed into *Agrobacterium tumefaciens* (verification of the compatibility of the vector with the agrobacterium strain is needed prior to transformation).
2. Sterilized LB or YEB media to grow Agrobacterium, depending on availability and the needs of the Agrobacterium strain.
3. Selected antibiotics for the bacterial strain used. (If using binary vectors, antibiotics are recommended for both the bacterial strain and the plasmids.)
4. Culture tubes.
5. Incubator (with shaker) at 28 °C or 30 °C.
6. Insulin or 3-mL disposable syringes.
7. Needles, preference for short and thin gauge (27G 1¼). Larger syringes and wider needle gauges may hurt the plant and decapitate the leaves.
8. Plants at the proper stage for inoculation (Arabidopsis with 14–16 true leaves in pots).

2.2 Biolistic Inoculation in Arabidopsis thaliana Col-0 or Manihot esculenta var Kibandameno

2.2.1 Preparation of the Gold or Tungsten Particles

1. Gold (Bio-Rad or another producer) or tungsten (Bio-Rad or another producer) particles.
2. Balance (capable of 60 mg weighting).
3. Centrifuge (for microfuge tubes).
4. Absolute ethanol (200 Proof).
5. Deionized water (sterile).
6. Sonicator (Fisher Sonic Bath 1.9 L) or equivalent.
7. Microcentrifuge tubes (1.5 mL).
8. Sterilized tips and micropipettes.

2.2.2 Particle Bombardment onto Plants

1. Aliquoted gold particles ready for biolistics.
2. Purified plasmids of the geminivirus infectious clones, preferably at high concentration (>500 ng/μL). Rolling circle

amplification products or total DNA from infected plants can also be used for this purpose, but not all viruses will be infectious in this way.

3. Sterilized Swinnex® filter holder (or equivalent), one per treatment. (Millipore).
4. Micropipettes and sterile tips.
5. Sonicator (for instance jewelry sonicator or Fisher Ultrasonic Bath, 1.9 L).
6. Vortex or shaker.
7. Microdrop sprayer (Ascefran, LLC), low pressure or another biolistic device (Bio-Rad Helios Gene Gun System).
8. Plants at the proper stage for inoculation. Arabidopsis: (We have inoculated many ecotypes with this method, and all have worked, with very few exceptions [14]). Plants are grown at 20 °C with 8 h of light. Arabidopsis can be inoculated from six true leaf stages all the way to flowering if the inoculation is performed with a low-pressure device (less than 100 psi). For high-pressure inoculations (higher than 100 psi and up to 900 psi), it is better to use older plants closer to the 16–18 true leaves stage. Cassava: Propagated stem cuttings grown to a 6–8 leaf stage of a susceptible cultivar or other cultivars/varieties work well with this protocol. The plants can be grown between 28 °C and 30 °C with a light regime of a 12-h light/dark cycle (Aimone et al., 2021).

3 Methods

Methods 3.1 and 3.2 have been successfully used for beet curly top virus (BCTV) and tomato yellow leaf curl virus (TYLCV) in addition to CabLCV in Arabidopsis.

3.1 Agroinoculation in *Arabidopsis thaliana* Plants (See Video S1)

1. Use *Arabidopsis thaliana* Col-0 with 16–18 true leaves (*see* **Note 1**), grown at 20 °C with 8 h of light. Trays containing 18 9 cm^2 (3.5 in^2) pots allow the plant to expand the leaves so that symptoms are clear, and the plants grow well. Other configurations should also work, but this size tray is recommended for optimal results.
2. Grow 3–5 mL LB cultures of Agrobacterium carrying CabLCV A and B components at 30 °C overnight (O.D. should be at least 1.0 or to saturation).
3. Mix an equal amount of each culture (1:1).

   A 25-gauge needle in a syringe is used to inoculate the plant by making ~10 holes per plant, close and around to, but avoiding the central meristem. Deposit a droplet of culture in each plant (*see* **Note 2**).

4. Keep the inoculated plants under a cover (lid) for at least 1 day; take the lid off after that.
5. Observe the development of responses to the bacterial inoculation. A yellowing will appear ~3 dpi, which is caused by the response to *Agrobacterium*; the plants will recover, and viral symptoms should appear later on.
6. Observe and record the viral symptoms. Viral symptoms could start to appear at 9 dpi, and by 12–14 dpi, all plants should have full symptomatic tissue (*see* **Note 3**).

3.2 Other Plants Can Be Inoculated with This Method (See Note 4)

1. Grow the Agrobacterium cultures to saturation and mix 1:1, if bipartite.
2. Using a 25-gauge syringe, inoculate the stem under the flag leaf by puncturing it 5–10 times and leaving a droplet of the culture on it (*see* **Note 5**).

3.3 Low-Pressure Geminivirus Bombardment

3.3.1 Stock Particle Preparation for Biolistics (Fig. 1)

1. Weigh 60 mg of gold particles or tungsten particles.
2. Add 2 mL of 99.5% ethanol (200 proof) and mix well in a vortex or shaker.
3. Centrifuge at 10,000 rpm for 2 min.
4. Decant supernatant and eliminate it, making sure the particle pellet stays intact.
5. Resuspend the gold pellet in 1 mL of deionized water (sterile).
6. Sonicate for 10 min in a water bath sonicator and mix continuously.
7. Prepare 250 μL aliquots and resuspend in 750 μL of sterile deionized water. Store aliquots at −20 °C.
8. Use 50 μL aliquots for bombarding (*see* **Note 6**).

3.3.2 Geminivirus Bombardment (Low-Pressure Microdrop Sprayer) (Fig. 2 and Video S2)

1. Add 5 μg of DNA from each component (*see* **Note 7**) to gold particles while vortexing (50 μL of prepared gold particles). This will use 10 μg per every 50 μL aliquot of particles, which will bombard six plants.
2. Add 50 μL $CaCl_2$, 2.5 M, while agitating.
3. Add 20 μL of 0.1 M spermidine, while agitating.
4. Mix for 3–5 min by vortexing.
5. Spin tubes at 10,000 rpm for 10 s. Remove and discard the supernatant.
6. Add 250 μL of 200-proof ethanol (100%) and resuspend pellets. Sonication may be required to resuspend; however, do not over-sonicate. This could cause breakage of the DNA.

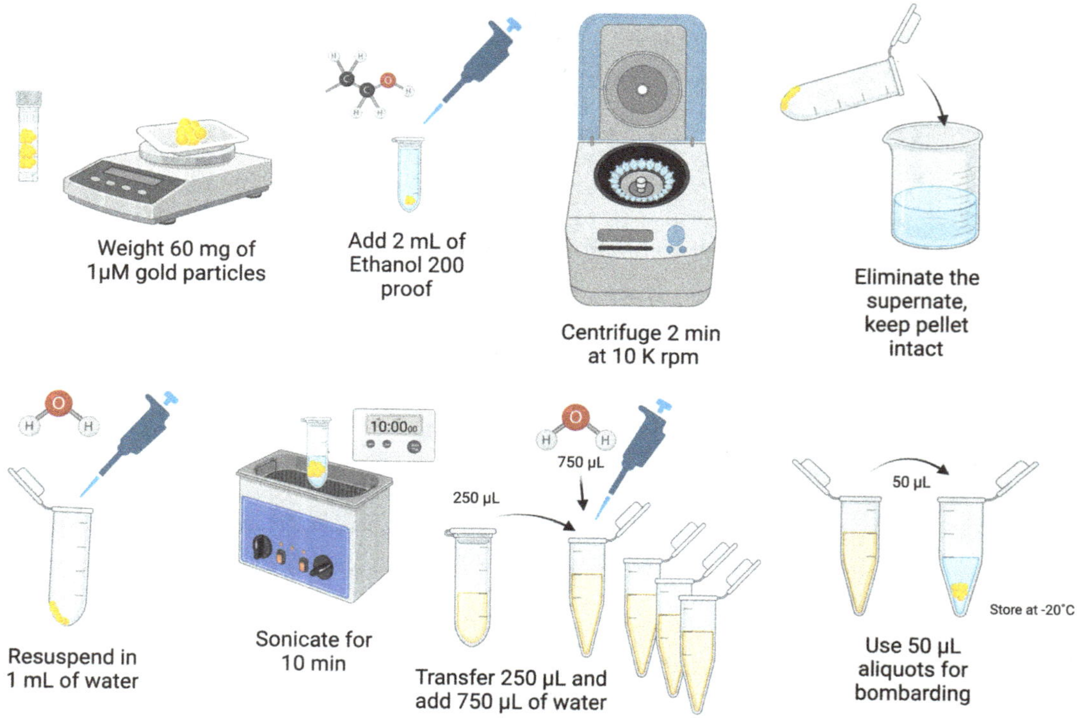

Fig. 1 Preparation of the particle's stocks. After weighing 60 mg of particles, absolute ethanol is added, followed by resuspension and short centrifugation at 10 K rpm. Once centrifuged, the supernatant is eliminated, and the pellet resuspended in 1 mL of water and then is placed in a bath sonicator for 10 min. After sonication and with the particles fully resuspended, 250 μL is transferred to a new tube, and 750 μL of water is added. While maintaining the particles resuspended, transfer 50 μL of the particle resuspension to a new microcentrifuge tube, where the particles can be used immediately or stored at −20 °C for long-term storage. (Figure created using BioRender.com)

7. Spin tubes at 10,000 rpm for 10 s. Remove and discard the supernatant.
8. Resuspend DNA pellets in 65 μL of 200-proof ethanol and keep the tubes on ice until use.
9. Use 10 μL to add to the center of the Swinnex® filter, close it, and bombard it. Alternatively, a pipette and tip can be used to directly add the volume onto the Swinnex® filter without opening it (*see* Fig. 3 and Video S3).

4 Notes

1. Younger plants can be inoculated, but the virus may kill them very quickly. Older plants can also be inoculated, but they are slightly more resistant to the infection, although eventually, they will be infected.

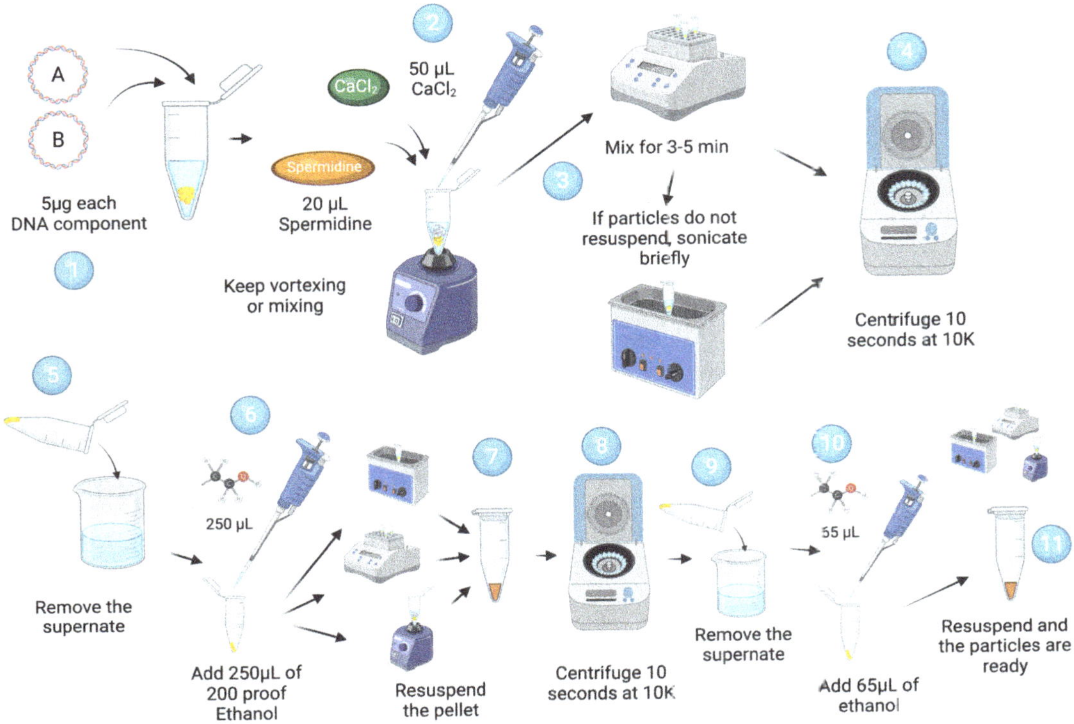

Fig. 2 Coating of the particles to be delivered with the DNA. (1) For 50 μL of particles, 5 μg of each DNA component (A and B for bipartite begomoviruses) is added (the volumes of this solution will vary depending on the concentration of the purified DNA). (2) 50 μL of $CaCl_2$ and 20 μL of spermidine are added to the continuously moving particles with DNA. (3) The particles can be placed in a shaker or sonicated, but sonication must be very brief to avoid breakage of the DNA. (4) A short centrifugation at 10 K rpm is followed by eliminating the supernatant, by decanting, or micropipette (5). (6) 250 μL of absolute ethanol is added. (7) Another resuspension is required (use a sonicator, shaker, or vortex). Centrifugation at 10 K rpm followed by supernatant removal (8 and 9). The last removal must be performed very carefully. The last steps involve adding 65 μL of absolute ethanol (10) and resuspending the particles with the DNA (11). The particles are ready to be used for delivery. (Figure created using BioRender.com)

2. The droplet can be deposited prior to poking.
3. For BCTV, symptom appearance is delayed compared to CabLCV and shows only after 21 dpi. For TYLCV, the symptoms are really mild and show after 28 dpi. Growth conditions, including temperature, light quality, and intensity, significantly affect symptom development.
4. For species different than Arabidopsis, agroinoculation can be used; however, not all the viruses will infect all the hosts (this is known to be effective in *Nicotiana benthamiana*, *Nicotiana tabacum*, *Nicandra physalodes*, *Datura stramonium*, and *Lycopersicon lycopersicon*). Successful inoculation by biolistics has been observed with the following viruses: CabLCV, BCTV,

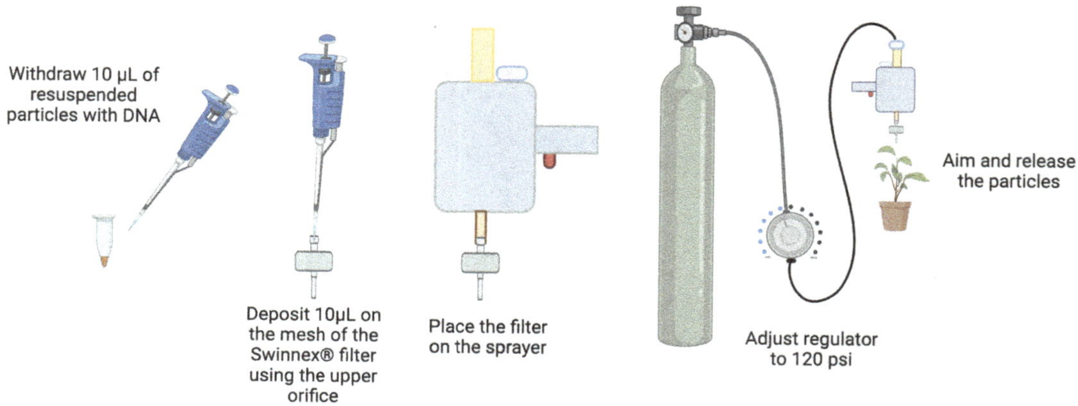

Fig. 3 Delivery of DNA-coated particles onto plants. From a fully resuspended particle suspension, remove 10 µL of particles and deposit them in the mesh of the Swinnex® filter or equivalent, by using the orifice at the top of the filter or unscrewing the top of the filter. Screw back the top if needed, and then screw the filter onto the mouth of the microdrop sprayer. Use your selected pressure for particle delivery; 120 psi is the recommended pressure for most applications. (Figure created using BioRender.com)

TYLCV, ToMoV, TGMV, PepGMV, PHYVV, ACMV, and EACMV and even CBSV-Ug (*Potyviridae*).

5. The best place to inoculate in this system is the stem, at the base of the upper leaf (younger). Other tissue may work; however, it will depend on the architecture of the plant and the ability to reach the phloem tissue easily.
6. The final concentration of particles will be 15 µg/µL. Store at −20 °C. This procedure has been modified from Cabrera-Ponce et al. 1997.
7. 5 µg is good, but 2.5 µg of cloned DNA can also work. For Arabidopsis, a low amount of 100 ng per component has worked. However, for cassava, it is strongly suggested to keep it at 10 µg of total DNA (5 µg per component).

Acknowledgments

This work was supported by the Bill & Melinda Gates Foundation Grant OPP1149990 and National Science Foundation Grant OISE-1545553. JTAI thanks Ayden Ferrell for helping prepare the manuscript.

References

1. Hanley-Bowdoin L, Bejarano ER, Roberts D, Mansoor S (2013) Geminiviruses: masters at redirecting and reprogramming plant processes. Nat Rev Microbiol 11(11):777–788
2. Jeske H (2009) Geminiviruses. In: de Villiers E-M, zur Hausen (eds) TT viruses: the still elusive human pathogens. Springer Verlag, Berlin Heidelberg

3. Rojas MR, Macedo MA, Maliano MR, Soto-Aguilar JOS, Briddon RW, Kenyon L, Rivera-Bustamante RF, Zerbini FM, Adkins S, Legg JP, Kvarnheden A, Wintermantel WM, Sudarshana MR, Peterschmitt M, Lapidot M, Inoue-Nagata AK, Gilbertson RL (2018) World management of geminiviruses. Annu Rev Phytopathol 56:637–677

4. Rentería-Canett I, Xoconostle-Cázares B, Ruiz-Medrano R, Rivera-Bustamante R (2011) Geminivirus mixed infection on pepper plants: synergistic interaction between PHYVV and PepGMV. Virol J 8:104

5. Fiallo-Olivé E, Lett J-M, Martin DP, Roumagnac P, Varsni A, Zerbini FM, Navas-Castillo J, ICTV Report Consortium (2021) ICTV virus taxonomy profile: geminiviridae 2021. J Gen Virol 102(12):001696

6. Brewer HC, Hird DL, Bailey AM, Seal SE, Foster GD (2018) A guide to the contained use of plant virus infectious clones. Plant Biotechnol J 16:832–843

7. Renukadevi P, Sangeetha B, Malathi VG, Nakkeeran S, Satya VK (2022) Enigmatic emergence of seed transmission of geminivirus. In: Gaur RK, Sharma P, Czosnek H (eds) Geminivirus: detection, diagnosis and management. Academic Press, Oxford

8. Dijkstra J, de Jager CP (1998) Practical plant virology. Springer, Heidelberg

9. Grimsley N, Hohn T, Davies JW, Hohn B (1987) Agrobacterium-mediated delivery of infectious maize streak virus into maize plants. Nature 325:177–179

10. Garzón-Tiznado JA, Torres-Pacheco I, Ascencio-Ibáñez JT, Herrera-Estrella L, Rivera-Bustamante R (1993) Inoculation of peppers with infectious clones of a new geminivirus by a biolistic procedure. Phytopathology 83(5): 514–521

11. Ascencio-Ibáñez JT, Settlage SB (2007) DNA abrasion onto plants is an effective method for geminivirus infection and virus-induced gene silencing. J Virol Meth 142:198–203

12. Guenoune-Gelbart D, Sufrin-Ringwald T, Capobianco H, Gaba V, Polston JE, Lapidot M (2010) Inoculation of plants with begomoviruses by particle bombardment without cloning: using rolling circle amplification of total DNA from infected plants and whiteflies. J Virol Meth 168:87–93

13. Rajabu CA, Kennedy GG, Ndunguru J, Ateka EM, Tairo F, Hanley-Bowdoin L, Ascencio-Ibáñez JT (2018) Lanai: a small, fast growing tomato variety is an excellent model system for studying geminiviruses. J Virol Meth 256:89–99

14. Reyes MI, Flores-Vergara MA, Guerra-Peraza O, Rajabu C, Desai J, Hiromoto-Ruiz YH, Ndunguru J, Hanley-Bowdoin L, Kjemtrup S, Ascencio-Ibáñez JT, Robertson D (2017) A VIGS screen identifies immunity in the Arabidopsis Pla-1 accession to viruses in two different genera of the Geminiviridae. Plant J 92 (5):796–807

Chapter 7

Absolute and Relative Quantification of Single-Stranded Positive-Sense RNA Viruses from Plant Tissue

Pinky Dutta and Kristiina Mäkinen

Abstract

Reverse transcription quantitative PCR (RT-qPCR) allows sensitive and specific measurement of mRNA transcripts from a given sample in a short period of time. Relative and absolute RT-qPCR are two strategies that could be used to quantify mRNA transcripts, based on the goal of the experiment. Here, we describe the protocol for the quantification of plant viral RNA transcripts from an infected sample using both strategies.

Key words Reverse transcription quantitative PCR, mRNA, Viral RNA, Relative gene expression, Absolute quantification

1 Introduction

Positive-sense plant RNA viruses play a prominent role in reducing crop yield and thus have a massive research interest [1, 2]. Since the prevention of viral disease incidence and spread is mainly dependent on preventive measures, early detection of infected plant materials is of utmost importance. Methods such as enzyme-linked immunosorbent assay (ELISA), reverse transcription polymerase chain reaction (RT-PCR), and reverse transcription quantitative PCR (RT-qPCR) are actively used for diagnosis and research purposes [3, 4]. With the advent of next-generation sequencing techniques, screening of several viruses can be done simultaneously; however, RT-qPCR remains a gold standard in the targeted detection of viruses [5].

RT-qPCR is a kinetic method and is based on the detection and quantification of a fluorescent reporter [6]. Fluorescent dye (e.g., SYBR Green), probes (e.g., TaqMan probe), and labeled oligonucleotide primers are different types of fluorescent reporters that are used in RT-qPCR [7]. SYBR green emits a fluorescent signal upon intercalating nonspecifically between double-stranded DNA, which

Elizabeth P. B. Fontes and Kristiina Mäkinen (eds.), *Plant-Virus Interactions*, Methods in Molecular Biology, vol. 2724, https://doi.org/10.1007/978-1-0716-3485-1_7,

is quantified by a quantitative PCR (qPCR) machine. Use of fluorescent probes provides higher specificity as intercalating dyes can bind and generate signal from non-specific products. RT-qPCR is a two-step process in which RNA molecules in a sample are reverse transcribed into complementary DNA (cDNA), which is further used as a template for qPCR. The assay is highly sensitive and specific, enabling detection and measurement from minute amount of nucleic acids allowing detection even from a single cell [8]. Apart from scrutinizing virus load and proliferation during pathogen infection, the assay is also used to study transcriptional changes in plants during different biotic and abiotic stress conditions [9–12].

Quantification of viral load can be determined by two strategies of RT-qPCR: relative and absolute quantification [13]. Relative gene expression analysis enables the quantification of a gene of interest relative to a reference gene. Comparative C_T (threshold cycle) method, also known as $2^{-\Delta\Delta C_T}$[14], is a widely used method to determine changes in expression of the gene of interest relative to a reference or a control gene. The $2^{-\Delta\Delta C_T}$ method relies on similar PCR efficiency of the target and reference gene. When primers with different PCR efficiency are used, Pfaffl method should be used to calculate relative gene expression [15]. The Pfaffl method accounts for the differences in PCR efficiency of both the reference and the target genes while calculating the ratio of deviation in threshold cycle of target gene over reference genes. Absolute expression of a gene of interest is quantified by calculating the copy number of a gene after transforming the data (C_T value) via a standard curve [16]. A standard curve to be used for viral RNA quantification can be constructed either using a known concentration of plasmid containing the viral genomic sequence or purified viral particles.

Implementation of relative or absolute quantification in a gene expression assay depends on the goal of the experiment. To ensure good experimental practice and generation of reliable and reproducible data, the MIQE guidelines for standardization of qPCR experiments should be followed [17, 18]. It states the minimum requirements required for publication of RT-qPCR data. Appropriate reference genes should be selected and validated for each experiment [14, 17, 18] to ensure their expression remains unaffected due to the treatment, and it is recommended that a minimum of two reference genes should be used [17, 18].

Here, we describe both the relative and absolute quantification of potato virus A RNA from *Nicotiana benthamiana*-infected plants. A layout of the RT-qPCR experimental design is in Fig. 1.

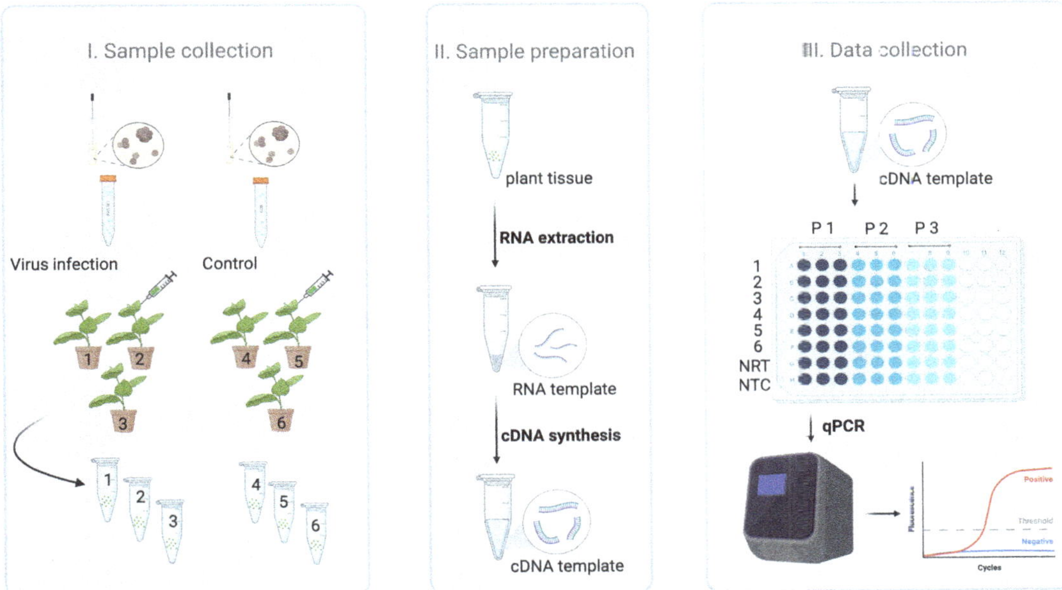

Fig. 1 Experimental set-up of viral RNA quantification in *N. benthamiana*. Numbers 1–3 and 4–6 represent biological replicates of infected and control samples. P1 = virus-specific primers, P2 and P3 = reference primers. NRT non-reverse transcriptase control, NTC non-template control. (Created with BioRender.com)

2 Materials

2.1 Initiation of Infection and Sample Collection

1. PVA viral construct derived from infectious complementary DNA (icDNA) of PVA strain B11 (GenBank Accession No. AJ296311) [19] and tagged with *Renilla* luciferase reporter rluc (35S-PVA: *rluc*) [20].
2. 35S-GUS construct [21] as a control.
3. *Agrobacterium tumefaciens* strain C58C1 containing the *vir* gene expression helper plasmid pGV2260 (*see* **Note 1**).
4. Four- to six-week-old *N. benthamiana* grown in a greenhouse with 16-h light at 22 °C and 8-h dark at 18 °C cycle.
5. LB + MACK: Luria–Bertani (LB) medium supplemented with 10 mM 4-morpholineethanesulfonic acid (MES) (pH 6.3), 20 μM acetosyringone (AS), carbenicillin (100 μg /mL), and kanamycin (50 μg/mL). To prepare 100 mL of LB + MACK medium, add 2 mL 0.5 M MES (pH 6.3), 13.33 μL 150 mM AS, 1 mL carbenicillin (100 μg/mL), and 500 μL kanamycin (50 μg/mL) to 96.48 mL of LB (*see* **Note 2**).
6. LB + agar plates: LB + agar medium supplemented with kanamycin (25 μg /mL) and carbenicillin (100 μg/mL). Add 1 mL carbenicillin (100 μg/mL) and 250 μL kanamycin (25 μg/mL) to 98.75 mL of LB agar cooled to 55 °C. Pour into petri dishes

and store at 4 °C for up to 1 month. 100 mL is enough to prepare five plates (*see* **Note 3**).

7. Infiltration or induction buffer: To make 100 mL of infiltration buffer, add 2 mL of 0.5 M MES pH 6.3, 1 mL of 1 M $MgCl_2$, and 100 μL of 150 μM AS to 96.9 mL nuclease-free water (*see* **Note 4**).
8. 1-mL syringe.
9. Cork borer.
10. 1.5-mL microcentrifuge tube.
11. 2-mm stainless steel beads.
12. Liquid nitrogen.

2.2 Two-Step RT-qPCR

1. Cryo-mill or mortar and pestle.
2. Heat block.
3. TRIzol (Invitrogen, USA) or total plant RNA extraction kit (Omega Bio-Tek, USA) or other similar product.
4. 100% chloroform.
5. 100% isopropanol.
6. 70% ethanol.
7. Nuclease-free water/MQ water.
8. Qubit RNA BR Assay Kit (Thermo Fisher Scientific).
9. Qubit Fluorometer/NanoDrop spectrophotometer.
10. 96-well/384-well plate.
11. 96-well/384-well plate centrifuge.
12. RQ1 DNase (Promega).
13. Gel-electrophoresis apparatus.
14. 1× TBE (Tris–borate–EDTA) buffer: Prepare 1 L of 1× TBE buffer by adding 100 mL of 10× TBE stock solution to 900 mL of nuclease-free water (*see* **Note 5**).
15. 1% agarose gel: Dissolve 1 g of agarose in 100 mL of 1× TBE buffer by heating in a microwave-safe flask. Cool down the agarose solution to about 50 °C and add any agarose gel stain [e.g., ethidium bromide (0.2–0.5 μg/mL) or any nucleic acid stain].
16. RevertAid cDNA synthesis kit (Thermo Scientific, Lithuania) or other similar product.
17. PVA CP specific primers: forward (5′- CATGCCCAGG TATGGTCTTC) and reverse (5′-ATCGGAGTGGTTGCAGT GAT).
18. Oligo-dt or random primers.

19. Primers for reference genes from *N. benthamiana*: PP2A, forward (5′-GACCCTGATGTTGATGTTCG) and reverse (5′-GAGGGATTTGAGAGAGATTTC); FBOX, forward (5′-GG CACTCACAAACGTCTATTTC) and reverse (5′-TGGGAGGCATCCTGCTTAT).
20. Maxima SYBR green qPCR master mix (Thermo Scientific) or other similar product.
21. Clear optical adhesive film.
22. PCR thermocycler.
23. qPCR thermocycler.

3 Methods

3.1 Initiation of PVA Infection in N. benthamiana via Agrobacterium-Mediated Infiltration

1. Revive *Agrobacterium* strain C58C1 harboring PVA and GUS control constructs from glycerol stock by streaking on a LB + agar medium and incubate at 28 °C for 2 days.
2. Inoculate a single colony of each construct to 5 mL LB+ MACK and incubate on a shaker with 220 rpm at 28 °C overnight.
3. Inoculate fresh LB+ MACK with the primary starter culture (1: 100) and incubate at 28 °C for 4 h.
4. Harvest bacterial cells by centrifuging at 5000*g* for 5 min at 20 °C and wash once with MQ water and induction buffer, sequentially.
5. Prepare an infiltration mix by adjusting bacterial OD_{600} to 0.3 with induction buffer and incubate for 1 h at room temperature (*see* **Note 6**).
6. Inject the infiltration mix into the abaxial side of two fully developed *N. benthamiana* leaves to initiate infection. Inoculate control plants with infiltration mix containing *Agrobacterium* carrying the 35S-GUS construct. At least three biological replicates (three plants) are required per treatment (three treatment and three control).
7. Maintain the infected and control plants in the greenhouse (16 h–22 °C light/8 h–18 °C dark cycle).

3.2 Total RNA Extraction

1. Collect leaf tissue (~100 mg) from inoculated leaves using a cork borer 3 days post-inoculation (dpi) and snap freeze in liquid nitrogen (*see* **Note 7**). Leaf tissue from an individual plant is collected into a single tube. Samples from multiple plants should not be pooled.
2. Pulverize leaf tissues in a cryo-mill with six 2-mm steel beads or a pre-cooled mortar pestle with liquid nitrogen (*see* **Note 8**).

3. Add 750 μL of TRIzol to each sample, invert the tube to mix, and incubate for 5 min at RT.
4. Add 150 μL of chloroform, mix thoroughly, and incubate for 3 min at RT.
5. Centrifuge the sample at 12,000*g* at 4 °C for 15 min to separate the sample into three distinct phases.
6. Transfer the upper aqueous phase to a separate microcentrifuge tube and add 500 μL of 100% isopropanol and incubate for 10 min at 4 °C.
7. Centrifuge at 12,000*g* at 4 °C for 10 min to precipitate the RNA.
8. Remove the supernatant and add 500 μL of cold 70% ethanol.
9. Centrifuge at 7500*g* at 4 °C for 5 min and discard the supernatant.
10. Air dry the RNA pellet for 5 min and resuspend in 25 μL of nuclease-free water.
11. Measure RNA concentration using Qubit or NanoDrop.

3.3 Removal of Contaminating DNA

1. Perform DNase treatment with RQ1 DNase (Promega) or a similar product, on 1 μg of total RNA to get rid of contaminating DNA.
2. Incubate the samples at 37 °C for 30 min.
3. Heat-inactivate DNase by adding DNase stop solution and incubating at 65 °C for 10 min. DNase stop solution comes along with RQ1 DNase.

3.4 Reverse Transcription of Total RNA

To generate a DNA template compatible for a two-step RT-qPCR, RNA is reverse-transcribed to cDNA using reverse transcriptase.

1. Run 300 ng of DNase-treated RNA on an 1% agarose gel in 1× TBE buffer to check RNA integrity prior to reverse transcription (*see* **Note 9**).
2. For a 20 μL PVA cDNA synthesis mix, add 0.2 μg/mL of random primers (1 μL) to 500 ng of DNase-treated RNA, to yield a total volume of 12 μL (*see* **Note 10**).
3. Incubate the samples at 65 °C for 5 min to get rid of RNA secondary structures.
4. Chill on ice and add RT mix containing 4 μL of 5× reaction buffer, 1 μL (20 U) of RiboLock RNase inhibitor, 1 μL (200 U) of reverse transcriptase, and 2 μL (1 mM) dNTP mix (*see* **Note 11**).
5. A negative control to check for genomic DNA contamination is included by omitting the reverse transcriptase (RT minus), while template RNA is omitted in another to assess

contamination of reagents (non-template control). These reactions are done in the same way as the actual test reactions in steps 2–4, but to reach a volume of 20 μL, the missing volume of omitted RT and template RNA is supplemented with nuclease-free water.

3.5 Relative Quantification RT-qPCR

Multiple reference genes must be used for normalization of qPCR assay, and these genes must be validated for each experiment separately [17, 18].

1. qPCRs are performed in triplicate per cDNA sample.
2. Three different master mixes are to be prepared: one each for virus-specific primers and two reference genes.
3. To make a qPCR master mix of 9 μL, add 5 μL of 2× SYBR green qPCR mix, 500 μM (0.5 μL) each of forward and reverse primer, and 3 μL of nuclease-free water.
4. Add the master mix into a 96 or 384-well plate and add 1 μL of cDNA into each well (*see* **Note 12**).
5. Cover and seal the plate with a clear optical adhesive film.
6. The qPCR plate is briefly centrifuged (~10 s at 1000 rpm).
7. Amplify the samples in a real-time PCR machine by subjecting it to a single cycle of initial denaturation at 95 °C for 30 s and subsequently 39 cycles each of denaturation at 95 °C for 10 s, annealing at 59 °C for 30 s, extension at 72 °C for 30 s.
8. Generate a melt curve by heating from 60 °C to 95 °C in increments of 0.5 °C/s programmed together with the RT-qPCR run (*see* **Note 13**).

3.6 Absolute Quantification

The general procedure followed for absolute quantification is similar to relative quantification. However, to determine the exact copy number of a gene in a sample, the inclusion of a linear standard curve in the qPCR run is necessary.

1. Calculate the copy number of a PVA icDNA plasmid using the following formula:

Copy number

$$= \frac{\text{Amount of DNA (ng)} \times 6.022 \times 10^{23}}{\text{Length of DNA (bp)} \times 1 \times 10^{9} \times \text{mass of DNA bp}}$$

- 6.022×10^{23}—Avogadro's constant.
- 1×10^{9}—A factor used to convert g to ng.
- Mass of DNA bp—The average mass of a DNA bp is either 660 (dsDNA) or 330 (ssDNA) g/mole. This value depends on what is selected as the type of DNA.

2. To prepare a standard of the range 10^8–10 molecules/μL, make a serial dilution of PVA plasmid. Mix 10 μL of PVA plasmid with 90 μL of nuclease-free water, vortex, and centrifuge it briefly. Repeat the steps until 10 PVA plasmids/μL are obtained (*see* **Note 14**).

3.7 Data Analysis for Relative Gene Expression

Different methods of calculating relative gene expression exist. When the primer efficiency is equal, relative gene expression is calculated using the $2^{-\Delta\Delta C_T}$ method [14], while the Pfaffl method [15] should be used when the differences in primer efficiencies exist.

1. Calculate the PCR efficiency of primer pairs, by generating a linear standard curve of the primers. The linear regression results in the equation of a line with the general form: $y = mx + c$, where m is the slope of the line and c is the y-intercept.

$$\%\text{Efficiency} = (E - 1) \times 100,$$

where

$E = 10^{-1/m}$.

The qPCR run information can be exported to Microsoft Excel spreadsheet. The data contain C_T, values of the samples, which can be fit into an equation to determine gene expression as fold change.

2. Comparative C_T method

$$\text{Relative fold change} = 2^{-\Delta\Delta C_T},$$

where

$\Delta\Delta C_T = \Delta C_T$ (infected sample) – ΔC_T (healthy sample).

ΔC_T (infected sample) = C_T (viral coat protein gene in the infected sample) – C_T (reference gene in the infected sample).

ΔC_T (healthy sample) = C_T (viral coat protein gene in the healthy sample) – C_T (reference gene in the healthy sample).

3. Pfaffl method

$$\text{Relative fold change} = \frac{(E\ \text{reference})^{\Delta C_T(\text{reference})}}{(E\ \text{target})^{\Delta C_T(\text{target})}}$$

where

ΔC_T (target) = C_T (target gene in the healthy sample) – ΔC_T (target gene in infected sample).

ΔC_T reference) = C_T (reference gene in healthy sample) − ΔC_T (reference gene in infected sample).

E = PCR efficiency.

3.8 Data Analysis for Absolute Quantification

The reliability of the results obtained in absolute quantification depends on the accuracy of the standard curve. Absolute counts of a gene copy are calculated by plotting the C_T value of the sample to the standard curve and extrapolating a value.

The qPCR machine calculates and returns the copy number of the genes in the infected and healthy samples based on the standard curve included in the run.

Student's t-test is performed to determine if the difference in relative/absolute quantification between the infected and the control samples is statistically significant.

4 Notes

1. Both the viral and control constructs were transformed into *Agrobacterium tumefaciens* strain C58C1 containing the *vir* gene expression helper plasmid pGV2260.
2. Prepare Luria–Bertani (LB) medium by dissolving 10 g tryptone, 5 g yeast extract, and 5 g NaCl in 1000 mL of nuclease-free water and autoclave. Prepare 0.5 M MES by dissolving 97.62 g MES in 750 mL MQ water, adjust the pH to 6.7 with 10 N NaOH, and make up the volume to 1000 mL with MQ water. Autoclave and store at 4 °C. Prepare 100 mM AS by dissolving 490 mg AS in 25 mL dimethyl sulfoxide (DMSO), make 100 μL aliquots, and store at −20 °C. Carbenicillin (10 mg/mL) and kanamycin (10 mg/mL) stock solution of 100 mL is prepared by dissolving 1 g of the required antibiotic in 100 mL MQ water and sterile filtered through a 0.2-mm filter and stored at −20 °C.
3. To prepare LB+ agar plates dissolve 9 g tryptone, 4.5 g yeast extract, 13.5 g agar, and 4.5 g NaCl in 900 mL of nuclease-free water, boil and autoclave, and store at 4 °C.
4. Dissolve 203.3 g $MgCl_2 \cdot 6H_2O$ in 1000 mL MQ water to obtain 1 M $MgCl_2$ stock solution, autoclave, and store this stock solution at room temperature.
5. Prepare 10× TBE stock solution by dissolving 121.1 g of Tris base, 61.83 g of boric acid, and 7.44 g EDTA in 1000 mL nuclease-free water, autoclave, and store at room temperature.
6. Bacterial OD_{600} can be adjusted according to experimental design.

7. Date of sample collection can be determined based on experimental design.
8. It is essential to retain the integrity of viral RNA, and hence, the samples should be kept frozen during pulverization until TRIzol is added. Addition of TRIzol stabilizes the RNA, maintaining RNA integrity.
9. Sharp ribosomal RNA bands indicate that RNA is still intact, while any smearing indicates degradation.
10. Depending on the utility of the synthesized cDNA, the primer type is to be chosen. Random hexamers synthesize cDNA by binding to random sites on the total RNA population (viral RNA, rRNA, and plant mRNA). Usage of oligo-dT primers selectively enriches cDNA for poly-A containing viral and plant mRNAs, as it binds to the poly-A tail of mRNA, while gene-specific primers enrich the cDNA pool for the gene of interest.
11. These reagents come with the cDNA synthesis kit.
12. Using a pipetting robot/liquid handler to pipette reagents helps to limit pipetting errors and human fatigue. If a pipetting robot is not available, the use of a multichannel pipette is advisable.
13. A melt curve analysis enables the determination of the specificity of qPCR product formed. A single peak indicates amplification of a single PCR product, hence highly specific.
14. A standard can also be prepared using purified viral particles.

Acknowledgements

We thank the Academy of Finland (Grant 1332950) for KM and Doctoral Programme in Plant Sciences, University of Helsinki, for PD for financial support.

References

1. Scholthof KB, Adkins S, Czosnek H, Palukaitis P, Jacquot E, Hohn T, Hohn B, Saunders K, Candresse T, Ahlquist P, Hemenway C, Foster GD (2011) Top 10 plant viruses in molecular plant pathology. Mol Plant Pathol 12:938–954
2. Rybicki EP (2015) A top ten list for economically important plant viruses. Arch Virol 160: 17–20
3. Mehetre GT, Leo VV, Singh G, Sorokan A, Maksimov I, Yadav MK, Upadhyaya K, Hashem A, Alsaleh AN, Dawoud TM, Almaary KS, Singh BP (2021) Current developments and challenges in plant viral diagnostics: a systematic review. Viruses 13:412
4. Rubio L, Galipienso L, Ferriol I (2020) Detection of plant viruses and disease management: relevance of genetic diversity and evolution. Front Plant Sci 11:1092
5. Santala J, Valkonen JPT (2018) Sensitivity of small RNA-based detection of plant viruses. Front Microbiol 9:939
6. Morrison TB, Weis JJ, Wittwer CT (1998) Quantification of low-copy transcripts by continuous SYBR Green I monitoring during amplification. BioTechniques 24:954–958, 960, 962

7. Valasek MA, Repa JJ (2005) The power of real-time PCR. Adv Physiol Educ 29:151–159
8. Suomalainen M, Greber UF (2021) Virus infection variability by single-cell profiling. Viruses 13:1568
9. Catto MA, Mugerwa H, Myers BK, Pandey S, Dutta B, Srinivasan R (2022) A review on transcriptional responses of interactions between insect vectors and plant viruses. Cells 11:693
10. Mantri NL, Ford R, Coram TE, Pang EC (2007) Transcriptional profiling of chickpea genes differentially regulated in response to high-salinity, cold and drought. BMC Genomics 8:303
11. Sathiyaraj G, Lee OR, Parvin S, Khorolragchaa A, Kim YJ, Yang DC (2011) Transcript profiling of antioxidant genes during biotic and abiotic stresses in Panax ginseng C. A. Meyer. Mol Biol Rep 38:2761–2769
12. Xu M, Chen J, Huang Y, Shen D, Sun P, Xu Y, Tao X (2020) Dynamic transcriptional profiles of arabidopsis thaliana infected by tomato spotted wilt virus. Phytopathology 110:153–163
13. Freeman WM, Walker SJ, Vrana KE (1999) Quantitative RT-PCR: pitfalls and potential. BioTechniques 26:112–122, 124–125
14. Livak KJ, Schmittgen TD (2001) Analysis of relative gene expression data using real-time quantitative PCR and the 2(-Delta Delta C (T)) method. Methods 25:402–408
15. Pfaffl MW (2001) A new mathematical model for relative quantification in real-time RT-PCR. Nucleic Acids Res 29:e45
16. Bustin SA (2000) Absolute quantification of mRNA using real-time reverse transcription polymerase chain reaction assays. J Mol Endocrinol 25:169–193
17. Bustin SA (2010) Why the need for qPCR publication guidelines? – The case for MIQE. Methods 50:217–226
18. Bustin SA, Benes V, Garson JA, Hellemans J, Huggett J, Kubista M, Mueller R, Nolan T, Pfaffl MW, Shipley GL, Vandesompele J, Wittwer CT (2009) The MIQE guidelines: minimum information for publication of quantitative real-time PCR experiments. Clin Chem 55:611–622
19. Puurand U, Makinen K, Paulin L, Saarma M (1994) The nucleotide sequence of potato virus A genomic RNA and its sequence similarities with other potyviruses. J Gen Virol 75 (Pt 2):457–461
20. Eskelin K, Suntio T, Hyvarinen S, Hafren A, Makinen K (2010) Renilla luciferase-based quantitation of Potato virus A infection initiated with Agrobacterium infiltration of N. benthamiana leaves. J Virol Methods 164:101–110
21. Eskelin K, Hafren A, Rantalainen KI, Makinen K (2011) Potyviral VPg enhances viral RNA translation and inhibits reporter mRNA translation in planta. J Virol 85:9210–9221

Chapter 8

Quantification of Virion-Sense and Complementary-Sense DNA Strands of Circular Single-Stranded DNA Viruses

Edgar A. Rodríguez-Negrete and Ana Grande-Pérez

Abstract

Circular ssDNA viruses are ubiquitous and can be found in both prokaryotes and eukaryotes. To understand the interaction of ssDNA viruses with their hosts, it is important to characterize the dynamics of viral sense (VS) and complementary-sense (CS) viral strands during the infection process. Here, we present a simple and rapid protocol that allows sensitive and accurate determination of the VS and CS strands generated during viral infection.

The method consists of a two-step qPCR in which the first step uses a strand-specific (CS or VS) labeled primer and T4 DNA polymerase that lacks strand displacement activity and makes a single copy per VS or CS strand. Next, the T4 DNA polymerase and unincorporated oligonucleotides are removed by a silica membrane spin column. Finally, the purified VS or CS strands are quantified by qPCR in a second step in which amplification uses a tag primer and a specific primer. Absolute quantification of VS and CS strands is obtained by extrapolating the Cq data to a standard curve of ssDNA, which can be generated by phagemid expression. Quantification of VS and CS strands of two geminiviruses in infections of *Solanum lycopersicum* (tomato) and *Nicotiana benthamiana* plants using this method is shown.

Key words Circular ssDNA viruses, CRESS DNA viruses, Strand-specific quantification, qPCR, Virus sense, Complementary sense, Geminiviruses, Phagemid

1 Introduction

Circular single-stranded DNA (ssDNA) viruses are a very abundant and ubiquitous group of viruses [1, 2] that are receiving increasing interest in recent years due to the discoveries of metagenomics [3–15] and meta-transcriptomics [16, 17]. Bacteria, archaea, and eukaryotes can be infected by circular ssDNA viruses [18, 19]. However, only those transmitted by arthropods that infect plants are pathogenic and cause serious losses to agriculture-based economies [20]. In addition, circular ssDNA viruses are endowed with high mutation and recombination rates, making them an important group of emerging viruses with enormous evolutionary potential [21–27].

Elizabeth P. B. Fontes and Kristiina Mäkinen (eds.), *Plant-Virus Interactions*, Methods in Molecular Biology, vol. 2724, https://doi.org/10.1007/978-1-0716-3485-1_8,

A growing list of circular ssDNA viruses is classified into 21 families. Prokaryotic viruses that infect bacteria are found in the *Inoviridae*, *Paulinoviridae*, *Plectoviridae*, *Microviridae*, and *Finnlakeviridae* families [28–30], whereas archaeal circular ssDNA viruses are classified in the *Pleolipoviridae* and *Spiraviridae* families [31, 32]. Eukaryotic circular ssDNA viruses belong to the families *Bacilladnaviridae*, which infect only diatoms; *Smacoviridae*, found in vertebrates and insect feces [33]; *Anelloviridae* and *Circoviridae*, which infect higher vertebrates [34, 35]; *Genomoviridae*, which groups fungal viruses and a series of viruses associated with vertebrate feces and insects [36]; *Redondoviridae*, identified from human samples [37]; the three new families, *Naryaviridae*, *Nenyaviridae*, and *Vilyaviridae*, of circular ssDNA viruses associated with protozoan parasites belonging to the genera *Entamoeba* and *Giardia* [38]; and finally, the circular ssDNA plant–virus families, which are the recently added *Metaxyviridae* family, and the *Geminiviridae* and *Nanoviridae* families, plus their associated circular ssDNA satellites, classified in the *Alphasatellitidae* and *Tolecusatellitidae* families [39–41]. Most viruses with a circular ssDNA genome encode the replication-associated protein Rep and replicate by a rolling circle replication (RCR) mechanism [42, 43]. All eukaryotic circular ssDNA viruses have been grouped into the so-called CRESS DNA viruses, i.e., circular Rep encoding single-stranded DNA viruses [44, 45]. They replicate in the nucleus after the VS (virion-sense) strand, which is the encapsidated strand, is transformed into a dsDNA replicative intermediate, or covalently closed replicative form (RF), by host DNA polymerases [42]. Then, the Rep protein cleavages the RF to start the RCR that will produce VS copies that will be recircularized and encapsidated into progeny virions or will enter a new RCR cycle. Expression of ORFs may occur from the VS or from the complementary-sense (CS) strands [46].

To understand the mechanisms involved in virus–host interaction, characterization of viral strand accumulation during the infection process is essential. Although some qPCR methods exist to quantify circular ssDNA viruses [47–54], they do not discern the type (VS or CS) of viral molecules. The protocol presented here allows sensitive and accurate determination of the VS and CS strands generated during viral infection and thus enables detailed analysis of the infection process of circular ssDNA viruses.

The method consists of a simple and rapid two-step qPCR in which the first step uses a tagged strand-specific primer and T4 DNA polymerase to copy the CS or the VS strand. T4 DNA polymerase lacks strand displacement activity, and when it performs the complete circular DNA copy, it cannot proceed beyond the 5′ end of the extended primer. Therefore, only one copy is generated per VS or CS strand. Next, the DNA polymerase and unincorporated oligonucleotides are removed using a silica membrane spin column. Finally, the purified VS or CS strands are quantified by qPCR in a second step in which amplification uses a tag primer and a specific primer. Absolute quantification of VS and CS strands is obtained by extrapolating the Cq data to a standard curve consisting of serial dilutions containing between 10^7 and 10^3 molecules of ssDNA (which can be generated by phagemid expression) or circular dsDNA.

This protocol has been used to quantify sensitively and accurately the VS and CS strands of two geminiviruses in infections of *Solanum lycopersicum* (tomato) and *Nicotiana benthamiana* plants. It revealed differences in VS/CS ratios between viruses in both hosts and determined that the majority of CS molecules were found in dsDNA intermediates [55].

2 Materials

2.1 Generation of Circular DNA Standards for Quantitative PCR (qPCR)

1. Tomato yellow leaf curl Sardinia virus (TYLCSV) (accession number L27708) and tomato yellow leaf curl virus (TYLCV) mild strain (accession number AF071228) full-length genome monomeric constructs pBTSvs and pBTScs for TYLCSV, and pBTYMvs and pBTYMcs for TYLCV, inserted in pBSK II (−) phagemid, in virion sense (VS) and complementary sense (CS) relative to fl phage origin, respectively (*see* **Note 1**).
2. *Escherichia coli* F′ strain (*see* **Note 2**).
3. M13KO7 bacteriophage helper virus suspension (10^{10} pfu/mL) (*see* **Note 3**).
4. TYP media: 1.6% w/v Bacto Tryptone, 1.6% w/v yeast extract, 8.5 mM NaCl, and 1 mM K_2HPO_4 (*see* **Note 4**). Store at room temperature.
5. Nuclease-free deionized H_2O (*see* **Note 5**). Store at room temperature.
6. Ampicillin 100 mg/mL and kanamycin 70 mg/mL (*see* **Note 6**). Store at −20 °C.
7. PEG solution: 20% w/v polyethylene glycol (PEG) 8000, 2.5 M NaCl (*see* **Note 7**). Store at room temperature.
8. TE buffer: Tris–EDTA pH 8.0 (*see* **Note 8**). Store at room temperature.
9. Phenol:chloroform 1:1 solution (*see* **Note 9**). Store the mixture in a clean brown glass bottle in a fume hood at room temperature.
10. 3 M sodium acetate ($NaCH_3COOH$) pH 5.2 (*see* **Note 10**). Store at room temperature.
11. 100% absolute ethanol. Store at −20 °C.

12. 70% ethanol. Store at −20 °C.
13. Agarose gel electrophoresis equipment: 1% agarose gel in TBE buffer stained with EtBr or red safe, electrophoresis horizontal chamber, molecular weight DNA ladder (100 bp–10 kb). Gel loading dye: 50% glycerol, 40 mM EDTA pH 8.0, 0.1% bromophenol blue w/v, 0.1% xylene cyanol w/v (*see* **Note 11**).

2.2 Systemic Geminivirus Infections of Nicotiana benthamiana and Solanum lycopersicum

1. Plants: *N. benthamiana* and tomato (*S. lycopersicum*) at 3–4 weeks of stage of development.
2. TYLCSV (accession number L27708) and TYLCV mild strain (accession number AF071228) agroinfectious clones (*see* **Note 12**).
3. Luria–Bertani (LB) media: 1% tryptone, 0.5% yeast extract, and 1% NaCl pH 7.0 (*see* **Note 13**).
4. Rifampicin (100 mg/mL), tetracycline (5 mg/mL), and kanamycin (50 mg/mL) (*see* **Note 14**).
5. Infiltration solution: 10 mM MES pH 5.6, 10 mM $MgCl_2$, and 100 μM acetosyringone (*see* **Note 15**).
6. 2× CTAB extraction buffer: 100 mM Tris–HCl, 1.4 M NaCl, 20 mM EDTA, 2% CTAB, and 0.2% 2-mercaptoethanol (*see* **Note 16**).

2.3 Two-Step qPCR Quantification of VS and CS Strands

1. Primers for strand-specific (VS or CS) DNA labeling carrying a TAG sequence (AGTTTAAGAACCCTTCCCGC) at the 5′ end and target-specific sequence, VS (GGACTTTA CATGGGCCTTCAC), and CS (GAAGGCTGAACTTCGA CAGC) at 3′ end (*see* **Note 17**).

   TAG-VS: 5′ AGTTTAAGAACCCTTCCCGC GGACTTTA CATGGGCCTTCAC 3′.

   TAG-CS: 5′ AGTTTAAGAACCCTTCCCGC GAAGGCT GAACTTCGACAGC 3′.
2. Primers for strand-specific (VS or CS) and total viral DNA amplification (*see* **Note 18**).

   TAG: 5′ AGTTTAAGAACCCTTCCCGC 3′.

   VS: 5′ GGACTTTACATGGGCCTTCAC 3′.

   CS: 5′ GAAGGCTGAACTTCGACAGC 3′.
3. Primers for qPCR data normalization (*see* **Note 19**).

   25S UNIV (+): 5′ ATAACCGCATCAGGTCTCCA 3′.

   25 UNIV (−): 5′ CCGAAGTTACGGATCCATTT 3′.
4. T4 DNA polymerase kit (*see* **Note 20**).
5. dNTPs nucleotide mix at 1.7 mM each.
6. Silica-based spin columns inclusion range 100 bp–10 kb (*see* **Note 21**).

7. Quantitative PCR (qPCR) Master Mix containing SYBR Green or Eva Green chemistries.
8. Quantitative PCR (qPCR) thermocycler.

3 Methods

3.1 Synthesis of Single-Stranded DNA Standards

To obtain single-stranded circular molecules, *Escherichia coli* F′ strain is first transformed with tomato TYLCSV or TYLCV full-length genome monomeric constructs (pBTSvs and pBTScs for TYLCSV, and pBTYMvs and pBTYMcs for TYLCV, inserted in pBSK II (−) phagemid, in virion sense (VS) and complementary sense (CS) relative to f1 phage origin, respectively, using standard chemical or electroporation transformation methods and then infected with M13KO7 bacteriophage helper virus according to the next protocol.

1. Inoculate 1.5 mL of TYP media containing 100 μg/mL ampicillin with a fresh colony containing the monomeric viral construct (VS or CS) and 1.5 μL of M13KO7 helper virus suspension 10^{10} pfu/mL (final concentration 10^7 pfu/mL) and grow with shaking at 225 rpm and 37 °C in an orbital shaker for 60 min.
2. Then, add 1.5 μL of 70 mg/mL kanamycin stock solution (final concentration 70 μg/mL) and grow for a further 14–18 h at 225 rpm and 37 °C in an orbital shaker.
3. Centrifuge each culture at 13,000 rpm for 10 min, transfer the supernatant to a new tube, and add 200 μL of 20% PEG solution and mix gently.
4. Centrifuge the mixture at 13,000 rpm for 10 min, discard the supernatant, air dry the pellet, and resuspend in 300 μL of TE buffer.
5. Add 300 μL of 1:1 phenol:chloroform solution, mix vigorously by vortexing, centrifuge the mixture at 13,000 rpm for 10 min, and carefully transfer the upper aqueous phase to a new tube. Repeat **step 5**.
6. Add to the aqueous phase (300 μL) 30 μL of 5 M Na-CH_3COOH and 700 μL of absolute ethanol, mix gently by inversion, and centrifuge the mixture at 13,000 rpm for 10 min to precipitate the DNA.
7. Discard the supernatant, wash the pellet with 1 mL of 70% ethanol, centrifuge the mixture at 13,000 rpm for 5 min, discard the supernatant, air dry the pellet, and resuspend in 20 μL of nuclease-free deionized H_2O.

8. Assess the integrity of the single-stranded circular molecules containing the VS and the CS strands by 1% agarose gel electrophoresis.
9. In parallel, evaluate the purity and yield of the single-stranded circular molecules by spectrophotometric assay according to the A260/A280 ratio (*see* **Note 22**).
10. Determine the number of VS or CS molecules from the mass of the single-stranded circular viral standards using the following formula:

$$\text{ssDNA copy number} = \text{moles of ssDNA} \times \text{Avogadros number}$$

where

Avogadro's number = 6.025×10^{23}.

moles of ssDNA = mass of ssDNA (g)/molecular weight of ssDNA (g/mol).

molecular weight of ssDNA = number of deoxynucleotide monophosphates of ssDNA × average of molecular weight of deoxynucleotide monophosphates (308.97 g/mol).

To determine the number of double-stranded circular viral standard molecules, the average molecular weight of a base pair = 617.96 g/mol is used to calculate moles of dsDNA (*see* **Note 23**).

3.2 Systemic TYLCSV Infections of Nicotiana benthamiana *and Tomato*

For inoculation of TYLCSV or TYLCV in *N. benthamiana* and *S. lycopersicum* plants, *A. tumefaciens* GV3101 strain harboring viral partial dimeric tandem repeat constructs are used (*see* **Note 24**). As mock inoculation control, *A. tumefaciens* GV3101 strain harboring binary empty vector is used.

1. *A. tumefaciens* GV3101 strains containing the viral clones and the empty binary vector are refreshed in LB solid medium supplemented with 100 μg/mL rifampicin, 5 μg/mL tetracycline, and 50 μg/mL kanamycin (final concentration), incubating the petri dishes for 3 days at 28 °C.
2. An individual colony of each strain is pricked with a sterile toothpick and inoculated into 50 mL of LB liquid medium supplemented with 100 μg/mL rifampicin, 5 μg/mL tetracycline, and 50 μg/mL kanamycin (final concentration) and incubated overnight at 28 °C with vigorous shaking.
3. A bacterial pellet is obtained by centrifugation at 4500 rpm for 15 min, the supernatant is discarded, and the pellet resuspended in infiltration solution (10 mM MES pH 5.6, 10 mM $MgCl_2$, 100 μM acetosyringone) to OD_{600} = 1.0, and the suspension incubated at room temperature overnight in the dark.

4. For *N. benthamiana* plants (3 plants), 1 mL of the corresponding suspension is infiltrated in two totally expanded leaves in the abaxial zone using a 1-mL syringe without a needle. For *S. lycopersicum* plants (three plants), the same procedure is performed on two fully expanded leaves, and in addition, 1 mL of suspension is injected directly into the axillary shoot using a 1-mL syringe with a needle.
5. Plants are maintained for 4 days in a bioclimatic chamber at 24 °C under long-day conditions (16/8 light/dark) and then transferred to the greenhouse.
6. Uninoculated new apical leaf samples are collected from groups of three plants at 7, 14, 28, and 42 days post-inoculation (dpi), and tissue is used for total DNA extraction using the CTAB-based method. Equal amounts of extracted DNA (1 μg) collected at each time point from the three samples are pooled, and 200 ng of the mixture is analyzed by standard or two-step qPCR protocols.

3.3 Absolute Quantification of VS and CS Strands by Two-Step qPCR

We have developed a protocol for sensitive and reliable quantification of virion-sense (VS) and complementary-sense (CS) strands generated during viral infection, allowing detailed analysis of the infection process of circular ssDNA viruses, including the monopartite begomoviruses (Family *Geminiviridae*) TYLCSV and TYLCV. The method consists of a simple and rapid two-step qPCR in which the first step uses a tagged strand-specific primer and T4 DNA polymerase (which lacks strand displacement activity) to copy the CS or VS strand. Unincorporated oligonucleotides are then removed using a silica membrane spin column. Finally, the purified VS or CS strands are quantified by qPCR in a second step in which amplification takes place using a tag primer and a specific primer. In addition, the total amount of viral DNA (VS + CS) is quantified using viral-specific primers (Fig. 1). Absolute quantification of VS, CS, and total viral DNA strands is obtained by substituting the Cq data into the y-value of the equation of the corresponding standard curve regression plot in a dynamic range from 10^7 to 10^3 viral molecules.

1. Prepare single-stranded (VS or CS) and double-stranded (VS + CS) viral DNA standards by mixing 10 μL of 10^8 molecules/μL stock with 90 μL of ultrapure H_2O to obtain the corresponding 10^7 molecules/μL dilution. Mix by vortex and perform a short spin. Repeat the serial dilution procedure until 10^3 viral molecules/μL are obtained.
2. For tagging single-stranded standards in the first step, mix 1 μL of 1μM TAG-VS or TAG-CS primers (for labeling of VS or CS standards, respectively), 1 μL of dNTPs mix at 1.7 mM of each nucleotide, 1 μL of T4 DNA Pol 10× buffer, 1 μL of the

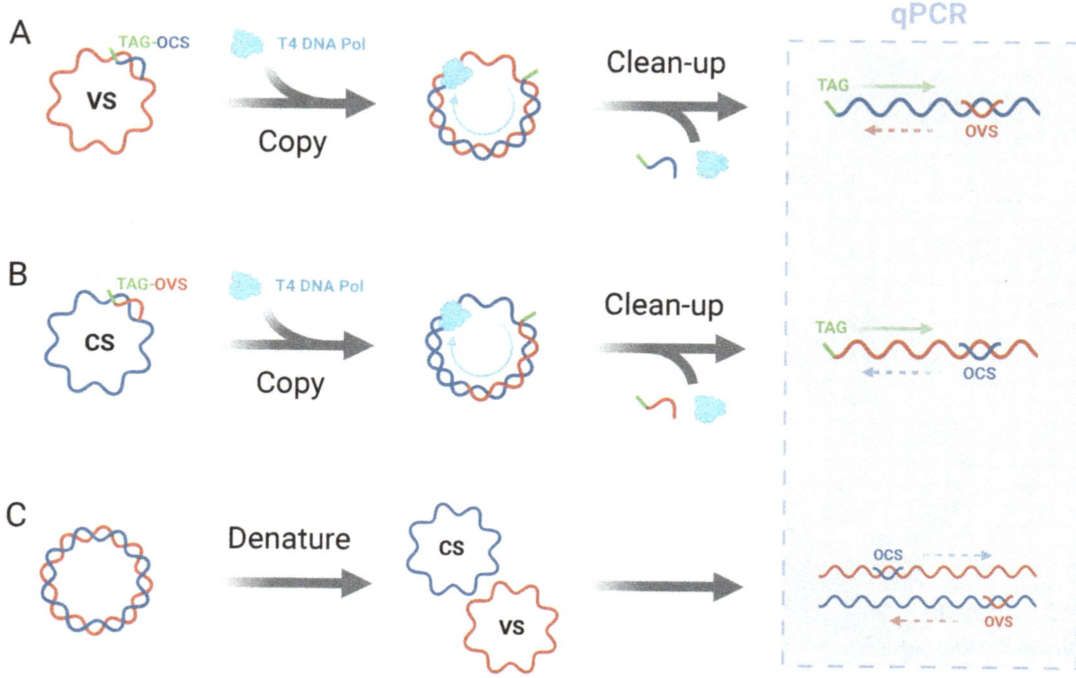

Fig. 1 Outline of the protocol for strand-specific quantification. Two-step quantitative PCR (qPCR) procedure for quantification of viral sense (VS) and complementary-sense (CS) DNA molecules. In the first step, a T4 DNA polymerase extension of (**a**) the VS strand using the OCS-TAG primer or of (**b**) the CS strand using the OVS-TAG primer is performed. Subsequently, qPCR amplification is carried out to quantify (**a**) the VS strand using primers TAG and OVS or (**b**) the CS strand using primers TAG and OCS. In both cases, the primers used for T4 polymerase extension are removed prior to qPCR using a silica column that allows recovery of DNA molecules larger than 100 bp and smaller than 10 kb. (**c**) Standard qPCR to quantify both the VS and CS strands using the OVS and OCS primers. (Created with BioRender.com)

corresponding single-stranded standard (concentration from 10^8 to 10^4), and 5 μL of ultrapure H_2O.

3. The reaction mixes are denatured at 95 °C for 10 min, cooled to room temperature, and incubated at 37 °C for 30 min with 1 unit T4 DNA polymerase (1 unit/μL).
4. Unincorporated oligonucleotides are removed using silica membrane spin columns (inclusion range 100 bp–10 kb), according to manufacturer's instructions, and the DNA is eluted in 50 μL of ultrapure H_2O.
5. For the second step of the qPCR procedure, 2 μL of the single-stranded tagging reactions or double-stranded standards are mixed with 10 μL of SYBR Green or Eva Green Master mix 2×, 1 μL of the corresponding 10 μM primer set (TAG + CS primers, or TAG + VS primers, or CS + VS primers, for quantification of VS, CS, or total viral strands, respectively), and 6 μL of ultrapure H_2O. Each standard dilution is analyzed by three technical replicates.

A

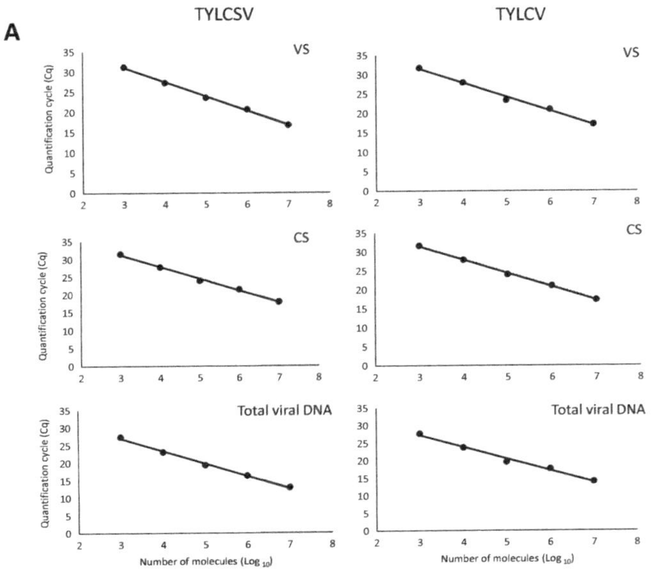

B

	TYLCSV	TYLCV
VS	$R^2 = 0.9962$	$R^2 = 0.98942$
	E = 95 %	E = 94 %
	y= -3.45x + 41.10	y = -3.4752x + 42.99
CS	$R^2 = 0.995$	$R^2 = 0.99854$
	E = 99 %	E = 96.7 %
	y = -3.3437x + 41.242	y = -3.4021x + 43.55
Total	$R^2 = 0.9955$	$R^2 = 0.9877$
	E = 97.5 %	E = 99 %
	y = -3.3829x + 37.30	y = -3.333x + 37.22

Fig. 2 Standard curves for total and strand-specific viral DNA quantification. (**a**) Serial dilutions ranging from 1×10^3 to 1×10^7 copies of tomato yellow leaf curl Sardinia virus (TYLCSV) or tomato yellow leaf curl virus (TYLCV) circular single-stranded (VS- or CS-specific) or double-stranded (total viral DNA) DNA phagemids are used as quantification standards. Quantification is performed according to the methods described in Fig. 1. Data represent the average of three independent qPCR replicates. (**b**) Correlation coefficients, qPCR efficiency values, and regression plot equations obtained

6. For qPCR, samples are subjected to one cycle at 95 °C for 30 s, 40 cycles at 95 °C for 10 s, and 60 °C for 15 s (*see* **Note 25**). Melting curves are obtained after a final step of 60 cycles of 10 s per cycle under a ramp rate of 0.5 °C every 10 s from 60 to 90 °C (*see* **Note 26**).
7. Standard curves for absolute quantification are obtained by linear regression analysis of the quantification cycle value (C_q) of each of the three technical replicates on the Log_{10} of DNA concentration (Fig. 2a). The linear regression results in the equation of a line with the general form: $y = mx + c$, where m is the slope of the line and c is the y-intercept. The efficiency (E) of qPCR is calculated as follows: $E = 10^{-1/m}$, and the percent efficiency % $E = (E - 1) \times 100$ (Fig. 2b) (*see* **Note 27**).
8. For time-course analysis of TYLCSV or TYLCV infections, equal amounts of extracted DNA (1 μg) from three samples collected at each time point are pooled, and the mixture is analyzed by two-step qPCR or standard qPCR to quantify strand-specific or total viral DNA molecules, respectively.
9. Two μL of DNA extracts (100 ng/μL) from infected *N. benthamiana* and *S. lycopersicum* plants are used for the first step, single-strand (VS or CS) tagging, following the procedure described above (**steps 2–4**), and the tagged DNA is

eluted in 20 μL of ultrapure H_2O (10 ng/μL, final concentration).

10. For ssDNA quantification, 2 μL (20 ng) of VS- or CS-tagged products is used, whereas for dsDNA quantification, 2 μL of unlabeled total DNA extracts (10 ng/μL) is used according to the qPCR protocol described above (**steps 5** and **6**). In addition, as a reference control for data normalization, the 25 S rRNA amplicon of each sample is amplified using 2 μL of the corresponding template using the same qPCR protocol.
11. Cq data obtained from VS, CS, or total viral DNA are normalized by subtracting the ΔC_q data of the corresponding 25S rRNA amplification.
12. Quantification of ssDNA (VS or CS) and dsDNA is obtained by extrapolation of the normalized C_q data with the corresponding standard curve.

    For TYLCSV:

    For VS quantification: VS copy number (Log_{10}) = (C_q − 41.10)/(−3.45).

    For CS quantification: CS copy number (Log_{10}) = (C_q − 41.242)/(−3.3437).

    For total viral DNA quantification: total viral DNA copy number (Log_{10}) = (C_q − 37.30)/(−3.3829).

    For TYLCV:

    For VS quantification: VS copy number (Log_{10}) = (C_q − 42.99)/(−3.4752).

    For CS quantification: CS copy number (Log_{10}) = (C_q − 43.55)/(−3.4021).

    For total viral DNA quantification: total viral DNA copy number (Log_{10}) = (C_q − 37.22)/(−3.333).
13. The resulting quantification data are plotted against the corresponding sampling time and expressed as Log_{10} of the copy number of viral DNA strands (Fig. 3).

4 Notes

1. In the presence of a phagemid bearing fl origin, the single-stranded DNA of the phagemid is packaged preferentially and secreted into the culture medium. This allows easy and efficient production of single-stranded DNA qPCR standards. For total viral DNA quantification, the plasmid constructs pBTSvs and pBTYMvs, containing double-stranded TYLCSV and TYLCV full-length genomes, respectively, are used as standard.

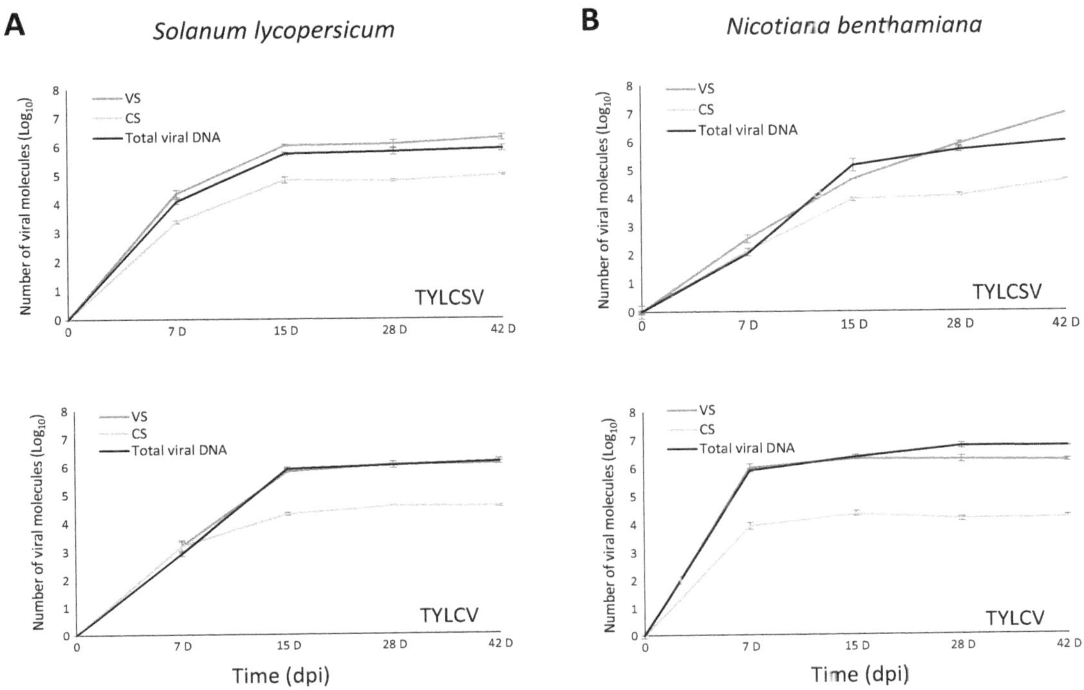

Fig. 3 Absolute quantification of virion-sense (VS), complementary-sense (CS), and total viral DNA molecules of tomato yellow leaf curl Sardinia virus (TYLCSV) and tomato yellow leaf curl virus (TYLCV) during systemic infections of tomato and *N. benthamiana*. Quantification of CS, VS, and total viral DNA molecules in a time-course experiment for TYLCSV (**a**) and TYLCV (**b**) was performed at 7, 15, 28, and 42 days post-inoculation (dpi). Equal amounts of extracted DNA (1 μg) collected at each time point from three samples were pooled, and the mixture was analyzed by two-step qPCR or standard qPCR to quantify strand-specific or total viral DNA molecules, respectively. Data represent the average of three independent qPCR replicates. Standard deviation (SD) is shown

2. *Escherichia coli* F′ (F-pilus encoding strain) is necessary for M13KO7 phage infection and single-stranded phagemid DNA production. *E. coli* XL1 Blue, or JM109 strains, harboring F′ factor is commercially available.
3. M13K07 is an M13 derivative with kanamycin resistance able to synthesize and package single-stranded phagemids containing M13 or fl origins. M13 derivatives are commercially available.
4. To prepare TYP media, add about 40 mL distilled water to a 250-mL glass beaker and sequentially dissolve 1.6 gr Bacto Tryptone, 1.6 gr yeast extract, 496 mg NaCl, and 174 mg K_2HPO_4. Make up to 100 mL with distilled water, autoclave in a storage glass bottle, and store at room temperature.
5. To prepare nuclease-free deionized H_2O, add 0.1% diethyl pyrocarbonate (DEPC) in deionized water, mix, and let sit at

room temperature overnight and autoclave. Store at room temperature.

6. To prepare stock solutions of ampicillin 100 mg/mL and kanamycin 70 mg/mL, dissolve 1 gr of sodium ampicillin or 0.7 gr of kanamycin monosulfate in sufficient deionized water to make a final volume of 10 mL in a 15-mL conical centrifuge tube. Filter the solutions through a 0.22-μm pore filter to sterilize. Aliquot and store at −20 °C.
7. Transfer 10 gr polyethylene glycol (PEG) 8000, 7.3 gr NaCl to a 50-mL conical centrifuge tube and add nuclease-free deionized water to 50 mL. Mix vigorously and continue mixing on a rotator until PEG 8000 is completely dissolved. Autoclave in a storage glass bottle and store at room temperature.
8. Mix in a 250-mL glass beaker 1 mL 1 M Tris-Cl pH 8.0 and 0.2 mL 0.5 M EDTA pH 8.0. Make up to 100 mL with deionized water, autoclave in a storage glass bottle, and store at room temperature.
9. Mix phenol (equilibrated with 8 hydroxyquinoline) with chloroform (1:1 v/v). Store the mixture in a clean brown glass bottle in a fume hood at room temperature.
10. Add about 80 mL of deionized water to a 250-mL glass beaker and dissolve 24.6 gr of sodium acetate. Adjust the pH to 5.2 with glacial acetic acid and make up to 100 mL with deionized water. Filter the solution through a 0.22-μm pore filter to sterilize it and store it in a glass bottle at room temperature.
11. To prepare the 1% agarose gel, mix 1 gr of ultrapure agarose and 100 mL of Tris–borate EDTA (TBE) buffer (50 mM Tris–HCl base, 50 mM H_3BO_3, 2.5 mM EDTA, pH 8.3) in a glass bottle. Melt the agarose in a microwave oven using 60-s pulses, stop, and swirl until complete dissolution of the agarose. Allow the agarose solution to cool to about 50 °C and add ethidium bromide (EtBr, 5 mg/mL stock solution) to a final concentration of 0.5 mg/mL or 10 μL Red safe (10,000×) to 1×. Pour the agarose into a gel tray with the comb in place and allow to cool until the gel has completely solidified. For sample loading, use 2–3 μL of 10× DNA gel loading dye.
12. TYLCSV and TYLCV agroinfectious clones harbor a partial dimeric tandem repeat of the cloned viral genomes in the binary vector pGreen 0029 transformed in *A. tumefaciens* strain GV3101.
13. To prepare LB media, combine 10 gr tryptone, 5 gr yeast extract, 10 gr NaCl, adjust pH to 7.0 with 1 N NaOH, and add 20 gr agar (for solid media only). Make up to 1 L with distilled water, autoclave in a glass storage bottle, and store at room temperature.
14. To prepare stock solutions of rifampicin 100 mg/mL, tetracycline 5 mg/mL, and kanamycin 100 mg/mL, dissolve 1 gr of sodium ampicillin, 50 mg of tetracycline hydrochloride, or 1 gr

of kanamycin monosulfate, in sufficient dimethyl sulfoxide (DMSO), absolute ethanol, or deionized water, respectively, to make a final volume of 10 mL in a 15-mL conical centrifuge tube. Sterilization is possible only for kanamycin solution by using 0.22-μm pore filters. DMSO and alcohol cause damage to filters, and sterilization of these solutions is not necessary. Aliquot and store at −20 °C.

15. To prepare 100 mL of infiltration solution, mix 1 mL of 1 M 2-(N-Morpholino) ethanesulfonic acid sodium salt (MES), pH 5.6, 1 mL of 1 M $MgCl_2$, 100 μL of 1 M 3′, 5′--dimethoxy-4′-hydroxyacetophenone (acetosyringone), in sufficient deionized water to make a final volume of 100 mL in a graduated cylinder. Make the infiltration solution just prior to use.
16. To prepare 20 mL of 2× CTAB extraction buffer, mix in a 15-mL conical centrifuge tube, 0.4 gr of CTAB, 2 mL of 1 M Tris–HCl pH 8.0, and 800 μL of 0.5 M EDTA pH 8.0. Make CTAB buffer just before extraction. It can be stored at room temperature for several days.
17. For DNA labeling primer design, TAG is a 20-nt sequence from the mammalian lymphocytic choriomeningitis virus. TYLCSV and TYLCV VS and CS primer sequences were designed against conserved regions of old-world (OW) begomovirus CP ORFs using the Primer Blast tool (https://www.ncbi.nlm.nih.gov/tools/primer-blast/).
18. For strand-specific amplification of TYLCSV and TYLCV strands, the combination of the primers TAG+VS or TAG +CS is required to detect VS and CS strands, respectively, using the corresponding labeled strand-specific template. For total viral DNA amplification, the combination of VS+CS primers is necessary, using the unlabeled viral dsDNA as template.
19. The tomato 25S nuclear ribosomal RNA gene (25S rRNA) (Accession Number X13557), highly conserved in the Solanaceae family, is used as a reference for qPCR normalization.
20. T4 DNA polymerase catalyzes DNA synthesis in the 5′–3′ direction in the presence of a template and primer. T4 DNA polymerase lacks 5′–3′ exonuclease and strand displacement activities, allowing single copying of a full-length circular DNA template.
21. After circular DNA labeling, a clean-up step is necessary to completely remove enzymes, salts, and primers. For this, it is important to use a silica-based spin column that only allows DNA fragments in the range of 100 bp–10 kb to be recovered.
22. Photometric quantification of DNA is based on the intrinsic absorption properties of nucleic acids. DNA absorbs at

260 nm, whereas contaminants, including phenolic compounds and proteins, absorb at 280 nm. The ratio of the readings at 260 nm and 280 nm (A_{260}/A_{280}) provides an estimate of DNA purity (pure DNA has an A_{260}/A_{280} ratio of 1.7–1.9). On the other hand, the DNA concentration is determined by measuring the absorbance at 260 nm (A_{260}) according to Beer–Lambert's law (an absorbance of 1 unit at 260 nm corresponds to 50 μg DNA per mL).

23. Alternatively, the number of molecules from the mass of ssDNA and dsDNA circular viral standards can be determined using the free online program http://cels.uri.edu/gsc/cndna.html.
24. Infectious clones of geminiviruses require more than one unit of the genome to produce a tandem repeat-containing two replication origins to achieve high infectivity rates. Partial dimeric constructs derived from geminiviruses (i.e., containing 1.n-mer of the viral genome such that they fulfill the necessary condition of the presence of two viral replication origins for effective Agrobacterium-guided artificial infection) are cloned into binary vectors and introduced into *A. tumefaciens* strains that enable the process of viral infection of the plant host.
25. For qPCR protocol, since a small amplicon of 188 bp is expected, a two-step protocol involving denaturation and annealing/synthesis is used.
26. A single peak produced in the melting curve analysis indicates a specific amplification product without PCR artifacts.
27. Linear regression of qPCR data and efficiency determination can be obtained from standard qPCR apparatus software or calculated manually with Microsoft Excel software.

Acknowledgements

This work was co-financed by the Programa Operativo FEDER 2014–2020 and Consejería de Economía y Conocimiento, Junta de Andalucía (Grant Number UMA18-FEDERJA-178 to A.G.-P.).

References

1. Shulman LM, Davidson I (2017) Viruses with circular single-stranded DNA genomes are everywhere! Annu Rev Virol 4(1):159–180
2. López-Bueno A, Tamames J, Velázquez D, Moya A, Quesada A, Alcamí A (2009) High diversity of the viral community from an Antarctic lake. Science 326(5954):858–861
3. Rosario K, Breitbart M (2011) Exploring the viral world through metagenomics. Curr Opin Virol 1(4):289–297
4. Rosario K, Duffy S, Breitbart M (2012) A field guide to eukaryotic circular single-stranded DNA viruses: insights gained from metagenomics. Arch Virol 157(10):1851–1871

5. Young JC, Chehoud C, Bittinger K, Bailey A, Diamond JM, Cantu E et al (2015) Viral metagenomics reveal blooms of anelloviruses in the respiratory tract of lung transplant recipients. Am J Transplant 15(1):200–209
6. Simmonds P, Adams MJ, Benkő M, Breitbart M, Brister JR, Carstens EB et al (2017) Consensus statement: virus taxonomy in the age of metagenomics. Nat Rev Microbiol 15(3):161–168
7. Rastrojo A, Alcamí A (2018) Viruses in polar lake and soil ecosystems. Adv Virus Res 101: 39–54
8. Wang H, Li S, Mahmood A, Yang S, Wang X, Shen Q et al (2018) Plasma virome of cattle from forest region revealed diverse small circular ssDNA viral genomes. Virol J 15(1):11
9. Yoshida M, Mochizuki T, Urayama SI, Yoshida-Takashima Y, Nishi S, Hirai M, et al. Quantitative viral community DNA analysis reveals the dominance of single-stranded DNA viruses in offshore upper bathyal sediment from Tohoku, Japan. Front Microbiol [Internet]. 2018 [cited 2022 Sep 19];9. Available from: https://www.frontiersin.org/articles/10.3389/fmicb.2018.00075
10. Fontenele RS, Lacorte C, Lamas NS, Schmidlin K, Varsani A, Ribeiro SG (2019) Single stranded DNA viruses associated with capybara faeces sampled in Brazil. Viruses 11(8):710
11. Liu Q, Wang H, Ling Y, Yang SX, Wang XC, Zhou R et al (2020) Viral metagenomics revealed diverse CRESS-DNA virus genomes in faeces of forest musk deer. Virol J 17(1):61
12. Kim OTP, Kagaya Y, Tran HS, Minei R, Tran TTH, Duong HTT et al (2020) A novel circular ssDNA virus of the phylum Cressdnaviricota discovered in metagenomic data from otter clams (Lutraria rhynchaena). Arch Virol 165(12):2921–2926
13. Benler S, Koonin EV (2021) Fishing for phages in metagenomes: what do we catch, what do we miss? Curr Opin Virol 49:142–150
14. Kinsella CM, Deijs M, Becker C, Broekhuizen P, van Gool T, Bart A et al (2022) Host prediction for disease-associated gastrointestinal cress DNA viruses. Virus Evol 8(2):veac087
15. Rosario K, Mettel KA, Benner BE, Johnson R, Scott C, Yusseff-Vanegas SZ et al (2018) Virus discovery in all three major lineages of terrestrial arthropods highlights the diversity of single-stranded DNA viruses associated with invertebrates. PeerJ 6:e5761
16. Porter AF, Shi M, Eden JS, Zhang YZ, Holmes EC (2019) Diversity and evolution of novel invertebrate DNA viruses revealed by meta-Transcriptomics. Viruses 11(12):1092
17. Shi M, Zhang YZ, Holmes EC (2018) Meta-transcriptomics and the evolutionary biology of RNA viruses. Virus Res 243:83–90
18. Malathi VG, Renuka DP (2019) ssDNA viruses: key players in global virome. Virus Dis 30(1):3–12
19. Gregory AC, Zayed AA, Conceição-Neto N, Temperton B, Bolduc B, Alberti A et al (2019) Marine DNA viral macro- and micro-diversity from pole to pole. Cells 177(5): 1109–1123.e14
20. Wang XW, Blanc S (2021) Insect transmission of plant single-stranded DNA viruses. Annu Rev Entomol 66(1):389–405
21. Duffy S, Shackelton LA, Holmes EC (2008) Rates of evolutionary change in viruses: patterns and determinants. Nat Rev Genet 9(4): 267–276
22. Grigoras I, Timchenko T, Grande-Pérez A, Katul L, Vetten HJ, Gronenborn B (2010) High variability and rapid evolution of a nanovirus. JVI 84(18):9105–9117
23. Sánchez-Campos S, Domínguez-Huerta G, Díaz-Martínez L, Tomás DM, Navas-Castillo J, Moriones E et al (2018) Differential shape of geminivirus mutant spectra across cultivated and wild hosts with invariant viral consensus sequences. Front Plant Sci 9:932
24. Kazlauskas D, Varsani A, Krupovic M (2018) Pervasive chimerism in the replication-associated proteins of uncultured single-stranded DNA viruses. Viruses 10(4):187
25. Juárez M, Rabadán MP, Martínez LD, Tayahi M, Grande-Pérez A, Gómez P (2019) Natural hosts and genetic diversity of the emerging tomato leaf curl New Delhi virus in Spain. Front Microbiol 10:140
26. García-Arenal F, Zerbini FM (2019) Life on the edge: Geminiviruses at the Interface between crops and wild plant hosts. Annu Rev Virol 6(1):411–433
27. Ortega-del Campo S, Grigoras I, Timchenko T, Gronenborn B, Grande-Pérez A (2021) Twenty years of evolution and diversification of digitaria streak virus in Digitaria setigera. Virus Evol 7(2):veab083
28. Mäntynen S, Laanto E, Sundberg LR, Poranen MM, Oksanen HM, ICTV Report Consortium (2020) ICTV virus taxonomy profile: Finnlakeviridae. J Gen Virol 101(9):894–895

29. Knezevic P, Adriaenssens EM, ICTV Report Consortium null (2021) ICTV virus taxonomy profile: inoviridae. J Gen Virol 102(7): 001614
30. Knezevic P, Adriaenssens EM, ICTV Report Consortium null (2021) ICTV virus taxonomy profile: plectroviridae. J Gen Virol 102(5): 001597
31. Prangishvili D, Mochizuki T, Krupovic M, ICTV Report Consortium null (2020) ICTV virus taxonomy profile: spiraviridae. J Gen Virol 101(3):240–241
32. Liu Y, Dyall-Smith M, Oksanen HM (2022) ICTV virus taxonomy profile: Pleolipoviridae 2022. J Gen Virol 103(11): 001793
33. Varsani A, Krupovic M (2018) Smacoviridae: a new family of animal-associated single-stranded DNA viruses. Arch Virol 163(7): 2005–2015
34. Varsani A, Opriessnig T, Celer V, Maggi F, Okamoto H, Blomström AL et al (2021) Taxonomic update for mammalian anelloviruses (family Anelloviridae). Arch Virol 166(10): 2943–2953
35. Breitbart M, Delwart E, Rosario K, Segalés J, Varsani A, ICTV Report ConsortiumYR (2017) ICTV virus taxonomy profile: Circoviridae. J Gen Virol 98(8):1997–1998
36. Varsani A, Krupovic M (2021) Family Genomoviridae: 2021 taxonomy update. Arch Virol 166(10):2911–2926
37. Abbas A, Taylor LJ, Collman RG, Bushman FD (2020) ICTV virus taxonomy profile: Redondoviridae. J Gen Virol 102(1):jgv001526
38. Krupovic M, Varsani A (2022) Naryaviridae, Nenyaviridae, and Vilyaviridae: three new families of single-stranded DNA viruses in the phylum Cressdnaviricota. Arch Virol 167(12): 2907–2921
39. Walker PJ, Siddell SG, Lefkowitz EJ, Mushegian AR, Adriaenssens EM, Alfenas-Zerbini P et al (2021) Changes to virus taxonomy and to the international code of virus classification and nomenclature ratified by the international committee on taxonomy of viruses (2021). Arch Virol 166(9):2633–2648
40. Fiallo-Olivé E, Lett JM, Martin DP, Roumagnac P, Varsani A, Zerbini FM et al (2021) ICTV virus taxonomy profile: Geminiviridae 2021. J Gen Virol 102(12):001696
41. Thomas JE, Gronenborn B, Harding RM, Mandal B, Grigoras I, Randles JW et al (2021) ICTV virus taxonomy profile: Nanoviridae. J Gen Virol 102(3):001544
42. Gutierrez C (1999) Geminivirus DNA replication. Cell Mol Life Sci 56(3–4):313–329
43. Rampersad S, Tennant P (2018) Replication and expression strategies of viruses. Viruses 2018:55–82
44. Zhao L, Rosario K, Breitbart M, Duffy S (2019) Eukaryotic circular rep-encoding single-stranded DNA (CRESS DNA) viruses: ubiquitous viruses with small genomes and a diverse host range. Adv Virus Res 103:71–133
45. Krupovic M, Varsani A, Kazlauskas D, Breitbart M, Delwart E, Rosario K et al (2020) Cressdnaviricota: a virus phylum unifying seven families of rep-encoding viruses with single-stranded, circular DNA genomes. J Virol 94(12):e00582–e00520
46. Hanley-Bowdoin L, Settlage SB, Orozco BM, Nagar S, Robertson D (2000) Geminiviruses: models for plant DNA replication, transcription, and cell cycle regulation. Crit Rev Biochem Mol Biol 35(2):105–140
47. Péréfarres F, Hoareau M, Chiroleu F, Reynaud B, Dintinger J, Lett JM (2011) A novel synthetic quantification standard including virus and internal report targets: application for the detection and quantification of emerging begomoviruses on tomato. Virol J 8(1):389
48. Abrahamian PE, Abou-Jawdah Y (2013) Detection and quantitation of the new world Squash leaf curl virus by TaqMan real-time PCR. J Virol Methods 191(1):76–81
49. Tyagi AK, Pradier A, Baumer O, Uppugunduri CRS, Huezo-Diaz P, Posfay-Barbe KM et al (2013) Validation of SYBR Green based quantification assay for the detection of human Torque Teno virus titers from plasma. Virol J 10(1):191
50. Noris E, Miozzi L (2015) Real-time PCR protocols for the quantification of the begomovirus tomato yellow leaf curl Sardinia virus in tomato plants and in its insect vector. Methods Mol Biol 1236:61–72
51. Noris E, Miozzi L (2015) Real-time PCR protocols for the quantification of the Begomovirus tomato yellow leaf curl Sardinia virus in tomato plants and in its insect vector. In: Uyeda I, Masuta C (eds) Plant virology protocols: new approaches to detect viruses and host responses [Internet]. Springer, New York. [cited 2022 Sep 19]. p. 61–72. (Methods in Molecular Biology). Available from: https://doi.org/10.1007/978-1-4939-1743-3_6
52. Henriques AM, Duarte M, Barros SC, Fagulha T, Ramos F, Luís T et al (2018) Development and validation of a real-time PCR for the detection and quantification of porcine circovirus type 2. Virus Dis 29(3):355–361

53. Tu T, Zehnder B, Qu B, Ni Y, Main N, Allweiss L et al (2020) A novel method to precisely quantify hepatitis B virus covalently closed circular (ccc)DNA formation and maintenance. Antivir Res 181:104865

54. Simón A, Ruiz L, Velasco L, Janssen D (2018) Absolute quantification of tomato leaf curl New Delhi virus Spain strain, ToLCNDV-ES: virus accumulation in a host-specific manner. Plant Dis 102(1):165–171

55. Rodríguez-Negrete EA, Sánchez-Campos S, Cañizares MC, Navas-Castillo J, Moriones E, Bejarano ER et al (2015) A sensitive method for the quantification of virion-sense and complementary-sense DNA strands of circular single-stranded DNA viruses. Sci Rep 4(1): 6438

Chapter 9

Replication Assay of Begomovirus in Arabidopsis Protoplasts

Laura Gonçalves Costa Martins and Elizabeth P. B. Fontes

Abstract

Protoplasts are isolated plant cells from which the cell walls have been removed by treatment with fungal cellulase and macerozyme enzymes, which degrade the primary components of the cell wall. The protoplasts are totipotent, sensitive, and versatile; thereby, they have been extensively used to study signal transduction pathways, cell-autonomous responses, and replication of plant viruses. This system has several advantages over the use of whole plants for viral replication, including a high percentage of infected cells and uncoupling virus movement from replication assays. Here, we describe a simple and efficient protocol for begomovirus transfection and replication in Arabidopsis protoplasts. The method consists of four steps, (i) protoplast isolation, (ii) PEG–calcium transfection of begomovirus infectious clones, (iii) elimination of input plasmid DNA by DpnI digestion, and (iv) quantification of the viral newly synthesized DNA by qPCR. The protoplasts can be transformed efficiently with begomovirus infectious clones, and virus replication can be monitored by the accumulation of nascent viral DNA in the infected protoplasts.

Key words Protoplast isolation, Protoplast transfection, *Begomovirus replication*, *CabLCV*, *Viral DNA accumulation*

1 Introduction

Protoplasts are isolated plant cells that have had their cell walls removed enzymatically. The versatility of protoplasts has been extensively exploited to study cell signaling and metabolic pathways, transcription regulatory mechanisms, gene expression, subcellular localization of proteins, protein—protein interactions, and virus replication. Therefore, they represent an efficient platform for virus–host interaction molecular studies. Several protoplast isolation protocols have been established for model plants and some crops [1]. Nevertheless, protoplasts prepared from leaves are often the preferential choice. High-yielding mesophyll protoplasts can be easily obtained from Arabidopsis, tobacco, common bean, maize, rice, and many other plant species [1–6]. These mesophyll protoplasts can be easily adapted to assay virus DNA replication. This

Elizabeth P. B. Fontes and Kristiina Mäkinen (eds.), *Plant-Virus Interactions*, Methods in Molecular Biology, vol. 2724, https://doi.org/10.1007/978-1-0716-3485-1_9,

described protocol has been recently used to ensure the lack of replication function of the host protein NISP, involved in begomovirus intracellular movement [7].

Begomovirus is the largest and most agronomically relevant genus of the *Geminiviridae* family and contains both monopartite and bipartite viruses with a single- or double-genomic configuration, respectively [8, 9]. The genomic components of bipartite begomoviruses are designated DNA-A and DNA-B; the single-genomic component organization of monopartite begomoviruses resembles the DNA-A of bipartite begomoviruses [8, 9]. DNA-A contains six functionally characterized open reading frames (ORFs), two in the virion sense (the capsid protein, *CP* or *AV1* and *AV2*) and four in the complementary sense (replication initiator protein, *Rep* or *AC1*; transcription activator protein, *Trap* or *AC2*; replication enhancer protein, *Ren* or *AC3;* and *AC4*). The DNA-B encodes two genes, the virion-sense *BV1*, which corresponds to *NSP* (nuclear shuttle protein), and the complementary-sense *BC1* or *MP* (movement protein).

Geminiviruses use the rolling circle mechanism of replication to replicate the single-stranded DNA (ssDNA) genome inside the nucleus of infected cells. The unpacked viral ssDNA is first converted into double-stranded DNA (dsDNA) by the host DNA polymerase α, which synthesizes the complementary strand to generate the viral dsDNA intermediate [10]. Then, the host DNA polymerase δ synthesizes new, nascent copies of the geminivirus ssDNA genome. In addition to replicating its genome in infected cells, for productive infection in plants, the begomovirus genome must move from the nucleus to the cytosol via nuclear pores and then to adjacent cells via the plasmodesmata [11, 12]. Therefore, the efficient accumulation of begomoviral DNA within plants results from the viral genome replication in the nucleus of infected cells, like other DNA viruses, and intercellular transport of viral DNA to adjacent, uninfected cells, as for any other plant virus [13, 14].

To understand the mechanisms involved in virus–host interaction and the role of virus and host proteins in mediating virus replication uncoupled from virus movement, we describe a protocol for assaying begomovirus replication in Arabidopsis protoplasts. Although this method can be easily modified to assay replication of any other plant viruses in mesophyll protoplasts of the more permissive host *Nicotiana benthamiana*, some peculiarities for ssDNA viruses that replicate in the nuclei of infected plants via dsDNA intermediates must be considered. For instance, input DNA of infectious clones that naturally persists as contaminants in protoplasts can be easily distinguished from newly synthesized viral DNA by Dpn I digestion, which recognizes methylated dam sites in prokaryotic DNA. By using DpnI site-flanking primers in the viral genome, the amplicon to be quantified will result from

DpnI-resistant genomes that have been replicated in eukaryotic cells. The method involves transfecting Arabidopsis protoplasts with infectious clones containing partial tandem repeats of cabbage leaf curl virus (CabLCV) DNA-A, extracting and digesting total DNA with Dpn1, and quantifying newly replicated vDNA by qPCR.

Several different methods to prepare protoplasts have been described [15, 16]. For instance, for replication assays of geminivirus genomes, we used to prepare the protoplasts from suspension cells, which involves tissue culture manipulation for obtaining and maintenance of viable cell suspension [17, 18]. In contrast, a rapid and efficient protocol for mesophyll protoplast preparation from mature Arabidopsis leaves has been established [6]. These mesophyll protoplasts will probably become more prevalent in virus–plant interaction studies as the mesophyll protoplast preparation achieves high efficiency and sensitivity and does not require complex tissue culture medium and techniques.

2 Materials

2.1 Protoplast Isolation and PEG–Calcium Transfection of Plasmid DNA (Infectious Clones)

1. Plants: Arabidopsis thaliana at a 4-week stage of development (*see* **Note 1**).
2. Cellulase R10 (Yakult Pharmaceutical Ind. Co., Ltd., Japan) and macerozyme R10 (Yakult Pharmaceutical Ind. Co., Ltd., Japan). Store at −20 °C (*see* **Note 2**).
3. PEG 4000 (*see* **Note 3**).
4. Stock solutions to be autoclaved or filter-sterilized using a 0.45-mm filter and stored at 4 °C: 0.2 M MES pH 5.7 (3.9 g-anhydrous basis / 100 mL ddH_2O), 0.8 M mannitol (29.14 g/200 mL ddH_2O), calcium chloride—1 M $CaCl_2$ (11.09 g / 100 mL ddH_2O), potassium chloride—2 M KCl (22.36 g/150 mL ddH_2O), magnesium chloride—2 M $MgCl_2 \cdot 6H_2O$ (8.13 g/20 mL ddH_2O), and sodium chloride—1 M NaCl (11.68 g/200 mL ddH_2O).
5. Sock solutions to be filter-sterilized using a 0.45-mm filter and stored at −20 °C: 10% (p/v) BSA (0.5 g/5 mL ddH_2O).
6. Petri dish (100 × 25 mm^2 for 10 mL enzyme solution).
7. Nylon sieve (75 mm).
8. 0.45-μm syringe sterilization filter.
9. Sharp razor blade.
10. Desiccator.
11. 5% calf serum.
12. Hemocytometer.

13. Glass-bottom dish.
14. Cover slips.
15. Light microscope.
16. Enzyme solution (digestion solution) (*see* **Note 4**): 20 mM MES (pH 5.7), 1.5% (w/v) cellulase R10, 0.4% (w/v) macerozyme R10, 0.4 M mannitol, 20 mM KCl, 10 mM $CaCl_2$, and 0.1% (w/v) BSA. For a 30 mL final volume (*see* **Note 5**), add 3 mL of 20 mM MES solution preheated at 70 °C (*see* **Note 6**), 15 mL of 0.8 M mannitol, and 300 μL KCl into a 100 mL Erlenmeyer. Cover the flask with parafilm and warm the mix at 55 °C for 10 min (*see* **Note 7**). Then, add 0.45 g of cellulase R10 and 0.12 g of macerozyme R10. Cool the solution to room temperature (25 °C); add 300 μL 10 mM $CaCl_2$ and 300 μL 0.1% (w/v) BSA; and complete the volume to 30 mL with sterilized ddH_2O. Filter the final enzyme solution (0.45-mm syringe filter) (*see* **Note 4**).
17. WI solution: 4 mM MES (pH 5.7), 0.5 M mannitol, and 20 mM KCl. For 20 mL WI solution, add 12.5 mL of 0.8 M mannitol, 400 μL of 20 mM MES, and 400 μL of 2 M KCl and complete the volume to 20 mL with ddH_2O. The WI solution can be stored at room temperature (22–25 °C).
18. W5 solution: 2 mM MES (pH 5.7), 154 mM NaCl, 125 mM $CaCl_2$, and 5 mM KCl. To prepare 150 mL solution, add 23.1 mL of 1 M NaCl, 18.75 mL of 1 M $CaCl_2$, 375 μL of 2 M KCl, and 1.5 mL of 0.2 M MES and complete the volume with sterilized ddH_2O to 150 mL. The W5 solution can be stored at room temperature.
19. MMG solution: 4 mM MES (pH 5.7) containing 0.4 M mannitol and 15 mM $MgCl_2$. To prepare 30 mL solution, add 15 mL of 0.8 M mannitol, 225 μl of 2 M $MgCl_2$, and 600 μL of 0.2 M MES and complete the volume with ddH_2O to 30 mL. The MMG solution can be stored at room temperature.
20. PEG–calcium transfection solution (*see* **Note 8**): 20 a 40% (w/v) PEG4000, 0.2 M mannitol, and 100 mM $CaCl_2$. To prepare 20 mL solution, add 4 g of PEG 4000, 2.5 mL of 0.8 M mannitol, and 1 mL of 1 M $CaCl_2$ and complete the volume with ddH_2O to 20 mL.
21. Infectious clones: CabLCV DNA-A (1.3–1.6 copies of DNA-A harboring duplicated replication origins) and plasmid DNA-negative control (*see* **Note 9**).

2.2 qPCR and PCR-Based Diagnostic of Viral DNA

1. Agarose gel electrophoresis equipment: 1.5% agarose gel in TBE buffer stained with EtBr, molecular weight DNA ladder, and gel loading dye.

2. Restriction *enzyme* DpnI (*see* **Note 10**).
3. Cetyltrimethyl ammonium bromide (CTAB) extraction buffer: 100 mM Tris–HCl (pH 8), 20 mM EDTA (pH 8), 2.0 M NaCl, 2% (w/v) CTAB, 1% (w/v) PVP.
4. Specific primers may be used to confirm the accumulation of CabLCV DNA-A by PCR. Make sure that the amplified fragment contains DpnI sites (*see* **Note 11**):

   Rep_CaLCuV_Fwd/ Begomovirus Rep (AC1): 5′ ATGCCACGAAACCCTAAA 3′.

   Rep_CaLCuV_Rvs/ Begomovirus Rep (AC1): 5′ GCAATCTTGTGCTGTGCT 3′′.
5. Primers for PCR data normalization—actin gene was employed as an endogenous control:

   AtActin2 Fwd /*Arabidopsis thaliana* actin:

   5′ ATGGCTGAGGCTGATGATATTCAACC 3′.

   AtActin2 Rvs /*Arabidopsis thaliana* actin:

   5′ CCGTACAGATCCTTCCTGATATCCACA 3′.
6. Master Mix Kit (Life Technologies) SYBR Green to quantitative PCR (qPCR).
7. Quantitative PCR (qPCR) thermocycler.
8. Primers for qPCR to amplify DNA CabCLV-A (*see* **Note 11**):

   RtCabLCVA1 Fwd /CaLCuV DNA-A: 5′ GGAAGTTGAAGTAGGCTTTGCGG 3′.

   RtCabLCVA1 Rvs / CaLCuV DNA-A: 5′ GGACCCGTTGTACACCAGAACC 3′.
9. Primers for qPCR data normalization as an internal standard:

   qRT_18SrDNA-Fwd /*Arabidopsis thaliana* 18SrDNA:

   5′ TTTGCGCGCCTGCTGCC 3′.

   qRT_18SrDNA-Rvs/*Arabidopsis thaliana* 18SrDNA:

   5′ TGTGCTGGCGACGCATCATT 3′.

3 Methods

3.1 Protoplast Isolation from Arabidopsis Leaves

Plant protoplasts without cell walls offer an efficient cell-based experimental system for replication assays of plant viruses. Molecules of CabLCV DNA-A (or any other Arabidopsis-infecting geminivirus) can be introduced in Arabidopsis protoplasts using the method PEG–calcium fusion.

Growth of the Arabidopsis Plants

1. Grow Arabidopsis plants on soil in a greenhouse or an environmental-controlled chamber. Try to maintain a stable environment as much as possible, to avoid environmental changes and stresses to the planta (*see* **Note 1**).

Preparation of Plasmid DNA

2. Grow CabLCV DNA-A (infectious clones) and LeNIK2-NS-pDONR207 (as an internal control) constructs in a 50-mL flask containing 10 mL Luria–Bertani (LB) liquid medium, supplemented with specific antibiotic, with shaking at 160 rpm at 37 °C overnight (*see* **Note 9**).
3. Extract and purify plasmid DNA from *E. coli* using the QIAprep® Spin Miniprep Kit, following the manufacturer's procedure. Adjust the plasmid DNA concentration to 1–1.5 μg/μL of plasmid DNA (*see* **Note 12**).

Protoplast Isolation

4. Select well-expanded and healthy leaves from 3- to 4-week-old plants before flowering (Fig. 1a, red stars) (*see* **Note 13**).
5. From the middle part of a selected leaf, cut 0.5–1.0 mm leaf strips on a clean white paper filter using a fresh sharp razor blade (Fig. 1b). Be careful not to crush the tissue at the cutting site. The paper filter facilitates the inspection of crushing at the cutting sites (*see* **Note 14**).
6. Quickly and gently transfer the leaf strips into the prepared enzyme solution (10 leaves in 5 mL) by completely submerging both sides of the strips (Fig. 1c, d) (*see* **Note 15**).
7. Vacuum infiltrate the petri dish containing the leaf strips using a desiccator for 20–30 min at room temperature (*see* **Note 16**).
8. Incubate the leaf strips in the enzymatic digestion solution without shaking in the dark for 2–3 h at room temperature. The enzyme solution should turn green, indicating the release of the protoplasts (Fig. 1e) (*see* **Note 17**).
9. Check the release of protoplasts in the solution using a light microscope. The released protoplasts look spherical, while the undigested cells appear more irregular (Fig. 1f) (*see* **Note 18**).
10. To complete the digestion of the leaf, dilute the enzyme/protoplast solution with 5 mL of W5 solution before filtration.
11. Filter the enzyme solution containing protoplasts gently passing through a clean 70 μm nylon mesh (*see* **Note 19**).

Fig. 1 A flowchart depicting the protoplast isolation procedure. (**a**) A representative sample of a healthy 4-week-old Arabidopsis plant will be used for the protoplast isolation. Red stars indicate healthy expanded leaves. (**b**) Inspection of cutting edges. Arabidopsis leaves are cut with a razor blade and placed in a clean white paper filter to inspect cross sections in slices. (**c**, **d**) Leaf slices placed (**c**) and submerged (**d**) in digestion solution. (**e**) Arabidopsis leaves cut after 3-h digestion in enzymatic solution. (**f**) Isolated Arabidopsis protoplasts visualized by light microscopy. (**g**) Counting cells under the microscope using a hemacytometer. (**h**) Transfected protoplast pellets after centrifugation at room temperature and carefully removing the supernatant. (**i**) Arabidopsis protoplasts were transfected and visualized by light microscopy after 14 h of incubation in the dark

12. Transfer the digested protoplast to a 15-mL round-bottom tube using a wide-end tip pipet. Centrifuge the flow-through at 100 g to pellet the protoplasts at room temperature for 2–4 min (*see* **Note 20**).

13. Remove slowly as much supernatant as possible without disturbing the protoplast pellet. Then gently resuspend the protoplast pellet at 2×10^5 mL^{-1} in cold W5 solution. Count cells under a microscope using a hemacytometer (Fig. 1g). Incubate the resuspended protoplast mixture on ice for 30 min (*see* **Note 21**).
14. This 30-min incubation time on ice allows the protoplast to settle at the bottom of the tube by gravity. Remove the W5 solution using a 1-mL pipette without disturbing the protoplast pellets. Resuspend the protoplast pellet gently in MMG solution (kept at room temperature) at 2×10^5 protoplast mL^{-1}.

3.2 PEG–Calcium-Mediated Protoplast Transfection

1. Add 100 μL of protoplasts (2×10^4 protoplasts) into 2-mL round-bottom tubes.
2. Add 10 or 20 μL DNA (10–15 μg of plasmid DNA, infectious DNA-A clone), mixing gently into each aliquot of protoplasts (*see* **Note 22**).
3. Add 110 μL of freshly prepared PEG-Calcium solution and then gently rotate to mix well (handle six to ten samples for each transfection) (*see* **Note 23**).
4. Incubate the transfection mixture at room temperature for up to 15 min.
5. Dilute the transfection mixture with 400–440 μL W5 solution at room temperature. Mix well by gently rocking or inverting the tube to stop the transfection process.
6. Centrifuge at 100 × *g* for 2–4 min at room temperature. After centrifugation, the supernatant is carefully removed using a pipette.
7. Resuspend protoplasts gently with 1 mL of WI solution.

3.3 Protoplast Incubation

1. Coat a 12-well tissue culture plate with 5% (v/v) calf serum for a short time (1 min) for each well to prevent sticking of the protoplasts to the plastic surface. After 1-min incubation, aspirate the coating solution.
2. Transfer the protoplasts from each tube to each well of the culture plate and cover the plate.
3. Incubate the protoplasts at room temperature for 12–18 h (time 0), 1.5 days, 2.5 days, 3.5 days, and 4.5 days in the dark.
4. Then, collect the protoplasts by transferring the protoplast solution to 2 mL round-bottom tubes and centrifuging at 100 × g for 2–4 min at room temperature.
5. Remove the supernatant (Fig. 1h) and freeze samples on dry ice (*see* **Note 24**).

6. Before freezing the samples on dry ice, transfer 10 μL of transfected protoplasts to a glass-bottom dish and cover with a glass coverslip to visualize using a light microscope (Fig. 1i) to assess the status of the protoplast, monitor the integrity of their membrane and their ability to develop over time.

3.4 qPCR and PCR-Based Diagnostic of Viral DNA

1. Resuspend the protoplast pellet in CTAB buffer for DNA extraction [19].
2. Eliminate RNA contamination in genomic DNA samples by incubating the samples with RNase (10 ng/μL) at 37 °C for 10 min.
3. Determine DNA concentration by NanoDrop spectrophotometer (Fig. 2a, b) and ensure an equal amount of inoculum at time zero in comparative DNA replication assays by qPCR (Fig. 2c).
4. Digest 1 μg of total DNA extracted from protoplasts with 10 U of restriction enzyme Dpn I for 1 h at 37 °C. Stop the reaction by heat inactivation (20 min at 80 °C).
5. For PCR-based diagnostic, use approximately 100 ng template DNA, 1 μL (10 μM) specific forward and reverse primers, and 12.5 μL PCR Master Mix (if 2×) to a final volume of 25 μL and separate the products by electrophoresis through an agarose gel (Fig. 2d, e).
6. Determine the viral DNA accumulation via quantitative PCR (Fig. 2f). Reaction mix: 100 ng of DNA extracts from transfected Arabidopsis protoplasts, 1 μL of the corresponding 10 μM gene-specific primer, 10 μL of 2× SYBR Green Master Mix (if not 2×, calculate the volume accordingly), and ultrapure H_2O to 20 μL (*see* **Note 25**).
7. Set up the real-time PCR system to the following thermal cycling conditions: an initial step at 95 °C for 10 min, followed by 40 cycles at 94 °C for 15 s and 60 °C for 1 min.
8. For absolute quantification of viral DNA, prepare standard curves using serial dilutions of DNA-A of CabLCV (100 to 10^9 copies of viral genome per reaction). Reaction mix for qPCR: 1 μL of each dilution of viral DNA, 4 μL of forward and reverse primers, and 5 μL of SYBR Green (if the master mix is 2×). Use three technical replicates of each serial dilution and the same thermal cycle described for protoplast DNA. Determine the linear regression equation of the quantification cycle value (CT) x copy number of the viral genome ($\log_{10}$).
9. For normalizing the data against an internal control, prepare standard curves of 18S rDNA, which accounts for template input variation, using the same qPCR protocol. For the 18S

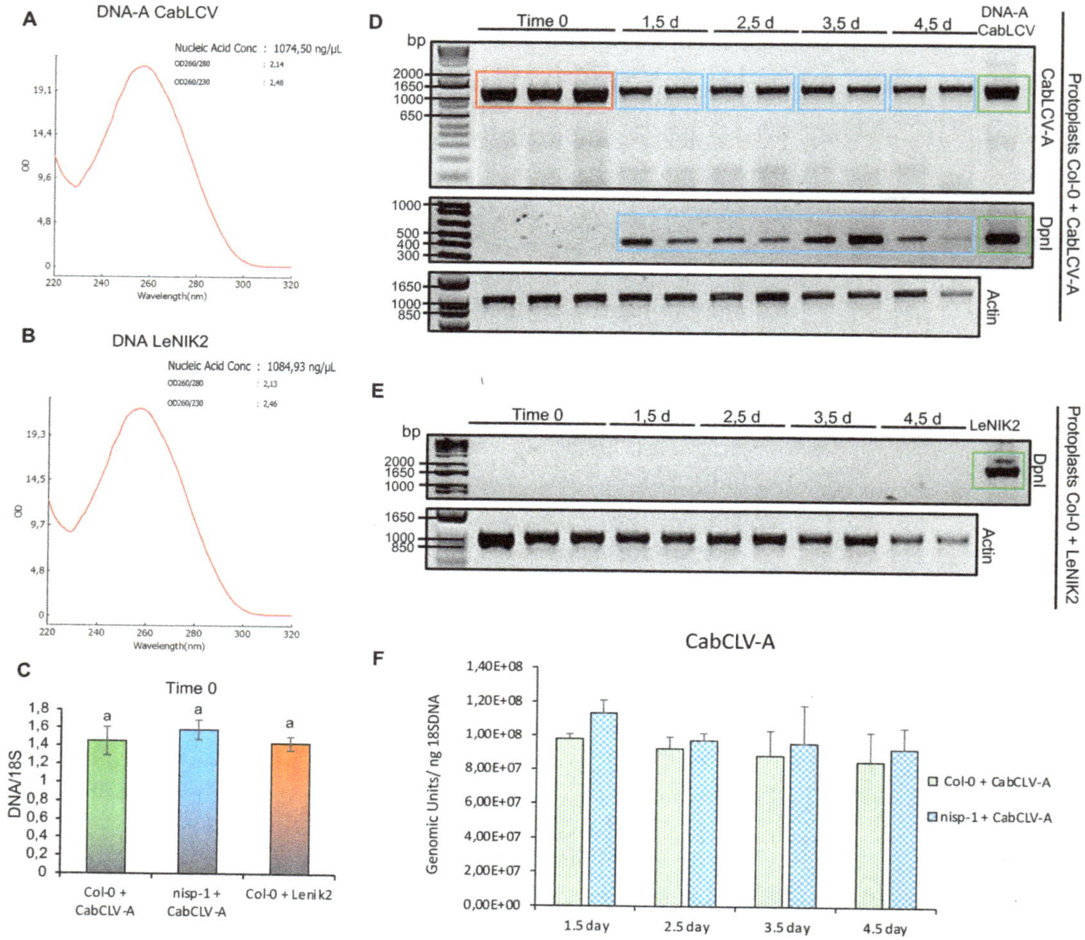

Fig. 2 Replication assay of CabLCV DNA-A in Arabidopsis protoplasts. (**a**) Quality control of the input CabLCV-DNA-A integrity by spectrophotometry. (**b**) Spectrophotometric analysis of input LeNIK2 control DNA. (**c**) Quantification of the DNA inoculum (CabLCV DNA-A infectious clone). Total DNA extracted zero-time post-transfection in *nisp*, and Col-0 protoplasts were quantified by qPCR and normalized to 18SrDNA. (**d**) Replication of CabLCV DNA-A in Col-0 protoplasts. Plasmids containing tandem copies repeats of CabLCV DNA-A were introduced into Col-0 protoplasts. Total DNA was isolated at the indicated days (d) post-transfection and either used directly as a template in PCR-based diagnostic of viral DNA (top gel, CabLCV DNA-A) or digested with DpnI prior to PCR (bottom gel, DpnI). In the Dpn1 agarose gel, the last lane corresponds to the amplification of input DNA not treated with DpnI, as a positive control. Newly synthesized viral DNA is represented by the DpnI-resistant amplicon. The early saturation of nascent viral DNA accumulation in protoplasts may be due to the low and limited number of transfected protoplasts prepared directly from leaves. Viable protoplasts were monitored via the amplification of an endogenous *Actin* gene. E. Negative control plasmid for the replication assay. The recombinant pDONR201 plasmid, containing a copy of an unrelated gene *LeNIK2,* was introduced into Col-0 by the polyethylene glycol method. Total DNA was extracted at different post-transfection intervals, digested with DpnI, and used as a template for the amplification of *LeNIK2*. No DpnI-resistant amplicon was detected. The last lane is a positive control using DpnI-untreated input DNA as a template. Actin was amplified from endogenous genomic DNA. F. Absolute quantification of newly synthesized viral DNA by qPCR. Total DNA extracted from transfected Col-0 and *nisp-1* protoplasts was digested with DpnI. The replicated DpnI-resistant amplicon was quantified by qPCR, using specific DpnI site-flanking primers in the viral DNA-A genome. Error bars indicate 95% confidence intervals based on replicated

rDNA standard curve, use an interval 75–800 ng/μL of genomic DNA.

10. Analyze the resulting quantification data using Tukey's test and the $2^{-\Delta CT}$ statistical method.
11. To calculate the genomic units/ ng 18S rDNA, convert the CT values using the standard curves for CabLCV plasmid DNA and 18S rDNA (Fig. 3a, b).

4 Notes

1. Arabidopsis plants are susceptible to environmental changes. Indeed, stress conditions may significantly alter the genetic profiles of plants and ultimately affect the integrity of the protoplast preparation. Therefore, maintain plant growth under a constant environment to increase the quality and transfection efficiency of the protoplasts.
2. Macerozyme R-10 and Cellulase "Onozuka R-10" are often used to isolate plant protoplasts. The original mesophyll protoplast isolation [6] recommends the enzymes from Yakult Pharmaceutical Ind. We never tested other providers.
3. PEG 4000 is critical to achieving high transfection efficiency; do not use PEG another molecular weight.
4. The enzyme solution should be prepared fresh; the final solution should be clear and light brown.
5. 5–10 mL of enzyme solution digest approximately 10–15 leaves, giving 0.5–1 × 10^6 protoplasts enough for more than 25 samples.
6. Before adding the enzyme powders, the MES solution must be preheated at 70 °C for 3–5 min for better dissolution of MES, and then, it should be cooled down to prepare the enzyme solution. Preheat an approximate volume to be used to prepare the fresh enzyme solution.
7. 55 °C for 10 min will inactivate DNase and proteases and increase the enzyme solubilization efficiency.
8. The PEG solution can be stored at RT for replication assays in mesophyll protoplasts. However, to increase transfection efficiency, use a freshly prepared PEG solution. The PEG solution should be prepared at least 1 h before transfection to be completely dissolved. Do not autoclave the PEG solution.

Fig. 2 (continued) samples from three independent experiments. (This figure was slightly modified from Ref. [7], license number 5481591346408)

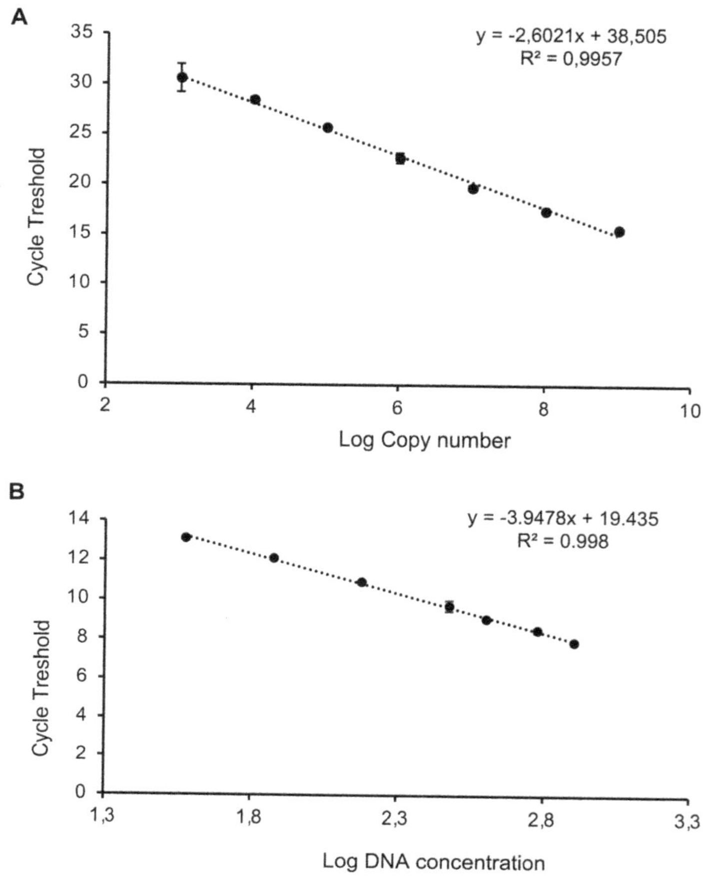

Fig. 3 Standard curve of SYBR green-based quantitative PCR amplification of plasmid CabCLV DNA-A and 18S rDNA. (**a**) The standard curve was plotted as the mean CT values obtained from each dilution of the standard plasmid CabCLV DNA-A vs the log_{10}DNA copy number (ten-fold dilutions ranged from 1.0×10^3 to 1.0×10^9 copies per μL). Values are the mean ± SD from three replicates. (**b**) Standard curve of 18SrDNA prepared from CabCLV-infected Arabidopsis protoplast. Each dilution was tested in triplicates. Values are the mean ± SD from three replicates

9. As an internal control, one may use any prokaryotic plasmid DNA, as it serves to confirm the efficiency of DpnI digestion. For the infectious DNA-A from begomoviruses, one should use a DNA construct containing tandem copy repeats (often 1.3–1.6 copies) of DNA-A and duplicated replication origins (*see* Chap. 4, Construction of geminivirus infectious clones for agroinoculation into plants). These constructs can be generated from cloned single copies of the virus genome (DNA-A) or can be obtained from different laboratories. Several infectious DNA-A from bipartite begomoviruses and infections

genome from monopartite begomoviruses and other geminiviruses are published and available upon request.

10. DpnI digests only dam methylated input DNA, not newly synthesized CabLCV DNA-A in the Arabidopsis protoplasts.
11. The primers designed to verify viral DNA accumulation in protoplasts by PCR or quantify (by qPCR) newly synthesized CabLCV-DNA-A in protoplast must amplify Dpn1 site-containing fragments of DNA-A. Digestion of the protoplast DNA with Dpn1 prior to qPCR will ensure that only Dpn1-resistant amplicons will be quantified. Those are derived from newly synthesized CabLCV DNA-A in Arabidopsis protoplasts and not from input DNA previously propagated in prokaryotic *E. coli*.
12. High concentrations of purified plasmid DNA are crucial for protoplast transfection.
13. The selection of healthy leaves at the proper stage of development is an essential step for mesophyll protoplast isolation. Plant stress responses are critical in steering the development of cells, tissues, and organs. Therefore, protoplasts prepared from healthy leaves without stress contribute to the high transfection efficiency of the protoplast preparation.
14. 10^6 protoplasts are enough for the experiment. Cut the leaves on a piece of clean white paper on a laboratory bench to provide proper support and easy inspection of the cut tissues.
15. For an enhanced protoplast yield, immediately dip and submerge the leaf strips in the enzyme solution to prevent them from drying on the paper during cutting.
16. Infiltration of leaf tissues is critical for mesophyll protoplasts' quick release and yield.
17. The efficiency of protoplast isolation differs significantly for different plant species, ecotypes, and genotypes, and each one requires optimization of the digestion time. However, a good time for starting would be 3 h of digestion, when most of the Arabidopsis protoplasts are released from the leaf.
18. Avoid over-digestion and long incubation times. The complete release of all the protoplasts from leaf strips is unnecessary.
19. This filtering process removes undigested leaf tissues and debris, while the released protoplasts pass through the mesh.
20. Set centrifuge acceleration and deceleration rates at the lowest setting to minimize protoplast damage during the centrifugation. Using the round-bottom tube and swinging-bucket rotors is highly recommended.

21. During protoplast resuspension, slowly add 1 mL of W5 solution along the side of the tube before adding the remaining 9 mL of W5 solution.
22. The size of the plasmid DNA construct impacts the transfection efficiency. For CabLCV DNA-A constructs with tandem repeat copies (1½ copies) of DNA-A, the transfection efficiency has been optimized for 10–15 μg of infectious DNA construct.
23. The concentration of PEG–calcium solution should be optimized empirically. For begomovirus replication assays, we use a 30% PEG final concentration.
24. If necessary, the samples can be stored at −80 °C until further analysis.
25. Each standard dilution is analyzed by three technical replicates.

Acknowledgments

This work was partly supported by the Brazilian research funding agencies CNPq, Fapemig, and Capes (Print Program).

References

1. Kanai R, Edwards GE (1973) Separation of mesophyll protoplasts and bundle sheath cells from maize leaves for photosynthetic studies. Plant Physiol 51:1133–1137. https://doi.org/10.2307/4263286
2. Nagata T, Takebe I (1971) Plating of isolated tobacco mesophyll protoplasts on agar medium. Planta 99:12–20. https://doi.org/10.1007/BF00392116
3. Nanjareddy K, Arthikala MK, Blanco L, Arellano ES, Lara M (2016) Protoplast isolation, transient transformation of leaf mesophyll protoplasts and improved Agrobacterium-mediated leaf disc infiltration of Phaseolus vulgaris: tools for rapid gene expression analysis. BMC Biotechnol 16:53. https://doi.org/10.1186/s12896-016-0283-8
4. Ren R, Gao J, Yin D et al (2021) Highly efficient leaf base protoplast isolation and transient expression systems for orchids and other important monocot crops. Front Plant Sci 12:626015. https://doi.org/10.3389/fpls.2021.626015
5. Toriyama HK (1985) Cell suspension and protoplast culture in rice. Plant Sci 41:179–183. https://doi.org/10.1016/0168-9452(85)90086-X
6. Yoo SD, Cho YH, Sheen J (2007) *Arabidopsis* mesophyll protoplasts: a versatile cell system for transient gene expression analysis. Nat Protoc 2:1565–1572. https://doi.org/10.1038/nprot.2007.199
7. Gouveia-Mageste BC, Martins LGC, Dal-Bianco M et al (2021) A plant-specific syntaxin-6 protein contributes to the intracytoplasmic route for the begomovirus CabLCV. Plant Physiol 187:158–173. https://doi.org/10.1093/plphys/kiab252
8. Chodon A, Karanthamalai J, Pandi G, Tennyson J (2022) The begomovirus–host interplay in viral replication and movement. In: Gaur RK, Sharma P, Czosnek H (Org) Geminivirus: detection, diagnosis and management, 1st edn. Academic Press, London, p 397–419. https://doi.org/10.1016/B978-0-323-90587-9.00001-8
9. Kumar RV (2019) Plant antiviral immunity against geminiviruses and viral counter-defense for survival. Front Microbiol 10:1460. https://doi.org/10.3389/fmicb.2019.01460
10. Wu M, Wei H, Tan H, Pan S, Liu Q, Bejarano ER, Lozano-Durán R (2021) Plant DNA polymerases α and δ mediate replication of geminiviruses. Nat Commun 2021(12):2780.

https://doi.org/10.1038/s41467-021-23013-2

11. Martins LGC, Raimundo GAS, Ribeiro NGA et al (2020) A Begomovirus nuclear shuttle protein-interacting immune hub: hijacking host transport activities and suppressing incompatible functions. Front Plant Sci 11: 398. https://doi.org/10.3389/fpls.2020.00398
12. Silva FA, Raimundo GAS., Fontes EBP (2022) Begomovirus:host protein-protein interactions in intracellular virus movement. In: Gaur RK, Sharma P, Czosnek H (Org) Geminivirus: detection, diagnosis and management. 1st edn. Academic Press, London, p. 347–356. https://doi.org/10.1016/B978-0-323-90587-9.00023-7
13. Wang A (2021) Cell-to-cell movement of plant viruses via plasmodesmata: a current perspective on potyviruses. Curr Opin Virol 48:10–16. https://doi.org/10.1007/s10265-014-0683-6
14. Kumar G, Dasgupta I (2021) Variability, functions and interactions of plant virus movement proteins: what do we know so far? Microorganisms 9:695. https://doi.org/10.3390/microorganisms9040695
15. Wu F-H, Shen S-C, Lee L-Y et al (2009) Tape-Arabidopsis sandwich – a simpler Arabidopsis protoplast isolation method. Plant Methods 5: 16. https://doi.org/10.1186/1746-4811-5-16
16. Priyadarshani SVGN, Hu B, Li W et al (2018) Simple protoplast isolation system for gene expression and protein interaction studies in pineapple (Ananas comosus L.). Plant Methods 14:95. https://doi.org/10.1186/s13007-018-0365-9
17. Fontes EPB, Eagle PA, Sipe PS, Luckow VA, Hanley-Bowdoin L (1994) Interaction between a geminivirus replication protein and origin DNA is essential for viral replication. J Biol Chem 269:8459–8465. https://doi.org/10.1016/S0021-9258(17)37216-2
18. Fontes EPB, Gladfelter HJ, Schaffer RL, Petty ITD, Hanley-Bowdoin L (1994) Geminivirus replication origins have a modular organization. Plant Cell 6:405–416. https://doi.org/10.1105/tpc.6.3.405
19. Doyle JJ, Doyle JL (1987) A rapid DNA isolation procedure for small quantities of fresh leaf tissue. Phytochem Bull 19:11–15

Chapter 10

Monitoring the Intracellular Trafficking of Virus-Induced Structures and Intercellular Spread of Viral Infection in Plants Using Endomembrane Trafficking Pathway-Specific Chemical Inhibitor and Organelle-Selective Fluorescence Dye

Yinzi Li and Aiming Wang

Abstract

Infection by positive-strand RNA viruses induces extensive remodeling of the host endomembrane system in favor of viral replication and movement. The integral membrane protein 6K2 of potyviruses induces the formation of membranous virus replication vesicles at the endoplasmic reticulum exit site (ERES). The intracellular trafficking of 6K2-induced vesicles along with microfilaments requires the vesicular transport pathway, actomyosin motility system, and possibly post-Golgi compartments such as endosomes as well. Recent studies have shown that endocytosis is essential for the intracellular movement of potyviruses from the site of viral genome replication/assembly site to plasmodesmata (PD) to enter neighboring cells. In this chapter, we describe a detailed protocol of how to use endomembrane trafficking pathway-specific chemical inhibitors and organelle-selective fluorescence dye to study the trafficking of potyviral proteins and potyvirus-induced vesicles and to unravel the role of endocytosis and the endocytic pathway in potyvirus infection in *Nicotiana benthamiana* plants.

Key words Potyvirus, Intracellular trafficking, Brefeldin A (BFA), Tyrphostin A23, Dynasore, FM4-64

1 Introduction

Upon infection, positive-sense, single-stranded RNA viruses induce extensive remodeling of the host endomembrane system in favor of the viral replication process [1]. The biogenesis of virus-induced membranous vesicles to anchor the viral replication complex (VRC) is a prerequisite for viral genome replication. Depending on the type of virus, a specific cytoplasmic organelle may be targeted for the induction of the virus-induced membranous vesicles. Such membranous structures provide specific cytoplasmic safeguarded

Elizabeth P. B. Fontes and Kristiina Mäkinen (eds.), *Plant-Virus Interactions*, Methods in Molecular Biology, vol. 2724, https://doi.org/10.1007/978-1-0716-3485-1_10,

microcompartments to protect virus genome replication from antiviral RNA silencing and other host defense mechanisms [2]. VRCs contain both viral and host components [3].

In plants, the endomembrane system includes the plasma membrane (PM), nuclear envelope, the endoplasmic reticulum (ER), the Golgi apparatus, trans-Golgi network (TGN), endosome, and vacuole [4]. The secretory and endocytic pathways are two major transport routes of the plant endomembrane system. In particular, the early secretory pathway between the ER and Golgi is mediated by COPII- and COPI-coated vesicles [5], whereas the transport between plasma membrane (PM), TGN, endosome, and vacuole, known as the late secretory pathway, is mostly dependent on clathrin-coated vesicles [6].

In the case of potyviruses, the potyviral integral membrane protein 6K2 induces the proliferation of ER membranes for the formation of viral vesicles to facilitate the assembly of viral replication complexes [7]. Previous studies revealed that the biogenesis of the potyviral 6K2 vesicles occurs at the COPII-accumulating ER exit sites (ERES) on the ER membrane [8]. The 6K2 virus-induced vesicles are mobile and align with microfilaments [9]. Many of them traffic from the ER to the periphery of chloroplasts [10, 11]. The early secretory pathway and actomyosin motility system are essential for these intracellular movements.

Endocytosis and endosomal trafficking regulate the internalization or uptake of membrane lipids, the proteins targeted to the plasma membrane, or extracellular materials into the cell. Endocytosis is mainly comprised of clathrin-mediated endocytosis (CME) and clathrin-independent endocytosis (CIE), based on the involvement of clathrin [12]. Many animal viruses have taken advantage of endocytosis to establish infection. The endocytic uptake and endosomal trafficking provide vesicular transport across the plasma membrane to allow viruses access to intracellular organelles [13]. In recent years, emerging evidence suggests that the endocytic pathway and endocytosis machinery have been hijacked by plant viruses for their infection [14–16]. The latest reports from our laboratory showed that potyviral 6K2-induced vesicles also target late endosomes during viral genome replication [17, 18].

The mechanism of how virus-induced vesicles traffic between endomembrane compartments remains largely unknown. To explore the involvement of the endomembrane system in intracellular viral transport, chemical inhibitors specific for certain endomembrane pathways and/or organelle-specific fluorescence dyes may serve as very useful tools.

Chemical inhibitors may transiently disrupt certain biological functions without the generation of genetic mutants. By using different inhibitors that target specific organelle structures and pathways, it becomes possible to affect secretory and trafficking pathways [19]. The perturbation effect may be optimized by adjusting treatment time and inhibitor concentration. Indeed, several chemical inhibitors, such as brefeldin A (BFA), tyrphostin A23, and dynasore have been used extensively in our research to investigate the function of the ER-Golgi transport and the endocytosis pathway associated with the trafficking of potyviral proteins and virus-induced vesicles (Table 1) [28].

BFA is an inhibitor that disrupts the ER-Golgi vesicular transport pathway (Robinson et al., 2008). To determine whether potyviral virus-induced vesicle is transported through ER-Golgi secretory pathway to chloroplasts, we examined the distribution of virus-induced 6K2 vesicles tagged with YFP (6K2-YFP) without or in the presence of BFA. Clearly, upon treatment of this inhibitor, some potyviral 6K2-induced vesicles were redistributed within the ER network, and targeting and trafficking of 6K2-induced vesicles to chloroplasts were affected [11] (Fig. 1).

Recently, we have reported that potyviruses recruit endocytosis dynamin-like proteins through interacting with several potyviral proteins including VPg and CI for viral replication and transport [17]. Two chemical inhibitors of endocytosis, dynasore and Tyr A23, were used in the study of investigating the role of the endocytosis pathway associated with potyvirus virus-induced vesicles and viral protein trafficking (Fig. 2).

Dynasore specifically targets dynamin to inhibit its activity and endocytosis by interfering with its GTPase activity [22]. Treatment with the dynamin-specific inhibitor dynasore disrupts endocytosis and impairs the transport of VPg and CI of the potyvirus turnip mosaic virus (TuMV) into endosomes, which results in the accumulation of VPg and CI on the PM. These findings suggest that the trafficking of TuMV VPg and CI is dependent on the endocytic pathway and further demonstrate that endocytosis and endosomal trafficking play an essential role in potyvirus infection.

Tyr A23 is also an endocytosis inhibitor that affects the recruitment of endocytic cargo into clathrin-coated vesicles at the plasma membrane [24]. Treatment with Tyr A23 did not affect TuMV intercellular movement (Fig. 2), indicating potyviruses may recruit a clathrin-independent endocytic pathway for intercellular movement.

In addition to chemical inhibitors, the technical advancement in live cell imaging and membrane-selective fluorescence dye is also instrumental for visualizing and tracking virus-induced vesicles and viral protein transport and intracellular architecture remodeling. The membrane-selective N-(3-triethylammoniumpropyl)-4-(p-diethylaminophenylhexatrienyl) pyridinium dibromide (FM4-64) is an amphiphilic styryl dye that is widely used to label the endosomal compartment and endocytic pathway. FM4-64 binds to the lipids of the outer leaflet of the plasma membrane and traffics from the plasma membrane through endosomal compartments and ultimately reaches the vacuole. It functions as an

Table 1
List of chemical inhibitors and fluorescence dye used in this protocol

Inhibitor or dye	Target	Effect	Reference
Brefeldin A (BFA)	Trans-Golgi network and early endosomes	Disruption of intracellular vesicle formation and protein trafficking from the ER to the Golgi apparatus	[20, 21]
Dynasore	Endocytosis	Inhibition of dynamin GTPase activity and clathrin-coated vesicle formation	[22, 23]
Tyrphostin A23	Endocytosis	Interfering the interactions between tyrosine motifs and the adaptor protein 2 (AP2) complex	[24, 25]
FM4-64	Endocytosis	Dye to label plasma membrane and endocytic organelle	[26, 27]

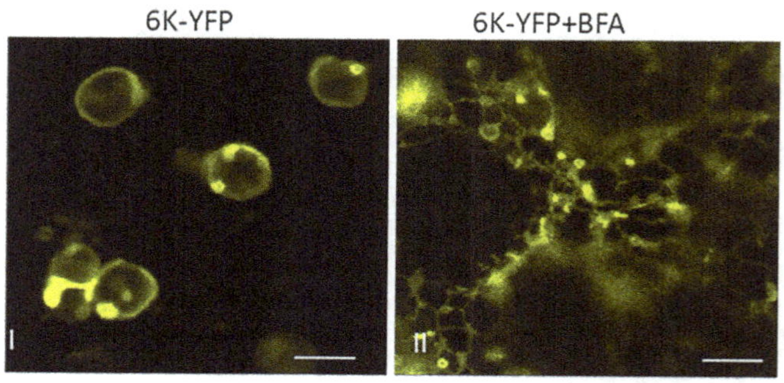

Fig. 1 The redistribution of potyviral 6K-YFP into the ER network by BFA. *Nicotiana benthamiana* epidermal cells transiently expressing TuMV 6K-YFP were treated with mock or 50 μg/mL BFA 24 h before observation under a confocal microscope. Without BFA, 6K-YFP targets the chloroplast (Panel I). Treatment of BFA (50 μg/mL) causes the redistribution of potyviral 6K-YFP into the ER network (Panel II). Scale bar = 8 μm. (Adapted from Ref. [11] with permission)

important and versatile staining tool to track the endocytic trafficking network and to monitor the virus-induced vesicle and viral protein trafficking in live plant cells [26, 27]. A colocalization study of the endocytic tracer FM4-64 with the CI protein of the potyvirus soybean mosaic virus (SMV) that was fused to YFP revealed that the CI colocalized with the punctate structures labeled by the FM4-64, suggesting its association with the endocytic pathway (Fig. 3).

Here, we describe a detailed protocol of how to use chemical inhibitors such as BFA, dynasore, and Tyr A23 to study the trafficking of potyvirus-induced vesicles and viral proteins in *Nicotiana*

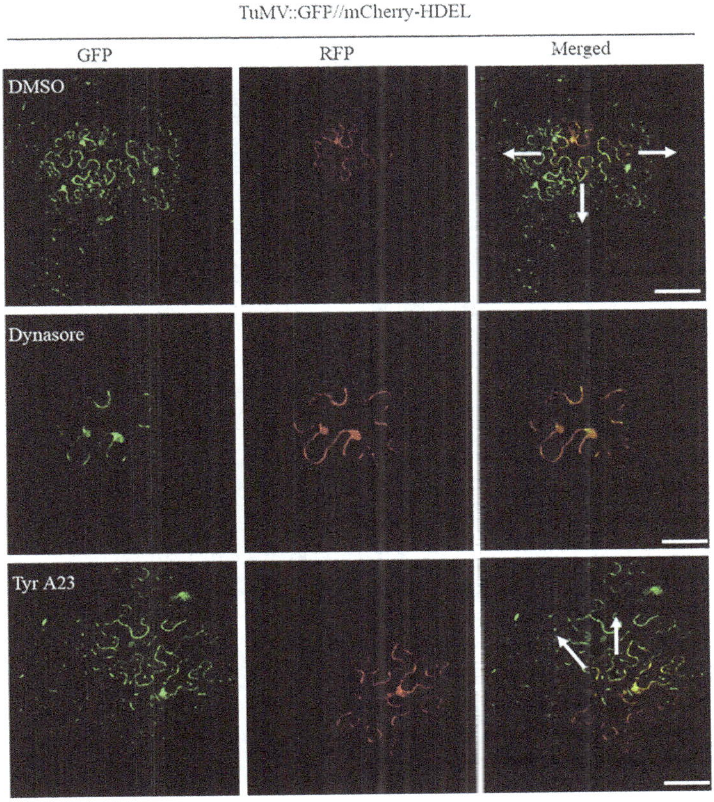

Fig. 2 TuMV intercellular movement is compromised by dynasore, a chemical inhibitor of endocytosis. Dynasore is known as a noncompetitive and reversible inhibitor which specifically targets dynamin to disrupt its GTPase activity. *Nicotiana benthamiana* epidermal cells were infiltrated of DMSO, 50 μM dynasore, or 150 μM Tyr A23 together with TuMV-GFP//mCherry-HDEL. Left panels, green fluorescence channel imaging GFP-tagged TuMV; middle panels, red fluorescence channel imaging mCherry-HDEL and right panels, merged images. Cells labeled by double fluorescence indicate primarily infected cells; secondarily infected cells emit only green fluorescence. Arrows point to the direction of viral movement. Scale bar = 100 μm. (Adapted from Ref. [17] with permission)

benthamiana leaf cells. We also provide a step-by-step guide for using the fluorescence dye FM4-64 to study the association of endocytosis and the endocytic pathway with potyvirus infection.

2 Materials

2.1 Plant Materials and Growth Conditions

1. Four-week-old *Nicotiana benthamiana* plants.
2. Controlled growth room settings: 22 °C, 60% humidity, and 16-h light/8-h dark cycle with 60–80 Lux light intensity.

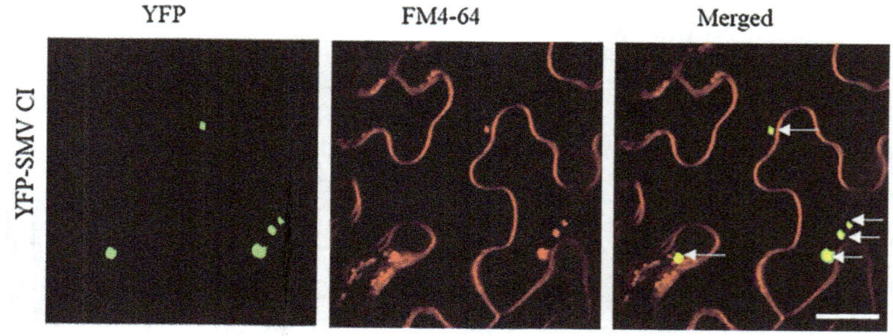

Fig. 3 Colocalization assay of YFP tagged SMV CI with fluorescent dye FM4-64 staining in *Nicotiana benthamiana* epidermal cells. Arrows indicate the colocalization of SMV CI protein with the endosome compartments labeled by FM4-64. Scale bar = 20 μm. (Adapted from Ref. [17] with permission)

2.2 Agrobacterium Culture and Infiltration Buffer

1. *Agrobacterium tumefaciens* GV3101 transformed with the binary vector containing the expression construct of interest.
2. Luria–Bertani broth (LB) medium.
3. Filter-sterilized antibiotics for agrobacterium selection (e.g., 100 μg/mL of kanamycin, 50 μg/mL of rifampicin, and 25 μg/mL of gentamicin).
4. Infiltration buffer: 10 mM $MgCl_2$, 10 mM 2-(N-morpholino) ethanesulfonic acid (MES), 150 μM acetosyringone, pH 5.6.

2.3 Chemical Inhibitor and Fluorescence Dye

1. Dimethyl sulfoxide (DMSO).
2. Brefeldin A (BFA): Stock solution is 10 mg/mL BFA in DMSO. Store aliquots at −20 °C.
3. Tyrphostin A23 (Tyr A23): Stock solution is 50 mg/mL Tyr A23 in DMSO. Store aliquots at −20 °C.
4. Dynasore: Stock solution is 10 mg/mL dynasore in DMSO. Store aliquots at −20 °C.
5. N-(3-Triethylammoniumpropyl)-4-(6-(4-(diethylamino)-phenyl)hexatrienyl) pyridinium dibromide (FM4-64): Stock solution is 2 mM FM4-64 in DMSO. Store aliquots at −20 °C.

2.4 Preparation of Stock Solutions and Working Solutions

1. Use 100% DMSO to make a stock solution to achieve full solubility of each chemical inhibitor and fluorescence dye (*see* **Note 1**). Mix the stock solutions well by vortexing. All stock solutions are kept at −20 °C.
2. Prepare working solutions freshly on the day of the experiment. Add the appropriate volume of the stock solution of inhibitor to the working solution and mix thoroughly by up-and-down pipetting (*see* **Note 2**).

- The working concentration of BFA is 50 μg/mL.
- The working concentration of Tyr A23 is 150 μM.
- The working concentration of Dynasore is 50 μM.

3. FM4-64 is a styryl dye that is widely used to image the endocytosis process (*see* **Note 3**). Prepare 2 mM stock solutions of FM4-64 by adding 100% DMSO.
4. Immediately before use, dilute the stock solution to prepare a working staining solution of 40 μM FM4-64. Keep this staining solution on ice.

2.5 Confocal Laser Scanning Microscopy

1. Confocal laser scanning microscope: Leica TCS SP2 inverted with 63× oil or 63× water-immersion objective in multitrack mode or other similar confocal microscopes.
2. Image analysis software: Leica Application Suite for Advanced Fluorescence (LAS AF; version 2.35) software (Leica Microsystem) or other compatible software.

2.6 Other Equipment

1. Temperature-controlled shaker and incubator.
2. Centrifuge.
3. Spectrophotometer.
4. Glass slide and cover glass.
5. Timer.
6. Aluminum foil.
7. Single-use syringe (1 mL) and syringe filter (0.22 μm).
8. Tweezers.

3 Methods

3.1 Plant Materials and Growth Conditions/Preparation of Plant Leaf Samples

Four-week-old *N. benthamiana* plants are used for agrobacterium-mediated transient expression. Grow the plants in a growth room at 22 °C with a 16-h photoperiod and relative humidity of 60%.

3.2 Transient Protein Expression

1. Transform each expression vector into *Agrobacterium tumefaciens* strain GV3101 by electroporation, respectively, and plate onto the LB-agar plate containing 100 μg/mL of kanamycin, 50 μg/mL of rifampicin, and 25 μg/mL of gentamicin and then incubate at 28 °C for 48 h.
2. Inoculate a single positive colony from the transformants into 4 mL of LB medium containing the appropriate antibiotics. Grow the agrobacterium culture overnight at 28 °C.

3. Harvest the agrobacterium cells by centrifugation and resuspend the pellet in infiltration buffer (10 mM $MgCl_2$, 10 mM MES, and 150 μM acetosyringone). Repeat three times to remove antibiotics and LB medium.
4. Incubate the agrobacterium culture at room temperature for 2 h before infiltration.
5. Adjust the agrobacterium culture to an optical density of 0.1 at 600 nm (OD600) for subcellular localization assay (*see* **Note 4**). Infiltrate the agrobacterium culture into leaf epidermal cells under gentle pressure using a syringe barrel against the underside of the leaves.
6. For co-agroinfiltration assays, mix two Agrobacterium cultures with equal volume, and infiltrate the mixed cultures into *N. benthamiana* leaf epidermal cells. Keep the total cell population of each mixture consistent for each infiltration experiment (*see* **Note 5**).
7. After agroinfiltration, maintain the plants under normal growth conditions for observation.

3.3 Chemical Inhibitor Treatment and FM Dye Staining

3.3.1 BFA Treatment

1. Infiltrate the leaf of *N. benthamiana* with 50 μg/mL BFA working solution. At 3 h post-BFA treatment, infiltrate the infiltrated leaf segments with the expression constructs.
2. After 24 h post-agroinfiltration, transfer the leaf sample to a microscope glass slide and cover it with a coverslip for observation under a confocal microscope and imaging (Fig. 1).

3.3.2 Tyr A23 and Dynasore Treatment

1. Co-infiltrate *N. benthamiana* leaves with either 150 μM Tyr A23 solution or DMSO (as a mock control) together with agrobacterium culture of TuMV-GFP//mCherry-HDEL expressing GFP-tagged recombinant TuMV and the mCherry-HDEL expression cassette at a low OD600 of 0.001.
2. Co-infiltrate *N. benthamiana* leaves with either 50 μM dynasore solution or DMSO (as a mock control) together with agrobacterium culture of TuMV-GFP//mCherry-HDEL at a low OD600 of 0.001.
3. At 92 h post-agroinfiltration, visualize the Tyr A23-, dynasore-, or DMSO-treated leaf sample under confocal microscopy (Fig. 2).

3.3.3 FM4-64 Dye Staining

1. Infiltrate *N. benthamiana* leaves with agrobacterium cultures expressing potyviral proteins tagged with YFP.
2. At 48 h post-agroinfiltration of the expression vector (*see* **Note 6**), infiltrate FM4-64 staining solution into the same leaf area for *in vivo* staining of the plasma membrane and endocytic organelles.

3. At 5 min post-dye-labeling, visualize the leaf sample for the stained cells by fluorescence and monitor for endocytosis and endosome localization using confocal microscopy (Fig. 3).

3.4 Confocal Microscopy and Imaging

1. Cut out a small section from the agroinfiltrated or dye staining area of *N. benthamiana* leaves.
2. Place the leaf sample in a microscope slip with the lower epidermis facing upward.
3. Mount the leaf sample with water or oil and then cover it with a coverslip.
4. Follow standard confocal microscopy procedure. Set the parameters for each fluorescent probe, the excitation, and emission wavelengths as follows (*see* **Note 7**):
   - For FM4-64: 543 nm (excitation) and 600–750 nm (emission).
   - For YFP: 514 nm (excitation) and 555–575 nm (emission).
   - For GFP: 488 nm (excitation) and 505–525 nm (emission).

4 Notes

1. Dimethyl sulfoxide (DMSO) is harmful, and personal protective equipment is needed.
2. To avoid repeated thawing and freezing cycles, aliquot the stock solution to smaller volumes and store at −20 °C.
3. FM4-64 is light sensitive. Wrap the tube with aluminum foil to minimize light exposure to ensure compound stability and prevent dye bleaching.
4. The low OD600 value of 0.001 is used for the agrobacterial culture expressing PCB301TuMV-GFP//mCherry-HDEL infectious clone to study viral intercellular movement, which allows distinguishing primary and secondary infections in a few epidermal cells by expressing two expression cassettes.
5. Keep the final OD600 of each agrobacterial culture as low as possible to avoid protein overexpression or overloading of the cellular environment that might result in false-positive fluorescence signals.
6. FM4-64 infiltration time point is important due to FM4-64 stability and the time required for transient expression. We recommend 48 h post-agroinfiltration of the expression vector.
7. Optimize laser power, gain, and offset for each separate fluorophore. Always use the sequential scanning mode for co-imaging multiple fluorescent signals to avoid crosstalk between different fluorophores.

Acknowledgments

The protocol described here is based on the work supported by grants from Agriculture and Agri-Food Canada and the Natural Sciences and Engineering Research Council of Canada to A.W.

References

1. Laliberté J-F, Sanfaçon H (2010) Cellular remodeling during plant virus infection. Annu Rev Phytopathol 48:69–91
2. Yang X, Li Y, Wang A (2021) Research advances in potyviruses: from the laboratory bench to the field. Annu Rev Phytopathol 59: 1–29
3. Wang A (2015) Dissecting the molecular network of virus-plant interactions: the complex roles of host factors. Annu Rev Phytopathol 53:45–66
4. Aniento F, de Medina S, Hernández V, Dagdas Y, Rojas-Pierce M, Russinova E (2022) Molecular mechanisms of endomembrane trafficking in plants. Plant Cell 34:146–173
5. Brandizzi F, Barlowe C (2013) Organization of the ER–Golgi interface for membrane traffic control. Nat Rev Mol Cell Biol 14:382–392
6. Reyes FC, Buono R, Otegui MS (2011) Plant endosomal trafficking pathways. Curr Opin Plant Biol 14:666–673
7. Grangeon R, Jiang J, Laliberté J-F (2012) Host endomembrane recruitment for plant RNA virus replication. Curr Opin Virol 2: 683–690
8. Wei T, Wang A (2008) Biogenesis of cytoplasmic membranous vesicles for plant potyvirus replication occurs at endoplasmic reticulum exit sites in a COPI- and COPII-dependent manner. J Virol 82:12252–12264
9. Cotton S, Grangeon R, Thivierge K, Mathieu I, Ide C et al (2009) Turnip mosaic virus RNA replication complex vesicles are mobile, align with microfilaments and are each derived from a single viral genome. J Virol 83:10460–10471
10. Wei T, Huang TS, McNeil J, Laliberté J-F, Hong J et al (2010) Sequential recruitment of the endoplasmic reticulum and chloroplasts for plant potyvirus replication. J Virol 84:799–809
11. Wei T, Zhang C, Hou X, Sanfaçon H, Wang A (2013) The SNARE protein Syp71 is essential for turnip mosaic virus infection by mediating fusion of virus-induced vesicles with chloroplasts. PLoS Pathog 9:e1003378
12. Paez Valencia J, Goodman K, Otegui MS (2016) Endocytosis and endosomal trafficking in plants. Annu Rev Plant Biol 67:309–335
13. Mercer J, Schelhaas M, Helenius A (2010) Virus entry by endocytosis. Annu Rev Biochem 79:803–833
14. Lewis JD, Lazarowitz SG (2010) Arabidopsis synaptotagmin SYTA regulates endocytosis and virus movement protein cell-to-cell transport. Proc Natl Acad Sci 107:2491–2496
15. Xu K, Nagy PD (2016) Enrichment of phosphatidylethanolamine in viral replication compartments via co-opting the endosomal Rab5 small GTPase by a positive-strand RNA virus. PLoS Biol 14:e2000128
16. Cabanillas DG, Jiang J, Movahed N, Germain H, Yamaji Y, Zheng H, Laliberté J-F (2018) Turnip mosaic virus uses the SNARE protein VTI11 in an unconventional route for replication vesicle trafficking. Plant Cell 30: 2594–2615
17. Wu G, Cui X, Chen H, Renaud JB, Yu K et al (2018) Dynamin-like proteins of endocytosis in plants are coopted by potyviruses to enhance virus infection. J Virol 92:e01320–e01318
18. Wu G, Cui X, Dai Z, He R, Li Y et al (2020) A plant RNA virus hijacks endocytic proteins to establish its infection in plants. Plant J 101: 384–400
19. Zhu D, Zhang M, Gao C, Shen J (2020) Protein trafficking in plant cells: tools and markers. Sci China Life Sci 63:343–363
20. Nebenfuhr A, Ritzenthaler C, Robinson DG (2002) Brefeldin a: deciphering an enigmatic inhibitor of secretion. Plant Physiol 130:1102–1108
21. Robinson DG, Langhans M, Saint-Jore-Dupas C, Hawes C (2008) BFA effects are tissue and not just plant specific. Trends Plant Sci 13:405–408
22. Macia E, Ehrlich M, Massol R, Boucrot E, Brunner C, Kirchhausen T (2006) Dynasore, a cell-permeable inhibitor of dynamin. Dev Cell 10:839–850
23. Sharfman M, Bar M, Ehrlich M, Schuster S, Melech-Bonfil S et al (2011) Endosomal

signaling of the tomato leucine-rich repeat receptor-like protein LeEix2. Plant J 68:413–423

24. Banbury DN, Oakley JD, Sessions RB, Banting G (2003) Tyrphostin A23 inhibits internalization of the transferrin receptor by perturbing the interaction between tyrosine motifs and the medium chain subunit of the AP-2 adaptor complex. J Biol Chem 278:12022–12028
25. Ortiz-Zapater E, Soriano-Ortega E, Marcote MJ, Ortiz-Masia D, Aniento F (2006) Trafficking of the human transferrin receptor in plant cells: effects of tyrphostin A23 and brefeldin A. Plant J 48:757–770
26. Bolte S, Talbot C, Boutte Y, Catrice O, Read ND, Satiat-Jeunemaitre B (2004) FM-dyes as experimental probes for dissecting vesicle trafficking in living plant cells. J Microsc 214:159–173
27. Rigal A, Doyle SM, Robert S (2015) Live cell imaging of FM4-64, a tool for tracing the endocytic pathways in Arabidopsis root cells. Methods Mol Biol 1242:93–103
28. Mishev K, Dejonghe W, Russinova E (2013) Small molecules for dissecting endomembrane trafficking: a cross-systems view. Chem Biol 20:475–486

Chapter 11

Ribosome Profiling of Plants

Jahanara Sonia, Pulkit Kanodia, Zachary Lozier, and W. Allen Miller

Abstract

Translation is a key step in control of gene expression, yet most analyses of global responses to a stimulus focus on transcription and the transcriptome. For RNA viruses in particular, which have no DNA-templated transcriptional control, control of viral and host translation is crucial. Here, we describe the method of ribosome profiling (ribo-seq) in plants, applied to virus infection. Ribo-seq is a deep sequencing technique that reveals the translatome by presenting a snapshot of the positions and relative amounts of translating ribosomes on all mRNAs in the cell. In contrast to RNA-seq, a crude cell extract is first digested with ribonuclease to degrade all mRNA not protected by a translating 80S ribosome. The resulting ribosome-protected fragments (RPFs) are deep sequenced. The number of reads mapping to a specific mRNA compared to the standard RNA-seq reads reveals the translational efficiency of that mRNA. Moreover, the precise positions of ribosome pause sites, previously unknown translatable open reading frames, and noncanonical translation events can be characterized quantitatively using ribo-seq. As this technique requires meticulous technique, here we present detailed step-by-step instructions for cell lysate preparation by flash freezing of samples, nuclease digestion of cell lysate, monosome collection by sucrose cushion ultracentrifugation, size-selective RNA extraction and rRNA depletion, library preparation for sequencing and finally quality control of sequenced data. These experimental methods apply to many plant systems, with minor nuclease digestion modifications depending on the plant tissue and species. This protocol should be valuable for studies of plant virus gene expression, and the global translational response to virus infection, or any other biotic or abiotic stress, by the host plant.

Key words RNA sequencing, Ribo-seq, translation, viral RNA, protein synthesis, translatome

1 Introduction

Translation, the key event in all living things in which the genetic code is decoded as proteins are synthesized, has been underappreciated as a regulatory stage of gene expression [1, 2]. Historically, gene expression monitoring has focused on the levels of mRNA (the transcriptome) determined by methods that continue to be more informative, such as the widely used high-throughput RNA sequencing (RNA-seq) [3]. However, monitoring RNA abundance alone is not sufficient to determine protein expression level, because, as with transcription, the translation efficiencies can

Elizabeth P. B. Fontes and Kristiina Mäkinen (eds.), *Plant-Virus Interactions*, Methods in Molecular Biology, vol. 2724, https://doi.org/10.1007/978-1-0716-3485-1_11,

change due to an external stimulus [4] or during development [5], and changes in mRNA level do not always correlate with changes in protein level [6]. Proteomics addresses this question, but measuring protein levels combined with RNA-seq provides only an indirect measure of translation efficiency and is affected by differences in protein stability.

Ribosome profiling (ribo-seq) has emerged as a powerful method that directly measures the translational efficiency of mRNAs [7, 8]. Ribo-seq reveals the positions and relative amounts of ribosomes on every mRNA in the cell at single nucleotide resolution. Thus, it can reveal mRNAs that are up- or down-translated in response to a change in environment or developmental cue. It also detects sites of ribosome pausing that occur during critical translational control events. Ribo-seq is a modified RNA-Seq-based technique, in which, instead of sequencing and counting total mRNA, only the mRNA tracts that are occupied by translating ribosomes are sequenced. Depending on the research objectives, ribo-seq can provide a panorama of the whole translation process [9]. In this method, a translation-arrested cell lysate is digested with RNase to degrade mRNA not protected by a translating ribosome. Then, the resulting ribosome-protected fragments (**RPFs**) are deep sequenced. The number of RPFs that map to each position on an mRNA indicates the ribosome density and thus its translational efficiency (**TE**) and also reveals ribosome pause sites (Fig. 1).

1.1 Application of Ribosome Profiling in Plant–Virus Interactions

In addition to identifying unexpected coding regions and translational control mechanisms in host mRNAs, ribo-seq is invaluable for characterizing the numerous noncanonical translation events that are a hallmark of many positive-strand RNA viral genomes. For example, many viruses use recoding events such as stop codon read-through or ribosomal frameshifting [10]. Ribo-seq is the most accurate way of measuring the exact rate of read-through or frameshifting because the proportion of RPFs obtained from the ORFs before and after the recoding event is proportional to the number of ribosomes translating each ORF. This is more accurate than measuring ratios of viral or reporter proteins translated via the viral read-through/frameshift sequence, because the protein products may differ in stability. Thus, their levels would not be proportional to the efficiency of translation of the ORFs that generated them. Also, viruses encode many small proteins from small ORFs that may not appear significant [11]. Ribo-seq reveals all translated ORFs, including those that do not initiate with a canonical AUG start codon. Application of ribosome profiling to virus-infected human cells has revealed such highly quantitative details of coding capacity and gene expression [12, 13].

With regard to plant–pathogen interactions, Ribo-seq has revealed an unexpected translational upregulation of many genes in the plant immune system in response to bacterial infection

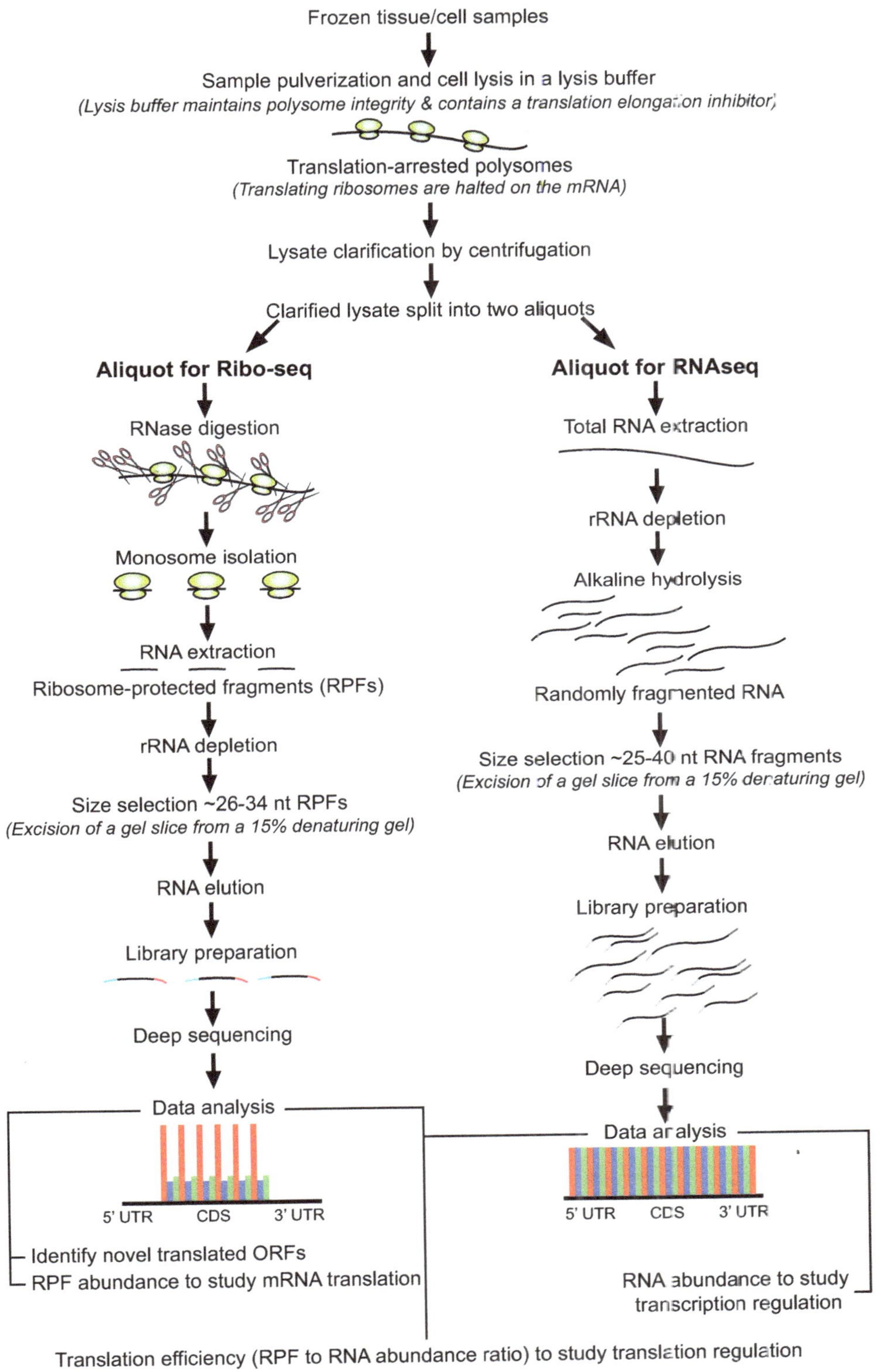

Fig. 1 Outline of ribosome profiling and RNA sequencing methodology. (Modified slightly from Fig. 2a of reference 20)

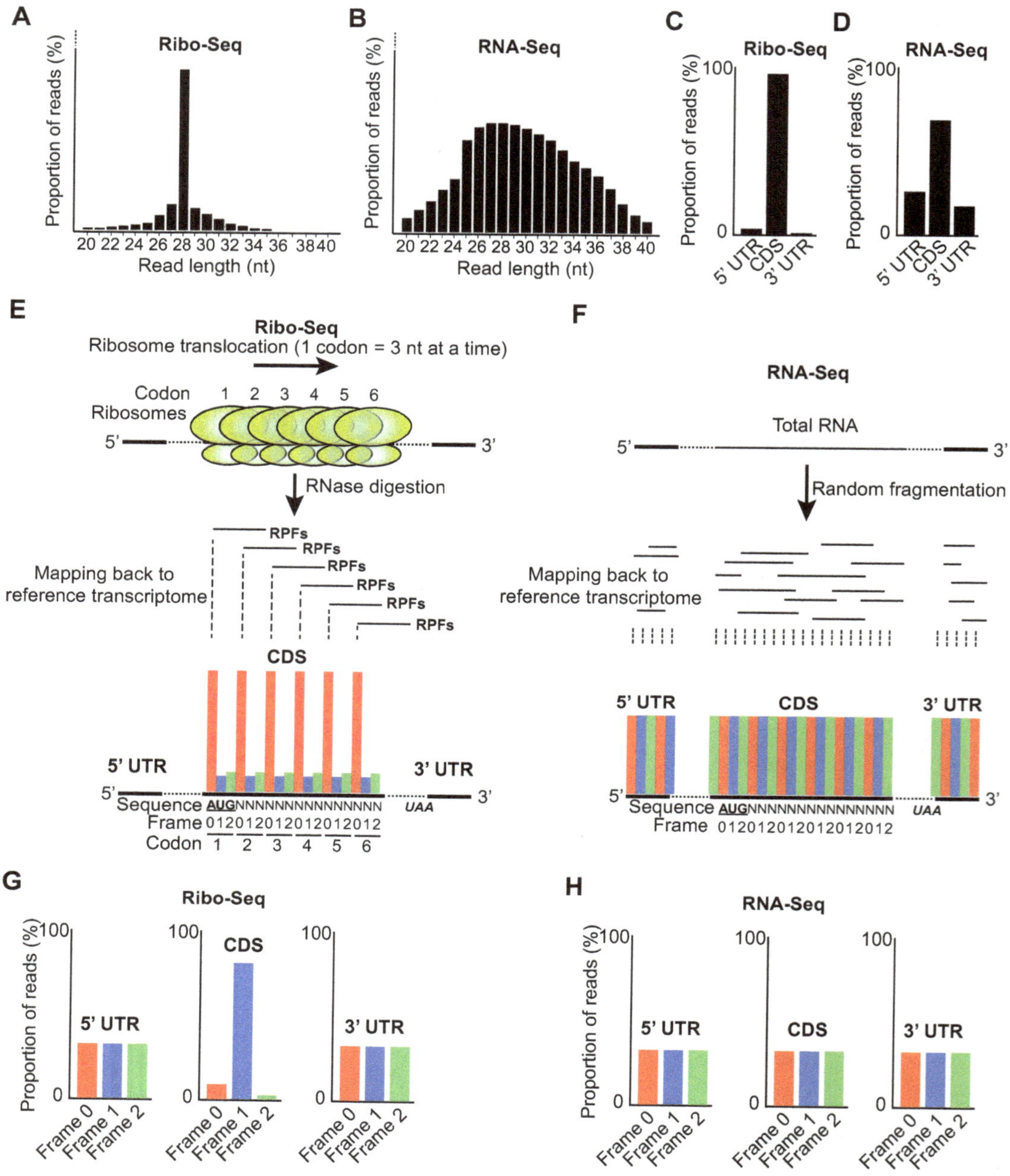

Fig. 2 Illustration of specific hallmarks of Ribo-seq data. (**a**) Read length distribution of Ribo-seq data shows a unimodal distribution. (**b**) Read length distribution of RNA-seq data shows a relatively uniform and broad distribution. (**c**) Ribo-seq data map predominantly to the CDS of all the mRNAs, whereas only a few reads map to the UTRs. (**d**) A substantial number of reads from RNA-seq data map to the UTRs and the CDS. (**e**) Metagene analysis representing the triplet periodicity of ribo-seq data. Because ribosomes move 3 nt (one codon) at a time during translation elongation, most RPFs map to every third nt in the reference transcriptome. (**f**) Metagene analysis showing the lack of triplet periodicity of RNA-seq data because it is obtained from random fragmentation of ribosome-free total RNAs. (**g**) Triplet periodicity of ribo-seq data is observed only for the reads that map to the CDS (true RPFs) and not for those that map to the UTRs. (**h**) No triplet periodicity is observed in RNA-seq data for reads that map either to the CDS or to the UTRs. The figure only includes

[14]. This later led to the identification of a signal transduction pathway in which phosphorylation of translation factors released specific innate immune system genes from a translationally inhibited state, upon detection of a bacterial effector [15]. A Ribo-seq study of sugarcane mosaic virus-infected maize found that photosynthetic and metabolic genes were both transcriptionally and translationally downregulated [16]. Given that many plant viral RNAs, such as the cap-independent translation elements of tombusvirids, and viral proteins, such as the genome-linked proteins (VPgs) of potyviruses and solemovirids, are known to bind key translation initiation factors [10, 17–19], these interactions might out-compete host mRNAs for the factors and alter their translation. Ribo-seq would be a powerful tool to answer these questions. Thus, we provide detailed methods here for plant virologists to employ Ribo-seq in the model plant Arabidopsis and the agronomic crop (and also a model organism) maize [20].

1.2 Bioinformatics

This chapter provides details only for preparing and sequencing RPFs at the laboratory bench. However, after that process, a substantial amount of sequence data analysis is necessary to confirm that the sequences represent true RPFs and to subsequently identify translational efficiencies, ribosome pause sites, and global control of translation. We provide a brief summary of data processing and analysis here. For more information on the processing and bionformatics of Ribo-seq data, we refer the readers to Kiniry et al. [21] and Xiao et al. [22]. Initial data treatment is the same for ribo-seq and RNA-seq data, starting by assessing the quality of raw data by tools such as FastQC [23], followed by processing via quality and adapter trimming by tools such as Cutadapt [24]. Bowtie2 [25] is used to align the processed reads to the organism's reference sequences for the rRNAs, tRNAs, and snoRNAs. The remaining reads are aligned to the reference transcriptome using Bowtie2 or to reference genome using splice-aware tools such as STAR [26]. There are two general approaches to counting reads: Consider only uniquely mapped reads or estimate the true location of multi-mapped reads. We recommend featureCounts [27] for counting only uniquely mapped reads and RSEM [28] for counting multi-mapped reads. Statistical analysis of counted reads can be performed with edgeR [29], DESeq2 [30], xtail [22], or riboVI [31]. The latter two are designed specifically for ribo-seq data. RiboVI is the most sensitive, as it identifies more genes as having statistically significant changes in translation efficiency (TE) with a lower false discovery rate than the other methods, but tends to give a lower change in TE for each gene. We advise working with an

Fig. 2 (continued) illustrations of high-quality ribo-seq data and does not represent real experimental data. Nucleotide (nt), ribosome-protected fragments (RPFs), coding DNA sequence (CDS), untranslated regions (UTRs). Reproduced from Ph.D dissertation of Pulkit Kanodia (2021) with permission of the author [38]

experienced statistician due to the complexity of the analyses. For RPF metagene analysis (discussed below), the authors recommend riboSeqR [32] if reference transcriptome mapping is used. If the RPFs are mapped to the reference genome, then RiboTaper [33] and RiboToolkit [34] are excellent tools for metagene analysis. Actively translating regions can be identified with RiboTaper [33]. Many of the above-listed programs and packages are used in a command-line environment and require some basic skills in scripting, R, and/or Python. Some web services perform all (RiboToolkit [34]) or many (RiboGalaxy [35]) of the above processes. Riboviz2 [36] and RiboVIEW [37] are command-line programs that perform many, but not all, of the steps above.

1.3 Hallmarks of High-Quality Ribo-seq Data

Ribo-seq data have specific hallmarks that are not present in RNA-seq data. However, some RNA secondary structures or ribonucleoprotein complexes not involved in translation may generate RPF-length fragments. Therefore, it is imperative to distinguish true RPFs from artifacts. The following features are hallmarks of high-quality Ribo-seq data with true RPFs:

(i) *Narrow Length Distribution.* Depending on variety of organisms, ribosome-protected fragment length may vary. Still ribosomes protect predominantly one specific size of RNA for a given organism (e.g., 28 nt for Arabidopsis [8] and 30 nt for maize [20]). In high-quality Ribo-seq data, the length distribution is mostly enriched to a single RPF length (Fig. 2a) along with very few other RPF lengths, whereas in RNA-seq data there is a broad read length distribution (Fig. 2b). RPFs with good triplet periodicity (described below) show a narrow length distribution [8].

(ii) *CDS Enrichment.* Almost all reads from ribo-seq data map to the coding sequence (CDS) because only the translating 80S ribosome protects mRNA from the RNase treatment used to obtain the RPFs. In contrast, RNA-seq data will always map to both CDS and UTRs (Fig. 2c–d).

(iii) *Triplet Periodicity.* As mentioned above, ribosomes generate more RPFs at pause sites. Because ribosomes pause briefly at each codon, this pausing can be detected in the form of more reads mapping to every third base in a CDS (Fig. 2e). This triplet periodicity is absent in RNA-seq data because RNA-seq is obtained from random fragmentation of total RNA in the absence of ribosomes (Fig. 2f). Both the CDS enrichment and triplet periodicity in ribo-seq data and lack thereof in RNA-seq data can be observed in the same metagene analysis plot that maps positions of reads relative to codons in all mRNAs combined (Fig. 2e–h).

1.4 Translation Efficiency

Once the quality of RPFs has been confirmed, the translation efficiency (TE) can be calculated for each mRNA using software such as Xtail [22]. TE is the ratio of RPF abundance to mRNA abundance; as the more ribosomes are translating an mRNA, the more RPFs will be captured by ribo-seq. Thus, ribo-seq, when compared to the RNA-seq that is performed on the same sample, allows one to measure precisely how much change in expression of a given gene is due to transcription and how much is due to translation, the ultimate goal of ribosome profiling.

2 Materials

2.1 Enzymes, Chemical Reagents, and Kits

1. Nuclease-free water.
2. TURBO DNase with 10x buffer (Invitrogen).
3. RNase I (Invitrogen).
4. SUPERase-In (Invitrogen).
5. Proteinase K (Thermo Scientific).
6. T4PNK kit (Thermo Scientific).
7. ATP solution.
8. Acid–phenol: chloroform, pH 4.5 (with IAA 125:24:1, Invitrogen).
9. Chloroform–isoamyl alcohol (24:1, Sigma-Aldrich).
10. 15 mg/mL GlycoBlue Coprecipitant (Invitrogen).
11. 3 M sodium acetate pH 5.5.
12. 1 M Tris–HCl pH 7.5, autoclave, store at room temperature.
13. 2 M KCl, autoclave, store at room temperature.
14. 1 M $MgCl_2$, autoclave, store at room temperature.
15. 10% (v/v) Triton X-100, autoclave, store at room temperature.
16. 10% (v/v) IGEPAL CA-630, autoclave, store at room temperature.
17. 2 M sucrose, 0.45-μm filter-sterilize, store at 4 °C.
18. 50 mg/mL cycloheximide in ethanol, store at −20 °C.
19. 50 mg/mL chloramphenicol in ethanol, store at −20 °C.
20. 1 M dithiothreitol (DTT), do not autoclave, store at −20 °C in dark.
21. EDTA-free protease inhibitor (Thermo Scientific).
22. 10% (w/v) SDS, 0.22-μm filter-sterilize, do not autoclave, and store at room temperature.
23. 0.5 M EDTA pH 8.0, autoclave, store at room temperature.

24. 1 M Na_2CO_3, 0.22-μm filter-sterilize, store at room temperature.
25. 1 M $NaHCO_3$, 0.22-μm filter-sterilize, store at room temperature.
26. Ice-cold 100% isopropanol.
27. Ice-cold 100% ethanol.
28. Ice-cold 80% ethanol.
29. 15% TBE-Urea Gel.
30. 2x RNA loading Dye (NEB).
31. 10x TBE buffer.
32. SYBR Gold Nucleic Acid Gel Stain (Invitrogen).
33. Zymo RNA Clean & Concentrator −5 Kit (Zymo Res.).
34. Ribo-Zero rRNA depletion kit for plant leaf (Illumina).
35. NEXTflex Small RNA-Seq Kit v3 (Perkin Elmer).
36. Qubit dsDNA HS Assay Kit (Invitrogen).
37. Liquid nitrogen.

2.2 Buffers

1. Polysome extraction buffer (PEB):; 20 mM Tris–HCl pH 7.5, 140 mM KCl, 5 mM $MgCl_2$, 1% (v/v) Triton X-100, 0.5% (v/v) IGEPAL CA 630, 146 M sucrose, 100 μg/mL cycloheximide, 100 μg/mL chloramphenicol, 0.5 mM DTT, 0.5 μL/mL Turbo DNase, and EDTA-free protease inhibitor. Store the PEB without TURBO DNase, DTT, cycloheximide, and chloramphenicol at −20 °C. Add these components to the PEB right before use.
2. Proteinase K buffer: 10 mM Tris–HCl pH 7.5, 1% SDS, and 200 μg/mL Proteinase K.
3. Sucrose cushion, 1.022 M sucrose, 20 mM Tris–HCl pH 7.5, 140 mM KCl, 5 mM $MgCl_2$, 100 μg/mL cycloheximide, 100 μg/mL chloramphenicol, 0.5 mM DTT, 5 μL/mL SUPERase-in. Prepare fresh.
4. 2x Fragmentation buffer: 2 mM EDTA pH 8.0, 12 mM Na_2CO_3, 88 mM $NaHCO_3$. Prepare fresh. Verify pH ~ 9.0 using a pH strip.
5. Fragmentation stop solution: 0.3 M sodium acetate pH 5.5, 53.6 μg/mL GlycoBlue. Prepare fresh.
6. RNA gel extraction buffer: 0.3 M sodium acetate pH 5.5, 1 mM EDTA pH 8.0, 10 mM Tris–HCl pH 7.5, 0.25% (w/v) SDS. Store at room temperature.

2.3 Equipment and Consumables

1. Sorvall Mini-Ultracentrifuge (Discovery M150).
2. S150-AT fixed angle rotor (Thermo Scientific).
3. 1 mL Mini-Ultracentrifuge tubes (Thermo Scientific).
4. TissueLyser (Eppendorf).
5. Thermo-mixer (Thermo Scientific).
6. Blue light transilluminator.
7. 50-mL falcon tubes.
8. 4.8-mm stainless steel metal beads (MidSci) (sterilize in an oven at 200 °C overnight).
9. 2 mm zirconium oxide beads (Fisher Sci) (sterilize in an oven at 200 °C overnight).
10. 0.22-μm Spin-X cellulose acetate filter columns (Sigma-Aldrich).
11. Qubit Fluorometer (Invitrogen).
12. Qubit assay tubes (Invitrogen).
13. NanoDrop Spectrophotometer.
14. Low-retention 0.5-mL, 1.5-mL, and 2-mL microfuge tubes (Fisher Sci).
15. Low-retention barrier/filter pipette tips (for use in entire protocol for pipetting).
16. Square petri dish (VWR).
17. Razor blades (VWR) (sterilize in an oven at 200 °C overnight).
18. Syringe filters.
19. 18G syringe needle.
20. 0.22-μm syringe filter.
21. 0.45-μm syringe filter.
22. Magnetic tube rack for PCR tubes or/and 1.5-mL microfuge tubes (for use in magnetic bead RNA cleanup steps in rRNA depletion and library preparation).
23. Agilent Bioanalyzer.
24. Amber-colored goggles and blue transilluminator.

2.4 Size Markers

1. 28 nt RNA oligo (Sigma-Aldrich, HPLC purified): *5′*AUGUA CACGGAGUCGACCCGCAACGCGA-*3′*.
2. 34 nt RNA oligo (Sigma-Aldrich, HPLC purified): *5′*AUGUA CACGGAGUCGAGCUCAACCCGCAACGCGA-*3′*.
3. 10/60 nt single-stranded DNA ladder (IDT).

3 Methods

3.1 Lysate Preparation

1. Collect the plant leaves in liquid nitrogen-chilled 50-mL falcon tubes containing two 4.8-mm beads (*see* **Notes 1** and **2**).
2. Coarsely pulverize the leaves by vigorously shaking the tube and transfer them to liquid nitrogen-chilled 1.5-mL microfuge tubes containing 8–10 2-mm zirconium oxide beads.
3. Subsequently, pulverize the tissues into fine powder using a TissueLyser (1500 rpm for 30 s, twice), followed by adding 800 μL of PEB (*see* **Note 3**).
4. Incubate the crude lysate in ice for 20 min on a rocker followed by clarification using two rounds of centrifugation (21,100 × *g*, 15 min, 4 °C).
5. Subsequently, use NanoDrop to adjust the absorbance of the lysate with PEB to A_{254} (Lysate – PEB) ≈ A_{260} (Lysate – PEB) ≈ 6 (*see* **Note 4**).
6. Save a 200 μL aliquot of lysate to be used for total RNA extraction (Subheading 3.3). Dispense 400 μL aliquots of remaining lysate in 1.5-mL microfuge tubes for RPF purification (Subheading 3.2). If necessary, these can be flash-frozen and stored at −80 °C for future extraction (*see* **Note 5**).

3.2 RNase 1 Digestion, Monosome Collection, and RPF RNA Purification

1. Thaw 400 μL of the absorbance-adjusted lysate in ice.
2. Centrifuge (21,100 × *g*, 5 min, 4 °C) to remove any remaining debris and transfer the supernatant to new 2-mL microfuge tubes.
3. Add 200U RNAse1, mix by inverting the tube, and spin down.
4. Incubate at 28 °C for 60 min in a thermomixer at 400 rpm.
5. Immediately transfer the tubes to ice.
6. Add 5 μL SUPERase-In to terminate the reaction.
7. To the 400 μL RNase-treated lysate, add 350 μL ice-cold PEB to make the total volume ~ 750 μL, and carefully layer it on 350 μL ice-cold sucrose cushion in mini-ultracentrifuge tubes (*see* **Note 6**).
8. Ultracentrifuge at 57,000 rpm (131,500 × *g*) for 90 min at 4 °C with slow acceleration and deceleration in a Sorvall mini-ultracentrifuge with 4 °C precooled S150-AT fixed angle rotor.
9. Carefully remove the supernatant by pipetting. The pellet consists of monosomes from which RPFs will be extracted below.
10. Use a 200-μL pipette to remove any residual supernatant.
11. Optional: Rinse the monosome pellet carefully with 500 μL nuclease-free water (*see* **Note 7**).

12. Add 600 μL Proteinase K buffer (prewarmed to 42 °C for 5 min) to the monosome pellet.
13. Incubate at room temperature for 5 min.
14. Resuspend the monosome pellet by pipetting and transfer to 2-mL microfuge tubes (*see* **Note 8**).
15. Incubate at 42 °C for 30 min followed by incubation at 65 °C for 5 min.
16. Immediately transfer the solution to 600 μL hot acid–phenol–chloroform that is pre-heated to 65 °C (*see* **Note 9**).
17. Mix by inverting the tubes.
18. Incubate the mix at 65 °C for 5 min (mix intermittently by inverting the tubes).
19. Spin down and incubate the tubes in ice for 5 min followed by centrifugation (21,100 × *g*, 2 min, 4 °C).
20. Transfer the ~600 μL aqueous phase to new 2-mL microfuge tubes.
21. Add 600 μL room temperature acid–phenol–chloroform to the aqueous phase and mix by inverting the tubes.
22. Incubate at room temperature for 5 min followed by centrifugation (21,100 × *g*, 2 min, 4 °C).
23. Transfer ~500 μL aqueous phase to new 2-mL microfuge tubes.
24. Add 500 μL chloroform–isoamyl alcohol and mix by inverting the tubes.
25. Incubate at room temperature for 1 min followed by centrifugation (21,100 × *g*, 2 min, 4 °C).
26. Transfer ~400 μL aqueous phase to new 1.5-mL tubes containing 2 μL GlycoBlue and 45 μL 3M sodium acetate pH 5.5.
27. Precipitate RPF RNA by the addition of ~450 μL ice-cold 100% isopropanol and overnight incubation at −80 °C.
28. Collect the RPF RNA pellet by centrifugation (21,100 × *g*, 45 min, 4 °C).
29. Wash the pellet by adding 1 mL ice-cold 80% ethanol (without disturbing the pellet), followed by centrifugation (21,100 × *g*, 10 min, 4 °C).
30. Wash the pellet for a total of two times.
31. Spin down and remove any residual ethanol using a 20-μL pipette tip.
32. Air-dry the pellet at room temperature for ~5 min.
33. Resuspend the pellet in 30 μL nuclease-free water followed by NanoDrop quantification. RNA can be stored at −80 °C at this stage.

3.3 Total RNA Extraction

1. Add 400 μL Proteinase K buffer (prewarmed at 42 °C for 5–10 min) to the 200 μL lysate aliquot that was set aside in Subheading 3.1, **step 6**.
2. Mix by inverting and spinning down.
3. Incubate at 42 °C for 30 min.
4. Immediately conduct RNA purification using the hot acid–phenol–chloroform method as follows (*see* **Note 9**).
5. Heat 600-μL acid–phenol–chloroform in 2-mL microfuge tubes to 65 °C.
6. Heat Proteinase K-treated lysate to 65 °C for 5 min and subsequently transfer to 600 μL hot acid–phenol–chloroform.
7. Mix by inverting the tubes.
8. Incubate the mix at 65 °C for 5 min (mix intermittently by inverting the tubes).
9. Spin down and incubate the tubes in ice for 5 min followed by centrifugation (21,100 × *g*, 2 min, 4 °C).
10. Transfer the ~600 μL aqueous phase to new 2-mL microfuge tubes.
11. Add 600 μL room temperature acid–phenol–chloroform to the aqueous phase and mix by inverting the tubes.
12. Incubate the mix at 65 °C for 5 min (mix intermittently by inverting the tubes).
13. Spin down and incubate the tubes in ice for 5 min followed by centrifugation (21,100 × *g*, 2 min, 4 °C).
14. Transfer the ~600 μL aqueous phase to new 2-mL microfuge tubes.
15. Add 600 μL room temperature acid–phenol–chloroform to the aqueous phase.
16. Mix by vortexing briefly.
17. Incubate at room temperature for 5 min followed by centrifugation (21,100 × *g*, 2 min, 4 °C).
18. Transfer ~500 μL aqueous phase to new 2-mL microfuge tubes.
19. Add 500 μL chloroform-isoamyl alcohol and mix by vortexing.
20. Incubate at room temperature for 1 min followed by centrifugation (21,100 × *g*, 2 min, 4 °C).
21. Transfer ~400 μL aqueous phase to new 1.5-mL tubes containing 2 μL GlycoBlue and 45 μL 3M sodium acetate pH 5.5.
22. Precipitate RNA by adding ~450 μL ice-cold 100% isopropanol and overnight incubation at −80 °C.

23. Collect the RNA pellet by centrifugation (21,100 × *g*, 45 min, 4 °C).
24. Wash the pellet by adding 1 mL ice-cold 80% ethanol (without disturbing the pellet), followed by centrifugation (21,100 × *g*, 10 min, 4 °C).
25. Wash the pellet for a total of two times
26. Spin down and remove any residual ethanol using a 20-μL pipette tip.
27. Air-dry the pellet at room temperature for ~5 min.
28. Resuspend the pellet in 20 μL nuclease-free water, followed by NanoDrop quantification (*see* **Note 10**).
29. RNA can be stored at −80 °C at this stage.

3.4 DNase Treatment of RNase-Digested RNA and Total RNA

1. This step removes any gDNA contamination in the RNA preparations because the subsequent rRNA depletion step is sensitive to the presence of gDNA.
2. Prepare 50 μL reactions in PCR tubes with 10 μg RNA, 1 μL TURBO DNase, and 5 μL 10x TURBO DNase buffer.
3. Incubate at 37 °C for 30 min.
4. Add an additional 1 μL TURBO DNase and mix.
5. Incubate at 37 °C for 30 min.
6. Purify the DNase-treated RNA using the Zymo RNA Clean & Concentrator 5 Kit following the manufacturer's protocol (*see* **Note 11**).
7. Quantify DNase-treated RNA using NanoDrop and can store at −80 °C.
8. Total RNA integrity can be verified by electrophoresis of denatured RNA in a 15% TBE-Urea gel at 120 V for 5 min and 200 V for 50 min (*see* **Note 12**).
9. The quality of RPFs can be verified by electrophoresis of denatured RNA in a 15% TBE-Urea gel at 120 V for 5 min and 200 V for 75 min (*see* **Note 13**).

3.5 rRNA Depletion

1. Use half a reaction of Ribo-Zero (Illumina) per 5 ug of DNase-treated RNA (*see* **Note 14**).
2. Follow the manufacturer's protocol, except with one modification of omitting the 50 °C incubation step, only for DNase-treated ribo-seq samples.
3. Subsequently, purify the rRNA-depleted RNA using Zymo RNA Clean & Concentrator −5 kit according to the protocol described in Subheading 4.
4. For the final step, elute the RNA in 11 μL nuclease-free water.

5. Re-elute using the eluate (flow-through) to enhance recovery.
6. rRNA-depleted RNA can be stored at −80 °C at this stage.

3.6 Fragmentation of rRNA-Depleted Total RNA by Alkaline Hydrolysis

1. Add 10 μL of 2x fragmentation buffer to 10 μL rRNA-depleted total RNA and mix by pipetting.
2. Incubate the mix at 95 °C for 20 min followed immediately by adding 280 μL of fragmentation stop solution.
3. Precipitate RNA by adding ~750 μL ice-cold 100% ethanol and overnight incubation at −80 °C.
4. Collect the RNA pellet by centrifugation (21,100 × *g*, 45 min, 4 °C).
5. Wash the pellet by adding 1 mL ice-cold 80% ethanol (without disturbing the pellet), followed by centrifugation (21,100 × *g*, 10 min, 4 °C).
6. Wash the pellet twice.
7. Spin down and remove any residual ethanol using a 20-μL pipette tip.
8. Air-dry the pellet at room temperature for ~5 min.
9. Add 10 μL nuclease-free water to the pellet.
10. Let the pellet soften in ice and then resuspend it slowly by vortexing.
11. RNA can be stored at −80 °C at this stage.

3.7 Size Selection for RNA-Seq Samples

1. 10 μL fragmented RNA was mixed with 10 μL 2x RNA loading dye.
2. For size markers, 50 ng of IDT 10/60 nt ssDNA ladder was mixed with water and 2x RNA loading dye to 20 uL.
3. Denature RNA at 70 °C for 5 min followed immediately by incubation in ice for 2 min.
4. Clean the 15% TBE-Urea gel wells with the running buffer (1x TBE) by pipetting.
5. Immediately load the samples in the gel (*see* **Note 15**).
6. Run the gel at 120 V for 5 min and 200 V for 50 min.
7. Open the gel cassette and stain gel in the lid of a square petri dish with ice-cold SYBR gold staining solution (2 μL SYBR Gold in 20 mL 1x TBE buffer) for 1 min.
8. Discard the staining solution and immediately add 20 mL ice-cold 1x TBE.
9. Discard the solution as much as possible.
10. Place the lid of the petri dish (with the gel in it) on a blue light transilluminator.

11. Wear the amber-colored goggles and in the dark and excise the gel slice corresponding to the region between the top of the 30 nt and top of the 40 nt band of IDT 10/60 nt ssDNA ladder (*see* **Note 16**).
12. Transfer the gel slice to a pierced 0.5-mL tube placed in a 2-mL microfuge tube (*see* **Note 17**).
13. Centrifuge (21,100 × *g*, 2 min, 4 °C) so that the gel slice passes through 0.5-mL tube into the 2-mL tube.
14. Discard the pierced 0.5-mL tube.
15. To the crushed gel pieces in the 2-mL tubes, add 500 μL of RNA gel extraction buffer.
16. Seal the tube with parafilm.
17. Incubate overnight at 4 °C on a shaker or nutator.
18. Subsequently, vortex the tubes vigorously and spin down them.
19. Decant gel–RNA mix to a 0.22 μm Spin-X cellulose acetate filter columns.
20. Use 100 μL RNA gel extraction buffer to retrieve any residual gel–RNA mix from the original tube (*see* **Note 18**).
21. Centrifuge (16,000 × *g*, 2 min, 4 °C).
22. Add 50 μL RNA gel extraction buffer to the column membrane to retrieve any residual RNA.
23. Centrifuge (16,000 × *g*, 1 min, 4 °C).
24. Transfer the flow-through to new 1.5-mL microfuge tubes.
25. Add 2 μL GlycoBlue.
26. Precipitate the size-selected RNA by adding an equal volume (~600 μL) of ice-cold 100% isopropanol and overnight incubation at −80 °C.
27. Collect the RNA pellet by centrifugation (21,100 × *g*, 45 min, 4 °C).
28. Wash the pellet by adding 1 mL ice-cold 80% ethanol (without disturbing the pellet) followed by centrifugation (21,100 × *g*, 15 min, 4 °C).
29. Wash the pellet for a total of two times.
30. Spin down and remove any residual ethanol using a 20-μL pipette tip.
31. Air-dry the pellet at room temperature for ~5 min.
32. Add 3.25 μL nuclease-free water to the pellet.
33. Let the pellet soften in ice and then resuspend it slowly by pipetting.
34. RNA can be stored at −80 °C at this stage.

3.8 Size Selection for Ribo-seq Samples (See Note 19)

1. 10 μL rRNA-depleted ribo-seq RNA was mixed with 10 μL 2x RNA loading dye.
2. For size markers, 2 pmol each of 28 nt and 34 nt RNA oligos were mixed with water and 2x RNA loading dye to 20 ul.
3. Denature RNA at 70 °C for 5 min followed immediately by incubation in ice for 2 min.
4. Clean the 15% TBE–Urea gel wells with the running buffer (1x TBE) by pipetting.
5. Immediately load the samples in the gel (*see* **Note 15**).
6. Run the gel at 120 V for 5 min and 200 V for 75 min.
7. Open the gel cassette and stain it in the lid of a square petri dish with ice-cold SYBR gold staining solution (2 μL SYBR Gold in 20 mL 1x TBE buffer) for 1 min.
8. Discard the staining solution and immediately add 20 mL ice-cold 1x TBE.
9. Discard the solution as much as possible.
10. Place the lid of the petri dish (with the gel in it) on a blue light transilluminator.
11. Wear the amber-colored goggles and in the dark and excise the gel slice corresponding to the region between the bottom of the 28 nt and the bottom of the 34 nt RNA oligos (Fig. 4) (*see* **Note 16**).
12. Transfer the gel slice to a pierced 0.5-mL tube placed in a 2-mL microfuge tube (*see* **Note 17**).
13. Centrifuge (21,100 × *g*, 2 min, 4 °C) so that the gel slice passes through 0.5-mL tube into the 2-mL tube.
14. The subsequent steps are exactly the same as in Subheading 3.7 (size selection of RNA-seq samples).
15. Add 3.25 μL nuclease-free water to the pellet.
16. Let the pellet soften in ice and then resuspend it slowly by pipetting.
17. RNA can be stored at −80 °C at this stage.

3.9 Sequencing Library Preparation

1. Prior to sequencing library preparation, modify the ends of the RNA fragments according to the following steps. (*See* **Note 20**)
   (a) Incubate 3.25 μL of size-selected RNA at 70 °C for 3 min.
   (b) Immediately transfer to ice.
   (c) Add 0.5 μL of 10x T4 PNK buffer A (without ATP), 0.25 μL of SUPERase-In, and 0.5 μL of T4PNK enzyme.
   (d) Incubate the reaction at 37 °C for 30 min.
   (e) Add 0.5 μL of 10 mM ATP.

(f) Incubate the reaction at 37 °C for 1 h.

(g) Add 5.5 μL nuclease-free water to the reaction.

(h) Incubate at 75 °C for 10 min to terminate the reaction.

(i) Transfer the tubes to ice.

2. Prepare cDNA libraries with NEXTflex Small RNA-seq kit according to the manufacturer's protocol with the following specifications (*See* **Note 21**):

   (a) Step A (3′ 4 N Adenylated Adapter Ligation): Dilute the 3′ 4 N adapter 1/three-fold (*see* **Note 22**) and conduct the incubation for ligation overnight.

   (b) Step B (Excess 3′ Adapter Removal) and Step C (Excess Adapter Inactivation): follow the manufacturer's protocol.

   (c) Step D (5′ 4 N Adapter Ligation): Dilute the 5' 4 N adapter 1/3-fold (*see* **Note 22**).

   (d) Step E (Reverse Transcription-First Strand Synthesis): Follow the manufacturer's protocol.

   (e) Step F (Bead Cleanup): Conduct bead cleanup for according to the "No size-selection" protocol (*see* **Note 23**).

   (f) Step G (PCR Amplification): Conduct 11 cycles of PCR with barcoded primers (*see* **Note 24**).

   (g) Step H2 (PAGE Size Selection & Cleanup): Follow the manufacturer's protocol.

3. Assess the quality of the libraries using Agilent High-Sensitivity DNA Assay Kit on a Bioanalyzer.

4. Quantify libraries using Qubit dsDNA HS Assay Kit.

5. Dilute and pool the libraries according to the sequencing platform and sequencing depth requirement.

6. Sequence the libraries on the Illumina NovaSeq 6000 platform to obtain at least 50-bp single-ended reads (*see* **Note 25**).

3.10 Ribo-seq and RNA-Seq Data Processing and Quality Assessment

1. Assess the quality of raw sequencing reads using FastQC [20].

2. Use Cutadapt [21] to remove the adapters from the raw sequencing reads using the following parameters: "-a TGGAATTCT CGGGTGCCAAGG—discard-untrimmed—minimum-length 23."

3. Use Cutadapt [21] with the parameters "-u 4 -u -4" to trim four nucleotides from the ends of ribo-seq and RNA-seq reads, which were added during library preparation.

4. Subsequently, retain the 27 to 32 nt ribo-seq reads and 25 to 40 nt RNA-seq reads and filter out the rest.

5. Use RiboToolkit [34] to assess ribo-seq-specific data characteristics, such as length distribution, triplet periodicity, frame enrichment, the distribution of reads over different mRNA features, etc.
6. Further data processing and analysis would largely depend on the experiment objectives and would include mapping the reads to ncRNA and using the ncRNA-unmapped reads to subsequently map to the organism's reference genome (*see* **Note 26**).

4 Notes

1. Preparing the lysate as soon as the plant tissue is collected is advisable. We recommend purifying RPF RNA from the lysate as early as possible. In case of storing at −80 °C, we recommend doing it after snap-freezing the lysate.
2. The amount of plant material collected would largely depend on the plant, the kind of tissues, and the age of the plant. Different amounts of tissues can be collected for different samples because, after lysate preparation, all the samples would be normalized to the same RNA/protein amounts using the absorbance at 254 and 260 nm.
3. The plant tissues should always be chilled in liquid nitrogen during tissue pulverization.
4. A_{254} and A_{260} refer to the absorbance measured at 254 and 260 nm wavelength, respectively, using NanoDrop. The absorbance of PEB is subtracted from the absorbance of lysate to give A_{254} (Lysate – PEB) or A_{260} (Lysate – PEB). If this is less than 6, it means more tissue is required. If it is more, it can be lowered by dilution with PEB. The absorbance is adjusted to 6 because RNase 1 digestion is optimized according to these values.
5. The PEB is optimized to preserve ribosome occupancy on an mRNA and encourage RNase digestion by RNase 1. Therefore, the total RNA integrity is not guaranteed to be maintained in this buffer. To overcome this limitation, the total RNA extraction must be carried out immediately after the adjustment of lysate absorbance to ensure the collection of full-length mRNAs. An aliquot of lysate can be stored at −80 °C for the extraction of RPFs at a later stage.
6. Sucrose cushion is a dense sucrose pad that allows separation of tightly coupled active ribosomes without any mechanical stress. During ultracentrifugation with sucrose cushion, ribosome footprint complex will travel through the highly dense sucrose pad with high sedimentation rate to form a pellet, while other

low molecular weight subunits (proteins/RNA) will remain in the supernatant. Though sucrose gradients can be used to separate ribosomes from other individual subunits according to their density, it may increase the chances of contamination with untranslated but protein-bound RNA molecules. Also, if the lysate volume is more than 400 μL, then add PEB accordingly to make the volume 750 μL before layering it on a sucrose cushion. Care must be taken when layering the lysate on a sucrose cushion to create a sharp junction between the two solutions with minimal to no mixing between them. Additionally, the time duration from sample layering to ultracentrifugation should be minimized. Furthermore, the pellet may be hard to see, and therefore, circle the area on the tube with a marker where the pellet is expected to form.

7. Rinsing the pellet with nuclease-free water is an optional step. Rinsing is done to wash away non-pelleted RNA and other contaminants. The risk of this step is that the pellet can be lost if it dissolves in the water or if it comes loose off the tube wall. Therefore, if rinsing is carried out, ice-cold nuclease-free water should be added on to the tube wall and removed by pipetting quickly but steadily on the tube wall opposite to the pellet side.
8. The pellet may not be completely dissolved at this stage. Therefore, care must be taken during transfer to make sure pellet pieces are not retained in the tube or in the pipette tip.
9. Acid–phenol–chloroform extraction must be carried in a fumehood. Because phenol–chloroform is heated, the tubes usually open and splatter. Therefore, the necessary personal protective equipment must be worn, including safety glasses.
10. At this stage, total RNA can either be subjected to DNase treatment, followed by RNA integrity assessment by denaturing polyacrylamide gel electrophoresis, or stored at −80 °C. DNase treatment and gel electrophoresis of total RNA and RPFs can be conducted simultaneously.
11. A greater volume of ethanol is added to RPF RNA samples to ensure precipitation of the small ~28 nt RPFs.
12. Mix 0.5 μL DNase-treated RNA with 4.5 μL water and 5 μL 2x RNA loading dye. Similarly, for size markers, (i) mix 1 pmol each of 28 nt and 34 nt RNA oligos with water and RNA loading dye to 10 ul, and (ii) mix 50 ng 10/60 nt IDT ssDNA ladder with water and RNA loading dye to 10 ul. Denature the samples at 70 °C for 5 min and immediately transfer the samples to ice. Clean the wells of 15% TBE–Urea gel with the running buffer by pipetting and immediately load the denatured samples. Run the gel as indicated, open the gel cassette, and subsequently stain the gel in a square petri dish

with ice-cold SYBR Gold staining solution (3 μL SYBR Gold in 30 μL 1x TBE buffer) for 2 min on a shaker. Cover the petri dish with aluminum foil to protect it from light. Discard the staining solution and add 30 mL ice-cold 1x TBE buffer. Immediately proceed to image the gel with a gel imager. For total RNA samples, the two larger rRNA bands should be present with minimal bands present between the RNA oligo size markers (Fig. 3a).

13. Mix 1 μL DNase-treated RNA with 4 μL water and 5 μL 2x RNA loading dye. Similarly, for size markers, (i) mix 2 pmol each of 28 nt and 34 nt RNA oligos with water and RNA loading dye to 10 ul and (ii) mix 50 ng 10/60 nt IDT ssDNA ladder with water and RNA loading dye to 10 ul. Denature the samples at 70 °C for 5 min and immediately transfer the samples to ice. Clean the 15% TBE-Urea gel wells with the running buffer (1x TBE) by pipetting and immediately load the denatured samples. Run the gel as indicated, open the gel cassette, and subsequently stain the gel in a square petri dish with ice-cold SYBR Gold staining solution (3 μL SYBR Gold in 30 μL 1x TBE buffer) for 2 min on a shaker. Cover the petri dish with aluminum foil to protect it from light. Discard the staining solution and add 30 mL ice-cold 1x TBE buffer. Immediately proceed to image the gel with a gel imager. One or a few sharp and distinct RPF bands should be observed between the RNA oligo size markers.

14. The sale of Illumina Ribo-Zero as a stand-alone kit has been discontinued by the time of writing this chapter. However, the components for rRNA depletion are included with the purchase of the Illumina TruSeq Library Preparation Kit. Alternatively, rRNA depletion kits from other vendors, such as the RiboMinus Plant Kit for RNA-seq from Thermo Fisher Scientific, may be used with appropriate modifications.

15. Load size markers on the gel after every two samples to ensure accurate and precise size selection.

16. Do not spend time visualizing the gel using a gel imager. As soon as the staining solution is added, the RNA starts diffusing out of the gel. Therefore, proceed to cut the gel immediately after staining. Firstly, cut the region of the samples horizontally and then vertically. Use a new baked blade (sterilized) for every cut. Then, using a new baked blade and a 10-μL tip, pick up the gel slice and transfer it to a pierced 0.5-mL tube in a 2-mL microfuge tube.

17. Using an 18G syringe needle, pierce a hole at the bottom of a 0.5-mL tube and then place this tube in a 2-mL microfuge tube. Upon centrifugation, the gel slice will be crushed into smaller pieces when it passes through the hole, getting then collected in the 2-mL tube.

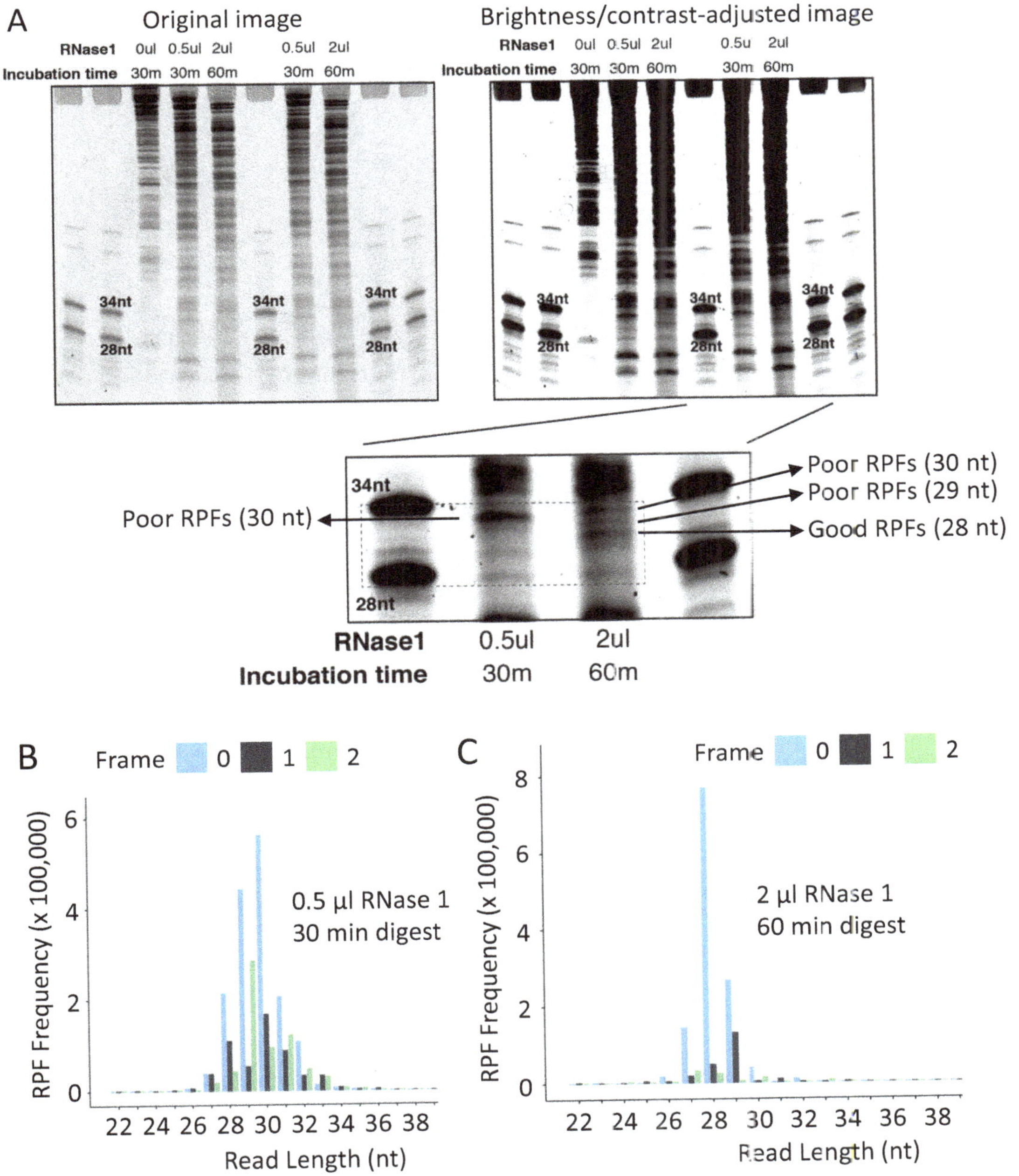

Fig. 3 Optimization of RNase 1 digestion. (**a**) Denaturing 15% PAGE gel shows distinct sharp bands (inset) under different digestion conditions that may or may not give good RPFs (**b** and **c**). RiboToolkit analysis of ribo-seq data with two different RNase digestion conditions that show (**b**) suboptimal triplet periodicity obtained by 0.5 μL RNase1 with 30 min digestion for reads of all sizes and (**c**) excellent triplet periodicity (vast majority of reads map to one reading frame) of the most abundant RPF size class (28 nt) obtained by digestion with 2 μL RNase1 for 60 min

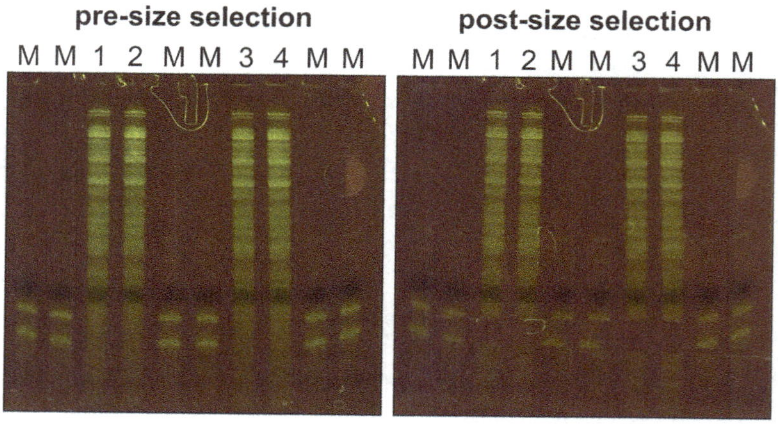

Fig. 4 Image of gel used for RNA size selection. 6% polyacrylamide gel in TBE–urea buffer. Lanes 1, 2, 3, and 4 show RNA from RNase1-treated lysate. M indicates 28 and 34 nt RNA markers. RNA is stained with SYBR Gold, visualized on a blue light box through orange filter (Subheading 3.8). Note the blank regions on gel at right, from which RNA bands in the desired size range have been excised

18. The gel in the gel–RNA mix may block the pipette tip. Therefore, using a sterile blade, cut off the "tip" portion of the pipette tip, and use this for pipetting gel–RNA mix.
19. Optimal RNAse1 digestion conditions may vary for different plant tissue, and re-optimization may be needed. For example, 600 μL of lysate from maize seedling roots was digested with 1500 U RNase1 at 28 °C for 30 min [20]. To optimize the RNase1 digestion condition, titrate lysate volume, units of RNase 1, digestion temperature, and duration (as shown in Fig. 3) and analyze the sharpness of the band between 28 nt and 34 nt. Subsequently, run a small RNA bioanalyzer using Agilent bioanalyzer small RNA assay kit to assess the quality of the bands (Fig. 3b).
20. Alkaline hydrolysis and RNase 1 digestion result in RNA fragments with 5′ hydroxyl and 3′ phosphate ends. However, adapter ligation for library preparation requires 5′ phosphate and 3′ hydroxyl ends. Treat the RNA with T4PNK in the absence of ATP dephosphorylates the 3′ end. Subsequently incubate with ATP phosphorylates the 5′ end.
21. For steps A, B, etc., refer to the Steps mentioned in the manufacturer's protocol.
22. The one-third dilution of adapters was determined empirically based on the RNA yield after size selection. Users may need to determine the adapter dilution based on their RNA recovery.
23. Because the input RNA for library preparation was already size-selected, the bead cleanup after reverse transcription can be

conducted with the modified, more straightforward protocol that does not need size selection. The modified protocol can be found on the manufacturer's website (perkinelmer-appliedgenomics.com/nextflex_small_rna_v3_no_size_selection_supplement-2/).

24. The optimal number of PCR cycles needs to be determined empirically. Because the reagents in the library preparation kit are very limited, we recommend the following approach to ensure optimal amplification for every library preparation. The bead cleanup after reverse transcription yields 18 μL of cDNA. 9 μL of this cDNA can be amplified (Step G) in a 12.5 μL PCR with 11 cycles, thereby consuming only half the volume of PCR reagents and the cDNA. Assessing the PCR product can be done by PAGE. If the libraries are overamplified, the remaining 9 μL cDNA can be amplified with 9 or 10 cycles. In contrast, if the PCR product is not overamplified but after cleanup, the quantity is insufficient for sequencing, then the remaining 9 μL cDNA can be amplified with 12 or 13 cycles. This approach ensures the libraries are optimally amplified and yield sufficient DNA for sequencing.
25. The ribo-seq libraries yield fewer reads than RNA-seq libraries, for analysis after data processing and filtering steps. Therefore, ribo-seq libraries must be sequenced at least twice the depth of that of RNA-Seq libraries. We recommend using Illumina HiSeq 3000/4000 or NovaSeq 6000 to obtain high sequencing depth. For example, in ref [38], the libraries were sequenced on the Illumina NovaSeq platform with S1 flow cell (ribo-seq) and SP flow cell (RNA-seq) to obtain ~100 M and ~ 50 M raw reads, respectively, per sample. After data processing, ncRNA depletion, etc., the yield per sample was ~20 M reads, each for ribo-seq and RNA-seq data, which mapped to the Arabidopsis reference genome.
26. As of this writing, use Bowtie v.1.2 (parameters: "-v 2") [25] to map the processed data to ncRNA (rRNA, snoRNA, tRNA) sequences (TAIR10). Use STAR v.2.5 (parameters: "--outFilterMismatchNmax 2 -- outFilterMultimapNmax 1") to map the ncRNA-unaligned reads to the Arabidopsis reference genome (TAIR10).

References

1. Sonenberg N, Hinnebusch AG (2009) Regulation of translation initiation in eukaryotes: mechanisms and biological targets. Cell 136: 731–745
2. Hershey JWB, Sonenberg N, Mathews MB (2019) Principles of translational control. Cold Spring Harb Perspect Biol 11
3. Wang Z, Gerstein M, Snyder M (2009) RNA-seq: a revolutionary tool for transcriptomics. Nat Rev Genet 10:57–63
4. Chung BYW, Balcerowicz M, Di Antonio M, Jaeger KE, Geng F, Franaszek K, Marriott P, Brierley I, Firth AE, Wigge PA (2020) An RNA thermoswitch regulates daytime growth in Arabidopsis. Nat Plants 6:522–532

5. Kage U, Powell JJ, Gardiner DM, Kazan K (2020) Ribosome profiling in plants: what is not lost in translation? J Exp Bot 71:5323–5332
6. Vogel C, Abreu Rde S, Ko D, Le SY, Shapiro BA, Burns SC, Sandhu D, Boutz DR, Marcotte EM, Penalva LO (2010) Sequence signatures and mRNA concentration can explain two-thirds of protein abundance variation in a human cell line. Mol Syst Biol 6:400
7. Ingolia NT, Ghaemmaghami S, Newman JR, Weissman JS (2009) Genome-wide analysis in vivo of translation with nucleotide resolution using ribosome profiling. Science 324:218–223
8. Hsu PY, Calviello L, Wu HL, Li FW, Rothfels CJ, Ohler U, Benfey PN (2016) Super-resolution ribosome profiling reveals unannotated translation events in Arabidopsis. Proc Natl Acad Sci U S A 113:E7126
9. Ingolia NT, Hussmann JA, Weissman JS (2019) Ribosome profiling: global views of translation. Cold Spring Harb Perspect Biol 11
10. Miras M, Miller WA, Truniger V, Aranda MA (2017) Non-canonical translation in plant RNA viruses. Front Plant Sci 8:494
11. Smirnova E, Firth AE, Miller WA, Scheidecker D, Brault V, Reinbold C, Rakotondrafara AM, Chung BY, Ziegler-Graff V (2015) Discovery of a small non-AUG-initiated ORF in poleroviruses and luteoviruses that is required for long-distance movement. PLoS Pathog 11:e1004868
12. Irigoyen N, Firth AE, Jones JD, Chung BY, Siddell SG, Brierley I (2016) High-resolution analysis of coronavirus gene expression by RNA sequencing and ribosome profiling. PLoS Pathog 12:e1005473
13. Stern-Ginossar N, Ingolia NT (2015) Ribosome profiling as a tool to decipher viral complexity. Annu Rev Virol 2:335–349
14. Xu G, Greene GH, Yoo H, Liu L, Marques J, Motley J, Dong X (2017) Global translational reprogramming is a fundamental layer of immune regulation in plants. Nature 545:487–490
15. Wang J, Zhang X, Greene GH, Xu G, Dong X (2022) PABP/purine-rich motif as an initiation module for cap-independent translation in pattern-triggered immunity. Cell 185(3186–3200):e3117
16. Xu T, Lei L, Shi J, Wang X, Chen J, Xue M, Sun S, Zhan B, Xia Z, Jiang N, Zhou T, Lai J, Fan Z (2019) Characterization of maize translational responses to sugarcane mosaic virus infection. Virus Res 259:97–107
17. Miller WA, Jackson J, Feng Y (2015) Cis- and trans-regulation of luteovirus gene expression by the 3′ end of the viral genome. Virus Res 206:37–45
18. Hebrard E, Poulicard N, Gerard C, Traore O, Wu HC, Albar L, Fargette D, Bessin Y, Vignols F (2010) Direct interaction between the Rice yellow mottle virus (RYMV) VPg and the central domain of the rice eIF(iso)4G1 factor correlates with rice susceptibility and RYMV virulence. Mol Plant-microbe Interact 23:1506–1513
19. Jiang J, Laliberté J-F (2011) The genome-linked protein VPg of plant viruses – a protein with many partners. Curr Opin Virol 1:347–354
20. Kanodia P, Vijayapalani P, Srivastava R, Bi R, Liu P, Miller WA, Howell SH (2020) Control of translation during the unfolded protein response in maize seedlings: life without PERKs. Plant Direct 4:e00241
21. Kiniry SJ, Michel AM, Baranov PV (2020) Computational methods for ribosome profiling data analysis. Wiley Interdiscip Rev RNA 11:e1577
22. Xiao Z, Zou Q, Liu Y, Yang X (2016) Genome-wide assessment of differential translations with ribosome profiling data. Nat Commun 7:11194
23. Andrews S (2010) FastQC: a quality control tool for high throughput sequence data. http://www.bioinformatics.babraham.ac.uk/projects/fastqc
24. Martin M (2015) Cutadapt removes adapter sequences from high-throughput sequencing reads. EMBnet J 17
25. Langmead B, Salzberg SL (2012) Fast gapped-read alignment with Bowtie 2. Nat Methods 9:357–359
26. Dobin A, Davis CA, Schlesinger F, Drenkow J, Zaleski C, Jha S, Batut P, Chaisson M, Gingeras TR (2012) STAR: ultrafast universal RNA-seq aligner. Bioinformatics 29:15–21
27. Liao Y, Smyth GK, Shi W (2014) featureCounts: an efficient general purpose program for assigning sequence reads to genomic features. Bioinformatics 30:923–930
28. Li B, Dewey CN (2011) RSEM: accurate transcript quantification from RNA-seq data with or without a reference genome. BMC Bioinform 12:323
29. Robinson MD, McCarthy DJ, Smyth GK (2009) edgeR: a Bioconductor package for differential expression analysis of digital gene expression data. Bioinformatics 26:139–140
30. Love MI, Huber W, Anders S (2014) Moderated estimation of fold change and dispersion

for RNA-seq data with DESeq2. Genome Biol 15:550

31. Walker DC, Lozier ZR, Bi R, Kanodia P, Miller WA, Liu P (2023) Variational inference for detecting differential translation in ribosome profiling studies. Front Genet 14:1163361
32. Chung BY, Hardcastle TJ, Jones JD, Irigoyen N, Firth AE, Baulcombe DC, Brierley I (2015) The use of duplex-specific nuclease in ribosome profiling and a user-friendly software package for Ribo-seq data analysis. RNA 21: 1731–1745
33. Calviello L, Mukherjee N, Wyler E, Zauber H, Hirsekorn A, Selbach M, Landthaler M, Obermayer B, Ohler U (2016) Detecting actively translated open reading frames in ribosome profiling data. Nat Methods 13:165–170
34. Liu Q, Shvarts T, Sliz P, Gregory RI (2020) RiboToolkit: an integrated platform for analysis and annotation of ribosome profiling data to decode mRNA translation at codon resolution. Nucleic Acids Res 48:W218–W229
35. Michel AM, Mullan JP, Velayudhan V, O'Connor PB, Donohue CA, Baranov PV (2016) RiboGalaxy: a browser based platform for the alignment, analysis and visualization of ribosome profiling data. RNA Biol 13:316–319
36. Cope AL, Anderson F, Favate J, Jackson M, Mok A, Kurowska A, Liu J, MacKenzie E, Shivakumar V, Tilton P, Winterbourne SM, Xue S, Kavoussanakis K, Lareau LF, Shah P, Wallace EWJ (2022) Riboviz 2: a flexible and robust ribosome profiling data analysis and visualization workflow. Bioinformatics 38: 2358–2360
37. Legrand C, Tuorto F (2020) RiboVIEW: a computational framework for visualization, quality control and statistical analysis of ribosome profiling data. Nucleic Acids Res 48:e7
38. Kanodia P (2021) Global effects of plant virus infection, viral noncoding RNAs, and unfolded protein response on plant gene expression. Ph. D. dissertation, Iowa State University

Chapter 12

Experimental Transmission of Plant Viruses by Aphids or Whiteflies

Irene Ontiveros, Juan Antonio Diaz-Pendón, and Juan José López-Moya

Abstract

Numerous species of plant viruses are naturally transmitted by insect vectors, mainly homopterans like aphids and whiteflies. Depending on the vector specificity and the mode of transmission, different durations of the periods for acquisition, retention, and inoculation are required for a successful transmission. Therefore, the experimental setup to perform controlled transmission experiments under laboratory conditions involves handling the vector organisms and managing the times for the different steps of the process to optimize and standardize the results. This chapter describes some basic procedures that can be applied to vector-mediated transmission experiments with selected viruses using aphids or whiteflies and different host plants.

Key words Non-persistent transmission, Semipersistent transmission, Circulative transmission, Non-viruliferous vector, Viruliferous vector, Acquisition, Retention, Inoculation

1 Introduction

Plant viruses often rely on vectors as a means of transmission to survive in nature. By using vectors, they can reach new susceptible hosts where they can continue acting as obligate intracellular parasites, making the transmission process one of the most critical steps in the viral cycle [1]. Among vectors, phytophagous arthropods play an important role in the dissemination of most plant viruses, with insect vectors being the most common [2]. Many plant viruses are spread from plant to plant by species belonging taxonomically to the *Homoptera*, specifically the *Aphididae* (aphids) and *Aleyrodidae* (whiteflies) families [3–6]. Examples of viruses transmitted by aphids are potyviruses, cucumoviruses, caulimoviruses, luteoviruses, poleroviruses, alfamoviruses, closteroviruses, and nanoviruses. In the case of whiteflies, examples are begomoviruses, criniviruses, ipomoviruses, carlaviruses, and torradoviruses. These two insects possess pierce-sucking mouthparts equipped with

Elizabeth P. B. Fontes and Kristiina Mäkinen (eds.), *Plant-Virus Interactions*, Methods in Molecular Biology, vol. 2724, https://doi.org/10.1007/978-1-0716-3485-1_12,

stylets that allow them to feed on plant tissues without causing significant damage, a characteristic that facilitates the establishment of new virus infections. The ability of these vectors to form large colonies and efficiently disperse themselves establishes them collectively as the most significant groups of vectors for plant viruses.

In addition to being an essential step in the cycle of numerous viruses, which certainly merits attention from the scientific community [7], there are other compelling reasons to establish protocols for conducting controlled vector transmission experiments. Currently, the rate of discovering new viruses [8] is steadily increasing, thanks to the application of powerful genomic tools that often yield complete genome sequence data from environmental samples. However, it is noteworthy that these sequences frequently lack relevant biological information [9]. For numerous newly discovered viruses, there is a lack of knowledge regarding their hosts or vector organisms. Consequently, identifying the vector of a new virus holds important implications for its precise taxonomical classification, as well as for effective management if it will become a potential causal agent of new plant diseases threatening crop production.

There are excellent recent reviews that can be consulted for details about the relationship between plant viruses and their insect vectors [10–13]. Also, in the case of whiteflies, basic management procedures for transmission have been described recently [14]. Mechanistically, the successful transmission of viruses by insect vectors requires several steps in the case of insect vectors: first, the acquisition of virions from an infected plant; second, the acquired virions might need to be retained in the vector, either at specific sites through binding to receptor-like elements in the digestive tract, or circulating from different anatomical structures, mainly from the gut to the salivary glands; and finally, delivery of virions is required, in many cases following salivation, allowing the virions to be deposited in a susceptible host plant to start the infection. The duration of these steps (acquisition, retention, and inoculation) can serve to classify the modes of transmission according to the length of the period when the vector keeps the virus infectivity. In non-persistent transmission, the insect can transmit the virus almost immediately after acquiring it from an infected source plant, and the capacity to transmit the virus is lost rapidly, often within a few minutes. In contrast, persistent transmission requires longer periods of acquisition, often followed by a period of retention before the vector finally becomes viruliferous and can transmit the virus. Semipersistent transmission is an additional category reserved for viruses with intermediate requirements for both acquisition and retention periods. A further refinement of the classification also considers the route followed by the virus within the insect vector, being either non-circulative or circulative transmission. In the non-circulative case, there is only a temporary and reversible association of the virus with the anterior tract of the digestive system (mouthparts or foregut). This type of transmission corresponds mainly with non-persistent and semipersistent transmission, in which the virus is not retained for long, and it can be immediately released and inoculated, completing the process only during short periods of time. Both acquisition and inoculation are thought to occur during short feeding probes. Since the non-persistent transmission occurs within such a narrow timeframe, it is indeed difficult to control the spread of these viruses using insecticide treatments targeted to kill the vectors. On the other hand, circulative transmission requires the passage of virions through the insect's body. They first pass through the gut barriers to reach the hemolymph and subsequently enter the salivary glands to be inoculated during salivation. In this case, the complete process might take days and often requires a period of latency since immediately after acquisition, the virus cannot be inoculated until the completion of its circulation within the vector. Additional categories can be established in this case when the virus replication only occurs in the plant hosts as non-propagative, being propagative if the virus is able to replicate as well in the vector. Interestingly, propagative viruses are parasites of both plants and insects, alternating between the two types of hosts to complete their cycles of dispersion, which could occasionally involve different development stages of the insect and even passage to the progenies.

In this chapter, our focus is on general procedures for testing the transmission of plant viruses using the two more common vector organisms: aphids and whiteflies. Alongside direct plant-to-plant transmission setups, we describe procedures based on artificial membrane feeding [15, 16], which played a crucial role in identifying auxiliary factor(s) involved in the transmission of certain plant viruses [17–19].

2 Materials

2.1 Biological Materials: Viruses, Host Plants, and Insect Colonies

Before starting the experimentation on vector transmission processes, it is essential to establish suitable conditions for obtaining viruses (*see* **Note 1**), host plants (*see* **Note 2**), and insect colonies (*see* **Note 3**) following fundamental procedures in virology, plant science, and entomology (*see* **Note 4**).

2.2 Materials for Containment and Handling of Vectors

In addition to the biological materials, various devices are necessary to confine vectors while feeding during the acquisition and inoculation processes. These devices include the following:

1. Insect-proof cages (sized appropriately to accommodate plants during the acquisition and inoculation periods).

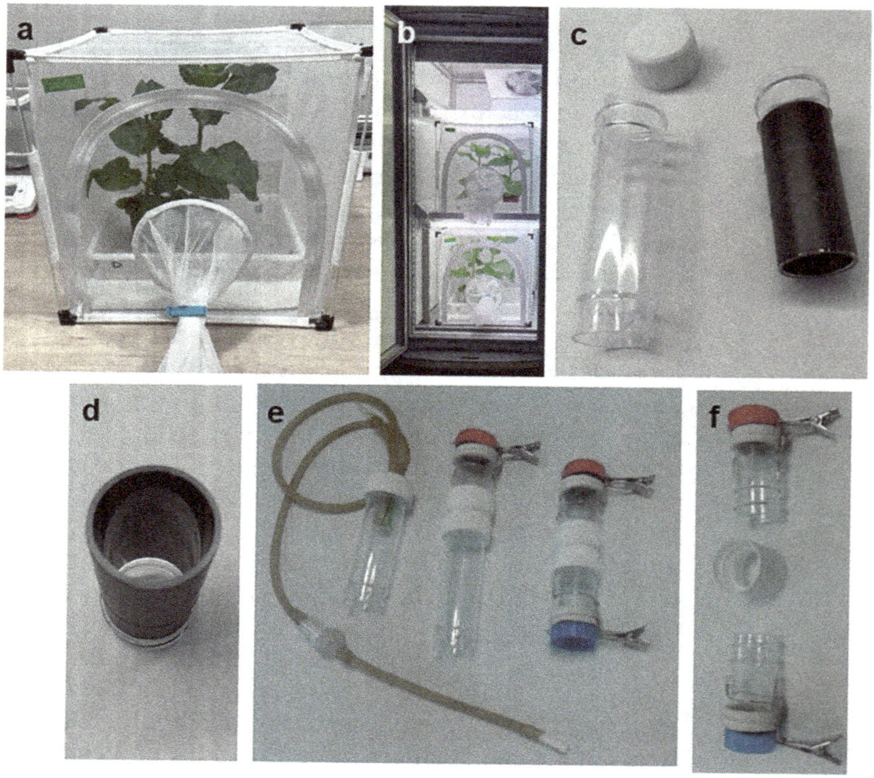

Fig. 1 Devices used for handling insect vectors in virus transmission experiments. The image shows a commercial portable insect-proof cage (BugDorm, measuring 47.5 × 47.5 × 47.5 cm) with aphid-infested tobacco plants inside (**a**). The cages are stackable and can be placed in growth chambers (**b**) for rearing different colonies of insects or conducting experiments that require vector confinement. To create small membrane feeding chambers, 30-mL tubes (polystyrene containers 25 × 90 mm with polypropylene caps) can be employed (**c**, left) by cutting the bottom and applying black adhesive tape for darkening (**c**, right). A closer look from above shows a cylinder (**d**) ready to be covered with a stretched parafilm membrane. The same tubes can be used to prepare various components for handling whiteflies. These include an aspiration device (**e**, left), clip cages that can be attached to tubes (**e**, center), or other clip cages (**e**, right, and disassembled in **f**) using two perforated caps glued together in opposing orientations to connect them

2. Containers and vials for capturing and immobilizing insects during fastening.
3. Paintbrush or aspiration device (*see* **Note 5**).
4. Clip cages (*see* **Note 6**).
5. Membrane feeding chambers prepared using stretched parafilm membranes and different feeding solutions (*see* **Note 7**).

Some examples are shown in Fig. 1.

2.3 Systems for Virus Detection

To determine the presence of the virus in individual plants following transmission, an appropriate diagnostic protocol might be needed in advance. The procedure for virus detection can encompass various methods, ranging from simple observations of symptoms to more advanced molecular techniques such as serological methods (ELISA or Western blot), hybridization using specific probes (dot-blot or tissue print), or methods based on PCR amplification (for DNA viruses) or RT-PCR (for RNA viruses).

3 Methods

All procedures will be conducted using plants and insects relocated to the designated area or room where the experiments will take place. In general, temperature and illumination should be maintained within ranges suitable for plant growth and insect rearing unless they need to be intentionally manipulated as variables within the experimental design.

The two primary methods for conducting transmission experiments are plant-to-plant transmission (Fig. 2) and artificial feeding (Fig. 3). These approaches are described in the following subsections.

3.1 Plant-to-Plant Transmission by Freely Moving Insect Vectors

This simplified method involves confining source and destination test plants together with potential vectors in a cage. The method can be used in studies to determine which insect species are capable of transmitting a specific virus, and it can be complemented by additional procedures to confirm unequivocally the identity of the vector and gather further information about the duration of the various stages of the transmission process.

1. Enclose infected source plants with test plants within a cage and introduce insect vectors at the desired density. The distance between the source and destination plants can be regarded as a parameter for the assay (*see* **Note 8**).
2. Allow unrestricted movement of vectors between plants, promoting mobility if required, and wait for the completion of the different periods for acquisition/inoculation.
3. Optionally, eliminate vectors through insecticide treatment (*see* **Note 9**).
4. Maintain the plants under suitable growing conditions, allowing sufficient time for virus infections to manifest.
5. Quantify the transmission (rate number of infected plants/total number of test plants) by directly observing symptoms or employing a detection system (*see* **Note 10**).

3.2 Plant-to-Plant Transmission Under Restricted Mobility Conditions of the Insect Vectors

1. Collect insects (*see* **Note 11**) in adequate containers and fasten them during a period determined experimentally (*see* **Note 12**).
2. Allow insects to access the infected plant (either complete or detached leaves) for the desired acquisition period (*see* **Note 13**). Alternatively, insects could be reared directly on virus-

Fig. 2 Procedure for conducting plant-to-plant aphid transmission experiments. Apterous adult aphids reared on infested plants (**a**) are carefully collected using a paintbrush (**b**) and placed in glass vials for a fasting period (**c**). The experimental arena for acquisition consists of a plate covered with wet filter paper and positioned inside a larger plate filled with soapy water (**c**). Small susceptible plants (**d**) are prepared in a tray to receive the virus, acquired from a leaf collected from an infected source plant (**e**). The aphids after fasting are grouped by tapping the vials, deposited on the detached infected leaf, and evenly distributed (**f**, **g**), with the acquisition time measured using a stopwatch (**h**). To prevent unwanted aphid visits, control plants (negative and positive: mock and mechanically inoculated, respectively) are covered with inverted sample tubes (**h**). Once the acquisition period is completed, the aphids are individually transferred to each test plant until the desired number of vectors per plant is reached and then covered (**i**). This process is repeated for all the test plants (**j**). After distributing the aphids, the controls are uncovered, dusted with carborundum (**l**), and either mock-inoculated or manually inoculated with a fresh extract of infected tissue (**m**, **n**), to be later rinsed and covered (**o**). Upon completion of the inoculation period, the plants are uncovered (**p**) and sprayed with insecticide (**q**) to eliminate the vectors. The percentage of transmission is calculated based on the number of infected plants, which are diagnosed once the positive control exhibits symptoms or tests positive

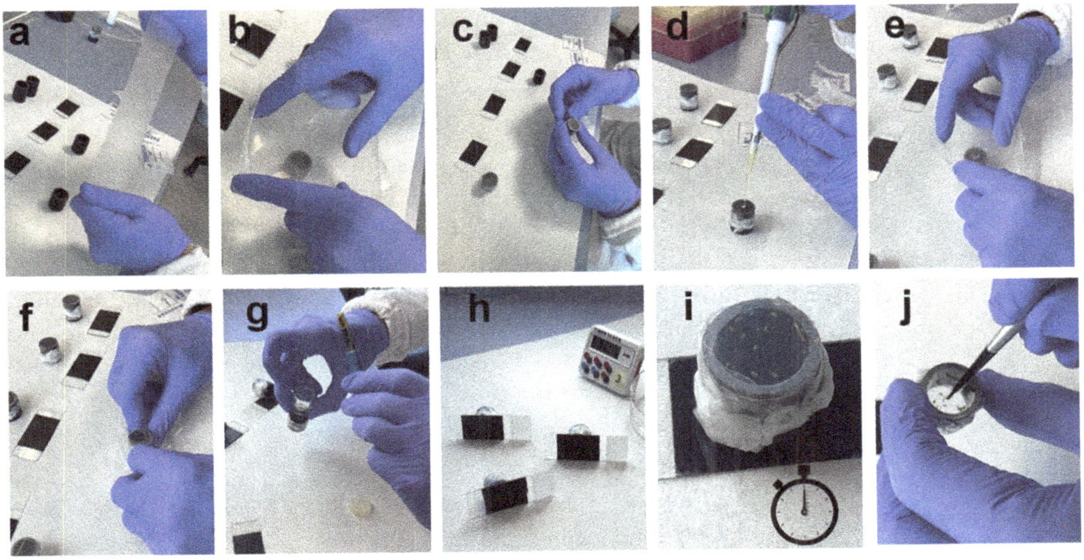

Fig. 3 Procedure for conducting membrane feeding transmission experiments. Apterous adult aphids are collected and placed for a fasting period, as shown in Fig. 2. In this case, chambers with artificial membranes are prepared using cylindrical section of rigid plastic tubes (polystyrene, diameter 2 cm, height 2.5 cm), and the exterior is darkened using black tape. Parafilm is then stretched in two dimensions and tightly wrapped around the top of the cylinder to form a membrane (**a**–**c**). A 15% sucrose solution is carefully placed on top of the membrane (**d**) and covered with an additional layer of stretched parafilm to create a sachet (**e**, **f**). The fasting aphids are introduced into the cylinder (**g**) and allowed to crawl until they reach the membrane attracted by the light. To cover the bottom of the chamber, a dark slide is used (**h**). The aphids can pierce the membrane with their stylets to access the sucrose solution, and the acquisition time is measured using a stopwatch (**i**). Once the acquisition period is completed, the cylinder is lifted, and the aphids are individually collected using a paintbrush (**j**). For the inoculation, the aphids are individually transferred to the test plants, as shown in Fig. 2

infected plants, therefore being already viruliferous at the time of collection, and consequently, this second step could be skipped.

3. Transfer insects to the test plants (*see* **Note 14**) and allow them to feed during the required inoculation period. For small plants, they can be individually covered with a containment cage such as an inverted beaker, while for larger plants, the restriction of mobility can be achieved using clip cages or by bagging the leaf or branch where the inoculation feeding will take place.
4. Uncover plants (or remove clip cages or bags) and eliminate vectors with insecticide treatment (optional).
5. Maintain the plants under suitable growing conditions, allowing sufficient time for virus infections to manifest.
6. Quantify the transmission (rate number of infected plants/ total number of test plants) by directly observing symptoms or employing a detection system.

3.3 Transmission After Virus Acquisition Through Artificial Feeding

1. Collect non-viruliferous insects in adequate containers and fasten them during a period determined experimentally.
2. Prepare artificial membranes with stretched parafilm (*see* **Note 15**) and the solutions to be tested (*see* **Note 16**).
3. Transfer the insects to the feeding chambers and allow them to feed through the parafilm on the solution containing the tested components. This can be done either in a single feeding, using individual components or mixes of different elements to be tested, or in sequential steps to provide each component in the desired order (see Subheading 3.4). The insects should have access to the virus preparation during the acquisition period as determined for the assay.
4. Transfer insects to the test plants and allow them to feed during the required inoculation period.
5. Uncover plants (or remove clip cages or bags, in case these were used) and eliminate vectors with insecticide treatment (optional).
6. Maintain the plants under suitable growing conditions, allowing sufficient time for virus infections to manifest.
7. Quantify the transmission (rate number of infected plants/ total number of test plants) by directly observing symptoms or employing a detection system.

3.4 Sequential Feeding Experiments

The processes of acquisition and inoculation can be divided into distinct sub-periods to carry out sequential feeding on various plant tissues and/or artificial membranes. This approach allows for refinement in the design of targeted experiments (*see* **Note 17**).

4 Notes

1. Before most types of transmission experiments, the virus infection might need to be initiated in hosts that will be used as a transmission source. This usually involves mechanical inoculation for certain viruses, rubbing manually an extract on leaves dusted with an abrasive (such as carborundum). Buffers to prepare crude extracts should be consulted in the relevant literature for each specific virus, but also examples for many viruses can be found in the dedicated database of "Description of plant viruses" available online at https://www.dpvweb.net/ [20]. For other viruses, live inocula might be needed and preserved through vegetative propagation or even through vector transmission. Finally, purified virions could be used in artificial membrane feeding for virus acquisition, either alone or assisted by auxiliary factors.

2. Virus-free seed stocks or other uninfected sources of propagules are a good system to initiate the growth of experimental plants. To exclude undesired visits of potential vectors, confinement in growth chambers or greenhouses will be desirable, or alternatively, the use of insect-proof cages (*see* Fig. 1) might be required during the whole cultivation cycle. Updated and valuable resources about handling plants for experiments with pathogens and pests can be found in the relevant literature and dedicated collections and databases, for instance at the European and Mediterranean Plant Pathology Organization (EPPO) website https://gd.eppo.int/ (Guidelines for the production of biological reference material. EPPO Bull, 51: 499–506. https://doi.org/10.1111/epp.12781).
3. Laboratory-established colonies (such as clonal parthenogenetic aphids and progenies populations of individually selected male and female whiteflies) are generally better than heterogeneous colonies or wild-captured individuals, which can carry other pathogens and might require to perform quarantine/acclimation periods before their use as vectors. Depending on the virus under study and the host range, a cautionary principle to avoid contaminations will be to rear insects in a non-host plant for the virus, although such a possibility might not always be available. As a general principle, different rooms or chamber facilities would serve to keep the insects separated from the experimental plants until the setting up of the transmission experiments. As in the case of biological materials, information on insect vectors can be found at EPPO, for instance, about whiteflies (*Bemisia tabaci*. EPPO Bulletin, 34: 281–288. https://doi.org/10.1111/j.1365-2338.2004.00729.x), and at a dedicated website of the French Institute of Agricultural Research INRAE (https://www6.inrae.fr/encyclopedie-pucerons_eng/).
4. Proper management of plant host and insect vectors is essential for successful transmission experiments. Both plants and insects are living organisms, and their susceptibility and performance can be influenced by environmental conditions and even circadian rhythms. Also, the influence of insect behavior needs to be considered [21]. As a rule, repetitions of the experiments must be performed at similar times of the day and in a comparable environment. If possible, controlled light and temperature conditions could be preferred to standardize the repetitions. In practical terms, a record of the conditions during each repetition should be kept for future reference and statistical treatment of the data.
5. To preserve their behavioral capacities, insect individuals must be handled carefully, keeping tools clean and minimizing the force required to remove them from the rearing plant. This involves gently collecting them using a fine paintbrush or

aspiration devices. An example of a home-made design for a collection device is shown in Fig. 1c, which uses a commercial sample tube (30 mL, cylindrical with conical bottom and screw cap) prepared using two silicon flexible thin hoses that pass through perforated holes in the cap of a tube and, therefore, can be screwed into different collecting tubes. One of the hoses is open and can be attached to a pipette tip to concentrate the aspiration, while the second hose should be connected to the vacuum source outside and covered with a fine mesh inside the device to avoid losing insects when the vacuum is applied. Holes for passing the two hoses through the container cap must be sealed to avoid leaks and consequently create a negative pressure in the tube for the suction of the insects.

6. Clip cages should be constructed using an openable hatch that could be clamped onto a leaf while maintaining moderate pressure to prevent insects from escaping, for example, using soft ring pads to avoid pressure damage on the leaf surface. To facilitate the transfer of insects, tubes of the same model as the ones used for the collection of vectors can be transformed. The construction involves first cutting the base of the tube to allow gluing and securing the hatch clamp, leaving the original screw-on cap on the other side of the cage to confine the insects, allowing ventilation through a mesh-covered opening. Additionally, this design can serve to transfer insects from the collection tube using a connector formed by two opposite caps, glued and perforated, to communicate two compatible tubes (Fig. 1f).
7. Under certain circumstances, purified virus particles and auxiliary factors can be supplied to the vectors using artificial feeding (*see* Fig. 3). For pierce-sucking insects, membranes can be prepared using stretched parafilm membranes placed above a darkened cylinder where the insects have been previously placed. The solution for feeding consists of an artificial diet placed either in a sandwich of two membranes forming a sachet (Fig. 3a–f), or above a single membrane with a spacer and a cover.
8. For the most straightforward transmission experiment confining together infected and test plants with the vector, negative control of a similar cage without including insects must be considered to rule out the possibility of other vector organisms acting regardless of the presence of the presumed vector. This is particularly important when testing viruses collected directly from field samples that might carry unidentified pests. Quarantine periods followed by detailed inspection and a few passages (if possible) under controlled laboratory conditions should be considered. While caging together source and test plants, it is advisable to avoid direct contact to preclude other types of

contact-driven transmission. The distances between plants must be adjusted previously to account for the flying capacity/other mobility conditions of the insects.

9. Insecticide treatment are performed once the period of inoculation is finalized, directly spraying on the plants (Fig. 2q) with a commercial product previously tested as effective on the same colony of insects. The treatment must be repeated, if necessary, especially in the case of whiteflies, to avoid the future risk of a new generation of vectors coming from eggs laid by females during the inoculation period. When the inoculation has been restricted to confined areas of leaves, one possibility to avoid infestation by next-generation whiteflies is to remove the leaves used for inoculation once the virus has been able to move to other parts of the plant through cell-to-cell and systemic movement.

10. Sampling at different intervals might be required to properly detect the presence of the virus after transmission. In case a mechanical inoculated control was included during the experiment (Fig. 2l, n), a good indication of the timing of appraisal for transmission rates can be deduced.

11. The stylets of sucking insects, when inserted into plant tissues, can be fragile and easily torn inadvertently during collection. In the case of actively feeding aphids, gently touching their abdomen with the paintbrush (Fig. 2a) activates a reflex response that rapidly retracts the stylet. This reflex can be observed through changes in the relative position of the antennae, which move from being parallel to the body to going forward and touching the surroundings in preparation for walking away. Once the antennae move and the aphid begins to ambulate, it becomes a safe moment to collect them without causing anatomical damage. Similarly, a convenient way to collect whiteflies is by attracting them to light sources, which can be achieved by shading a portion of the rearing boxes and leaving only the front part illuminated, or using yellow cards near the rearing plant. Both systems can stimulate their flight and facilitate their aspiration away from feeding areas on the plant. Once collected, whiteflies can be transferred between the different devices, connecting containers, and even the feeding clip cages or artificial membranes using gentle shaking, tapping, and swift movements of the sample tubes by a trained experimenter. Alternatively, connecting the different components prepared with compatible tube caps can be adopted using glued double connectors made with two perforated caps. Small clip cages with different designs that can be secured to leaves are good systems to allow the feeding of insects only in the selected part of the plant where they are confined. In the cases of whiteflies feeding for sufficiently long periods, eggs can be laid by females

on the leaves, and labeling of the temporarily infested leaves might be required to facilitate their later removal before the hatching without affecting the results.

12. Keeping the insects isolated from plants for a fasting period (around 1 h is often sufficient) might increase their focus on feeding when released and might have other positive effects like clearance of blocking stuff from hypothetical virus-specific receptors to favor acquisition. The containers used for fasting must be dry and with an adequate volume for the number of individuals or alternatively sufficiently ventilated to avoid suffocation in case of too crowded groups; for instance, glass vials of about 2–4 mL with tight caps can be used for groups of up to 50–100 aphids, which can be collected individually using a fine paintbrush (Fig. 2b, c). In the case of whiteflies, since they are attracted to light sources, covering the cage with an opaque fabric material would help move them to a section of the cage and facilitate their collection using a device with gentle aspiration. Once collected inside the tubes, the insects can be moved to the bottom by gently tapping the vials on the bench or table surface. For whiteflies, placing the tubes in cold conditions (such as on ice) for a few seconds before opening them helps to avoid the risk of escape.

13. Rearing viruliferous vectors (on virus-infected plants) could simplify the experimental procedures to proceed directly to the inoculation step. In other cases, the acquisition requires placing the insect on the virus source. This can be done by providing access of insects to the infected plant tissue: In the case of wingless aphids, this can be done on detached leaves on a petri dish over a wet filter paper. The petri dish can be placed inside a larger plate or tray surrounded with soapy water to better confine them and avoid escapes (*see* Fig. 2h). Aphids can also be placed directly on whole infected plants and recovered manually later. Since whiteflies fly actively, the acquisition can be forced on small, clipped cages on infected leaves or released inside cages containing the acquisition source. Another system adequate for whiteflies could be to connect the clip cage used for acquisition with a second clip cage for inoculation, creating a tunneled bridge that the whiteflies can use to move from infected to test plants and trimming the leaf section can make it easier to dry out and favor the migration of insects. In either case, it is important to allow individuals to feed on the infected plants for the duration of the acquisition access period. In the case of non-persistent viruses and aphid vectors, a stopwatch can be used to measure the time (Fig. 1j). Direct observation with a magnifying glass or a binocular microscope could reveal their behavior during feeding (*see* comments on **Note 10**). To further investigate feeding behavior, entomological equipment

such as Y-tube olfactometers and colored sticky cards can be used to monitor vectors [22]. Additionally, the implementation of techniques such as wiring individual insects and recording their feeding patterns using an electrical penetration graph (EPG) system could provide valuable insights [21].

14. Control of temperature can be used to reduce the mobility of insects for a while, but with certain time limits to avoid undesired mortality. The collection tubes containing insects can be placed momentarily on an ice bucket or left in a cold room for a few minutes, right before transferring them to other tubes, clip cages, or the membrane feeding chambers.
15. Artificial diets containing sugar are more attractive to insects. The percentage of sucrose (usually from 10% to 20%) can be experimentally optimized to maximize transmission rates in preliminary experiments. The diet can be contained in between two stretched layers of parafilm forming a sachet and placed at one end of a cylindrical chamber made with a short tube section (Fig. 3a–f), or for short feeding periods, the solution on the parafilm membrane can be covered with a glass coverslip, using a plastic flat washer as a spacer. In the case of collected whiteflies, they can be released through the other end of the tube, which would be covered by a ventilated lid (such as a perforated cap covered with a fine mesh). In all cases, zenithal illumination helps the movement of the insects toward the membrane.
16. The purification of virions and auxiliary factors allows combining them into artificial diets. The concentrations of the different components would need to be established experimentally, with expected linearity for a range of the concentrations chosen. Under certain conditions, the transmission percentages of serial dilutions can be used to titer the activity of a component better than using a concentration measurement.
17. Sequential feeding serves to identify the requirements of auxiliary factors for transmission [18]. The versatile system could even combine steps for the acquisition of components transiently expressed in plant tissues and later supplemented with membrane feeding on purified virions [23].

Acknowledgments

The authors' work at CRAG and IHSM has received funding from the European Union's Horizon 2020 Research and Innovation Program under Grant Agreement No. 101000570, Project VIRTIGATION, and from Grants PID2019-105692RB-100 and CEX2019-000902-S funded by MCIN/AEI/10.13039/

501100011033, and by the CERCA Programme/Generalitat de Catalunya. I.O. was supported by grant BES-2017-080808 funded by MCIN/AEI/ 10.13039/501100011033 and by "ESF Investing in your future." The authors thank Clara Ontañón for her assistance in capturing the photographs featured in the figures.

References

1. Ng JC, Perry KL (2004) Transmission of plant viruses by aphid vectors. Mol Plant Pathol 5: 505–511. https://doi.org/10.1111/j.1364-3703.2004.00240.x
2. Hull R (2014) Plant virology, 5th edn. Academic Press., ISBN 9780123848710. https://doi.org/10.1016/B978-0-12-384871-0.00023-6
3. Blackman RL, Eastop VF (2000) Aphids on the world's crops, an identification and information guide, 2nd edn. Wiley, Chichester, p 466. ISBN: 978-0-471-85191-2
4. Hullé M, Chaubet B, Turpeau E, Simon JC (2020) Encyclop'Aphid: a website on aphids and their natural enemies. Entomol Gen. https://doi.org/10.1127/entomologia/2019/0867
5. Stansly PA, Naranjo SE (2010) Bemisia: bionomics and management of a global pest. Springer., ISBN 978-90-481-2459-6, p 540. https://doi.org/10.1007/978-90-481-2460-2
6. Wang XW, Li P, Liu SS (2017) Whitefly interactions with plants. Curr Opin Insect Sci 19: 70–75. https://doi.org/10.1016/j.cois.2017.02.001
7. Chase O, Ferriol I, López-Moya JJ (2021) Control of plant pathogenic viruses through interference with insect transmission. In: Gaur RK, Paul Khurana SM, Sharma P, Hohn T (eds) Plant virus-host interaction: molecular approaches and viral evolution, 2nd edn, pp 359–381. https://doi.org/10.1016/B978-0-12-821629-3.00019-1
8. Roossinck MJ, Martin DP, Roumagnac P (2015) Plant virus metagenomics: advances in virus discovery. Phytopathology 105:716–727. https://doi.org/10.1094/PHYTO-12-14-0356-RVW
9. Hou W, Li S, Massart S (2020) Is there a "biological desert" with the discovery of new plant viruses? A retrospective analysis for new fruit tree viruses. Front Microbiol 11:592816. https://doi.org/10.3389/fmicb.2020.592816. https://doi.org/10.1016/B978-0-12-327550-9.50018-5
10. Ng JC, Zhou JS (2015) Insect vector-plant virus interactions associated with non-circulative, semi-persistent transmission: current perspectives and future challenges. Curr Opin Virol 15:48–55. https://doi.org/10.1016/j.coviro.2015.07.006
11. Gallet R, Michalakis Y, Blanc S (2018) Vector-transmission of plant viruses and constraints imposed by virus-vector interactions. Curr Opin Virol 33:144–150. https://doi.org/10.1016/j.coviro.2018.08.005
12. Jia D, Chen Q, Mao Q, Zhang X, Wu W, Chen H, Yu X, Wang Z, Wei T (2018) Vector mediated transmission of persistently transmitted plant viruses. Curr Opin Virol 28:127–132. https://doi.org/10.1016/j.coviro.2017.12.004
13. Wang XW, Blanc S (2021) Insect transmission of plant single-stranded DNA viruses. Annu Rev Entomol 66:389–405. https://doi.org/10.1146/annurev-ento-060920-094531
14. Polston JE, Capobianco H (2013) Transmitting plant viruses using whiteflies. J Vis Exp JoVE 81:e4332. https://doi.org/10.3791/4332
15. Pirone T (1964) Discussion and preliminary reports. Aphid transmission of a purified stylet-borne virus acquired through a membrane. Virology 23:107–108
16. Kunkel H (1977) Membrane feeding systems in aphid research. In: Harris KF, Maramorosch K (eds) Aphids as virus vectors. Academic Press, pp 311–338. ISBN: 9781483273884
17. Govier DA, Kassanis B (1974) Evidence that a component other than the virus particle is needed for aphid transmission of potato virus Y. Virology 57:285–286. https://doi.org/10.1016/0042-6822(74)90129-9
18. Pirone TP, Blanc S (1996) Helper-dependent vector transmission of plant viruses. Annu Rev Phytopathol 34:227–247. https://doi.org/10.1146/annurev.phyto.34.1.227
19. Leh V, Jacquot E, Geldreich A, Hermann T, Leclerc D, Cerutti M, Yot P, Keller M, Blanc S (1999) Aphid transmission of cauliflower mosaic virus requires the viral PIII protein.

EMBO J 18:7077–7085. https://doi.org/10.1093/emboj/18.24.7077

20. Adams MJ, Antoniw JF (2006) DPVweb: a comprehensive database of plant and fungal virus genes and genomes. Nucleic Acids Res 34(Database issue):D382–D385. https://doi.org/10.1093/nar/gkj023
21. Fereres A, Moreno A (2009) Behavioural aspects influencing plant virus transmission by homopteran insects. Virus Res 141:158–168. https://doi.org/10.1016/j.virusres.2008.10.020
22. Ontiveros I, López-Moya JJ, Díaz-Pendón JA (2022) Coinfection of tomato plants with Tomato yellow leaf curl virus and Tomato chlorosis virus affects the interaction with host and whiteflies. Phytopathology 112:944–952. https://doi.org/10.1094/PHYTO-08-21-0341-R
23. Goytia E, Fernández-Calvino L, Martínez-García B, López-Abella D, López-Moya JJ (2006) Production of plum pox virus HC-Pro functionally active for aphid transmission in a transient-expression system. J Gen Virol 87:3413–3423. https://doi.org/10.1099/vir.0.82301-0

Chapter 13

Quantification of Plant Virus Seed Transmission Rate in *Arabidopsis thaliana*

Miriam Gil-Valle, Cristina Sáez, Nuria Montes, and Israel Pagán

Abstract

More than 25% of all known plant viruses are transmitted through seeds, which makes this mode of dispersal of great importance for plant virus epidemics. Virus detection in seed stocks remains the most frequent approach for seed health testing, but current methods are not always standardized and/or do not allow analyzing large numbers of seeds. Here, we describe a high-throughput method to quantify plant virus seed transmission rate based on classical grow-out tests, which can be applied to widely different viruses and host species.

Key words Seed transmission, Plant virus, Grow-out test, RT-qPCR virus detection, *Arabidopsis thaliana*

1 Introduction

More than 25% of all known plant viruses are vertically transmitted from parents to offspring through seeds [1]. This transmission mode has a great impact on the epidemiology of these pathogens as it allows long-term survival and long-distance dispersal of viruses favored by the global trade of seeds. It is also a concern for germplasm banks, as virus infection may compromise seed viability [2]. In the laboratory, virus vertical transmission can be of great relevance, as it has been recently shown by [3] that seed-transmitted viruses are widespread in laboratory seed stocks of *Arabidopsis thaliana* (from now on, Arabidopsis), with unpredictable consequences in experimental settings. Consequently, the methods to detect viral presence in seeds and seedlings, the evaluation of vertical transmission through gametes and embryos and the strategies to obtain plant material certification are gaining increasing importance [4–6].

Conventional techniques have been focused on testing viral presence in the whole seed or in seedling batches through

Elizabeth P. B. Fontes and Kristiina Mäkinen (eds.), *Plant-Virus Interactions*, Methods in Molecular Biology, vol. 2724, https://doi.org/10.1007/978-1-0716-3485-1_13,

bioassays: Seed and seedling extracts have been used to inoculate indicator hosts, and grow-out tests after germination of contaminated seeds have been performed to check the development of viral symptoms in the seedlings [7]. These approaches present low specificity, are time- and space-consuming, and require trained specialists to perform inoculation and phenotyping. Imaging techniques (for instance, electron, fluorescence, or confocal laser scanning microscopy) also allow virus detection in infecting male or female gametes, pollen grains, embryos, seeds, and leaf tissues of the offspring [8]. Although these methods provide a very detailed image of the virus, the high cost of the specialized equipment required limits the use of these strategies. In addition, viral titer in seeds and seedlings is usually lower than that of parental plant tissues [9], increasing the difficulty of virus visualization. Serological-based strategies, such as the enzyme-linked immunosorbent assay (ELISA), have also been commonly used for virus detection in seeds/seedlings. ELISA test has high sensitivity and can be easily performed in large-scale tests with economical materials, which makes it a versatile technology [1]. However, it requires expensive specific antibodies for each viral species or isolate that must avoid cross-reaction with similar proteins of other pathogens or of the host plant. Due to the limitations of other methods, molecular techniques based on the detection of viral nucleic acids have become the most generalized method to assess vertical transmission of plant viruses since they provide the highest detection sensitivity and specificity combined with a relatively fast, cheap and easy procedure [10].

Although molecular techniques based on the hybridization of nucleic acid probes have been used to detect viruses in seeds and seedlings, the high detection potential of polymerase chain reaction (PCR)-based molecular methodologies is prevailing in studies evaluating seed transmission of plant viruses. (RT)-PCR allows the amplification of low levels of viral RNA/DNA, typical of seeds and seedlings, and usually involves simple and fast protocols [11]. Moreover, different variants and modifications of conventional PCR have been developed to detect several viruses at the same time in a single reaction (multiplex-PCR) or quantitatively determine the viral presence/absence in seeds or seedlings (quantitative PCR, qPCR) [12]. These methods require previous knowledge of the viruses to be detected in order to design specific primers. Although other, more recent, technologies, such as next-generation sequencing (NGS), offer massive identification of known and novel viral genomes, they are still highly expensive, specialized computational analysis of sequences is required, and validation methods need to be improved [13]. Thus, PCR-based methods are currently the preferred choice.

Detection techniques for test seed-transmitted viruses must combine effectiveness, accuracy, sensitivity, and reliability along with the simplicity, rapidity, inexpensive cost, and reproducibility to be routinely used [4, 14]. Although PCR-based methods generally combine these characteristics, standardized procedures to regulate seed trade and screening seed-transmitted viruses are not abundant. The International Seed Testing Association has released a few protocols concerning seed-transmitted virus detection [15], but not all of them are intended to be used with seedlings (virus detection in seeds does not necessarily entail transmission to the progeny) and/or are based on molecular methods. Here, we describe a rapid and affordable protocol based on RT-qPCR for the accurate detection of seed-transmitted plant viruses in seedlings. This method has been developed to detect plant virus transmission in Arabidopsis seeds but can be easily adapted to assess a wide range of plant species (indeed, we have successfully used it in tomato seeds) and virus species.

2 Materials

Prepare all solutions using ultrapure water (prepared by purifying deionized water to attain a sensitivity of 18.2 MΩ-cm at 25 °C) and analytical grade reagents.

2.1 Seed Germination

1. Seed sterilization solution: 3 mL HCl (37%) per every 100 mL bleach (3.78% sodium hypochlorite).
2. Vacuum desiccator.
3. Vaseline.
4. Sterilized paper (cut with the diameter of a Petri dish).
5. Sterilized wooden toothpick.
6. MS seed germination medium: Weigh 2,15 g of Murashige and Skoog medium. Add sucrose (final concentration: 1% w/v) and 0,5 g morpholinoethane sulfonic acid, monohydrate (MES). Add water to a volume of 900 mL. Mix and adjust pH to 5.6 with 2 M KOH (*see* **Note 1**). Make up to 1 L with water. Sterilize the medium in an autoclave and serve in Petri dishes within a horizontal laminar flow cabinet.
7. 3M Micropore® tape.
8. Plant growth chamber.
9. Dissecting forceps.
10. 1.5-mL tubes.
11. Liquid nitrogen.

2.2 Plant Total RNA Extraction

1. Plastic micropestle (*see* Fig. 1) or glass beads and a shaker.
2. Phenol-based reagent. Both NZYol® (Nzytech) and TRIzol® (Sigma) have been used in this protocol and perform well. Others will likely work equally well.
3. Chloroform.
4. 2-propanol.
5. 70% ethanol.

2.3 Virus Detection in RNA Extracts

1. One-step RT-qPCR kit. Conditions shown in the methods section have been set up for Brilliant III Ultra-Fast SYBR Green QRT-PCR Master Mix® (AGILENT).
2. Specific primers for the virus(es) to be detected.
3. 100 mM dithiothreitol (DTT).
4. Ribonuclease inhibitor (40 U/μL).
5. Thermocycler (384-well plates are advised for high-throughput detection).
6. Internal standards (optional, *see* Subheading 3.3).
7. Agarose D1 (low EEO for higher band definition).

3 Methods

3.1 Seed Germination and Seedling Growth

1. Put the desired number of seeds to be sterilized in a 1.5-mL tube with the lid opened (*see* **Note 2**). Then, place the tube within a vacuum desiccator. We use a borosilicate glass desiccator with a diameter of 26 cm. Next to the seeds, place an open bottle with the bleach. Add the HCl and immediately close the lid of the desiccator. Put Vaseline in the border of the lid to help maintain the vacuum. Allow the sterilization to proceed for 3.5 h. Afterward, use the vacuum to extract the gasses. Open the desiccator lid and close the tubes (Fig. 1a).
2. In a horizontal laminar flow cabinet, place seeds in a piece of sterilized paper, such that they can be easily transferred to the MS medium (Fig. 1b). Then, move them to Petri dishes containing MS seed germination medium. Handle seeds with a sterilized wooden toothpick to place them in the desired position. To do so, wet the tip of the toothpick in the MS medium to help the seed attach to the toothpick (Fig. 1c). Placing around 50–100 seeds per dish allows enough separation between seedlings after germination to facilitate collection. Then, close the plates with 3M Micropore® tape (Fig. 1d).
3. Cover the plates with an aluminum foil to maintain seeds in darkness (Fig. 1e) and place them at 4 °C for 4–7 days to allow seed stratification (*see* **Note 3**).

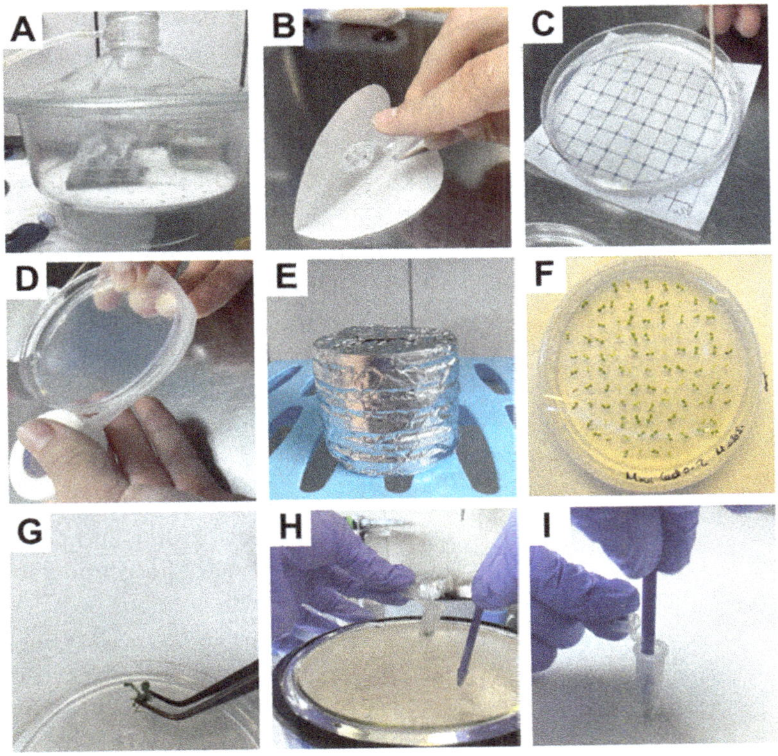

Fig. 1 Step-by-step summary of the protocol. (**a**) Seed sterilization. (**b**) Handling of seeds in the flow cabinet. (**c**) Sowing of seeds in MS medium. (**d**) Preparation of Petri plates containing seeds in MS medium for stratification. (**e**) Seed stratification. (**f**) Seed germination (picture was taken 5 days post-stratification). (**g**) Collection of seedlings from MS medium. (**h**) Sample freezing in liquid nitrogen. (**i**) Seedling grinding for RNA extraction

4. Remove the aluminum foil and move the Petri dishes to a plant growth chamber at 22 °C and 70% relative humidity under 16 h of light (intensity: 120–150 mol · m^{-2} · s).
5. After 7–10 days, the plateau of seed germination is reached in most Arabidopsis ecotypes (Fig. 1f) (*see* also **Note 4**). Collect seedlings at this point with dissecting forceps and place them in 1.5-mL tubes (Fig. 1g). Seedlings can be collected in pools that are further analyzed as a single sample to save time and resources. Place all seedlings of the same pool in the same tube (*see* **Note 5** for indications about pool size) and freeze samples in liquid nitrogen (Fig. 1h). Other, more complex pooling strategies can also be utilized in some circumstances (*see* **Note 6** for the description of alternative strategies). The process can be stopped at this point if necessary. When doing so, samples can be kept at −80 °C with no loss in total plant RNA yield. Otherwise, move to Subheading 3.2.

3.2 Plant Total RNA Extraction

1. Grind plant tissue to a powder using a plastic micropestle or glass beads and a shaker (Fig. 1i). Then, add 600 μL of phenol-based reagent (e.g., NZYol®, TRIzol®) and shake for 15 s. This volume of phenol-based reagent ensures extracting enough total plant RNA for virus detection regardless of the number of plants in the pool. Leave for 5 min at room temperature.
2. Add 200 μL of chloroform and shake for 15 s. Leave for 5 min at room temperature. After this point, samples must be kept in ice as much as possible to prevent RNA degradation.
3. Centrifuge the samples for 15 min at 12,000 g and 4 °C. Move the supernatant to a clean tube.
4. Add 2.5 volumes (here, 500 μL) of 2-propanol to the supernatant and keep 20–30 min at −20 °C. Then, centrifuge 10 min at 12,000 g and 4 °C.
5. Discard the supernatant and add 600 μL of 70% ethanol. Then, centrifuge 5 min at 7,500 g and 4 °C, and discard the supernatant. Let the pellet dry for 5 min and then resuspend in 30 μL of deionized water.
6. Use 1 μL to quantify RNA concentration in a spectrophotometer. Usual yield is about 200 ng/μL. Run 200 ng of RNA (typically, 1 μL) in a 1.2% agarose gel to check RNA integrity (*see* Fig. 2 for an example of RNA extracts obtained with this protocol).

3.3 Virus Detection in RNA Extracts

1. Use 50 ng of total plant RNA for virus detection. To do so, dilute samples to a concentration of 25 ng/μL and use 2 μL. Prepare sample duplicates (*see* **Note 7**).
2. Together with 2 μL of total plant RNA, add (per sample):
   - 5 μL of q-PCR kit AGILENT Brilliant III Ultra-Fast SYBR Green QRT-PCR Master Mix to perform the RT and the qPCR in one step. The protocol also works with Takara One Step TB Green™ PrimeScript™ RT-PCR Kit II master mix (Takara Bio Inc.) and can be easily adapted to similar kits from other brands.
   - 0.1 μL of DTT (100 mM).
   - 20 U of the ribonuclease inhibitor.
   - Specific primers at a final concentration of 0.2 mM (*see* **Note 8** on primers design).
3. Perform the amplification in a thermocycler. We typically use a LightCycler 480 II real-time PCR system (Roche) (*see* **Note 9**).
4. To attain virus RNA quantification, add internal standards (*see* **Note 10**).

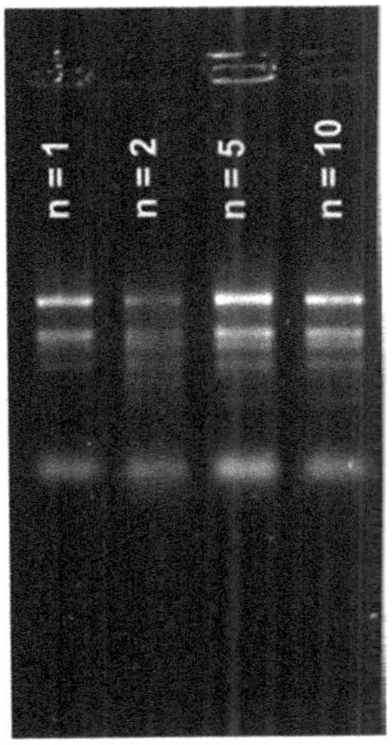

Fig. 2 Agarose gel (2%) with examples of RNA extracts from seedlings of different pool sizes ($n = 1, 2, 5$, and 10) obtained with the protocol described here. In all cases, 200 ng of total plant RNA was loaded into the gel (stained with Redsafe™)

5. Run RT-qPCR products in a 2% agarose gel at 90–100 V using 2 μL of the amplification product of purified viral RNA to check that an amplicon of the expected size is obtained.

4 Notes

1. If components of the medium are correctly weighted, pH will be between 5 and 5.2, so minor adjustments will be needed. If seeds are expected to be difficult to germinate (because of genotype-specific characteristics or age), gibberellic acid (GA4 + 7) at a final concentration of 5 μM can be added to the MS medium.
2. To successfully sterilize the total amount of seeds, the tube volume filled by seeds must not exceed one-tenth of the total tube volume.
3. Best results are obtained using 3- to 6-month-old seeds. Younger seeds may still retain some primary dormancy that prevents germination. This problem rarely occurs with typical "laboratory" genotypes such as Col-0 or *Ler*, but is much more common in other Arabidopsis genotypes, particularly in those collected in the field or derived from field-collected plants (e.g., NASC Iberian collection). Stratification helps to break dormancy and synchronize germination. If fresh seeds are used and the germination rate is low, increasing the stratification time may increase the germination rate.

4. In short-lived Arabidopsis genotypes, the maximum germination rate is reached after 5–6 days. However, some accessions, particularly long-lived ones, may need up to 10 days to reach the germination plateau. In Fig. 3, a time course of germination rates between 1 and 10 days post-stratification is represented for seeds of 10 different Arabidopsis genotypes, which include short-lived ones such as An-1, Cen-1, Fei-0, *Ler* and Ver-5 and long-lived ones such as Cad-0, Cdm-0, Cum-0, Kas-0, and Ll-0. Data on the life span of these plant genotypes can be found in [16, 17].
5. When samples are pooled, detection results reflect the number of positive pools rather than the number of seedlings testing positive. Therefore, pooling does not allow a direct quantification of virus seed transmission rate, and a likelihood function must be applied to estimate the number of positive seedlings per pool. As seed transmission is a binomial event, and in most Arabidopsis genotypes it happens at a low frequency [18], the likelihood of more than one seedling of the same pool being infected can be inferred using a Poisson distribution [19]. Hence, the seed transmission rate can be estimated following the equation: $p = 1 - (1 - Y/N)^{1/k}$, where p is the probability of virus transmission by a single seed, y is the number of positive samples (positive pools), n is the total number of samples assayed (number of pools), and k is the number of seedlings per sample (pool). It is important to note that the accuracy of the estimate, particularly at higher seed transmission rates, depends on the following: (i) the number of analyzed seedlings, (ii) the number of seedlings per pool, and (iii) the seed transmission rate of the virus. Figure 4 shows simulations of seed transmission rates estimated using the equation defined above as a function of the pool size and the observed seed transmission rate (that is, the seed transmission rate calculated by analyzing each seedling independently: pool size = 1). In the simulations, the total number of analyzed seedlings ($n = 100$) was maintained constant as this is enough for a meaningful estimate. Simulations show that seed transmission estimates are accurate regardless of the pool size as far as seed transmission remains low (below 20%), and reducing pool size to increase accuracy is only advisable if higher seed transmission rates are expected. At high transmission rates, accuracy is highly reduced. For instance, for a pool size of 5, seed transmission rates between 60% and 100% cannot be distinguished. In this case, reducing pool size is mandatory.
6. When a large number of seedlings needs to be analyzed, a two-dimensional pooling strategy can be applied to reduce the time and cost of the process [20]. Briefly, RNA extracts are arranged into 12 by 12 matrices and then pooled by row

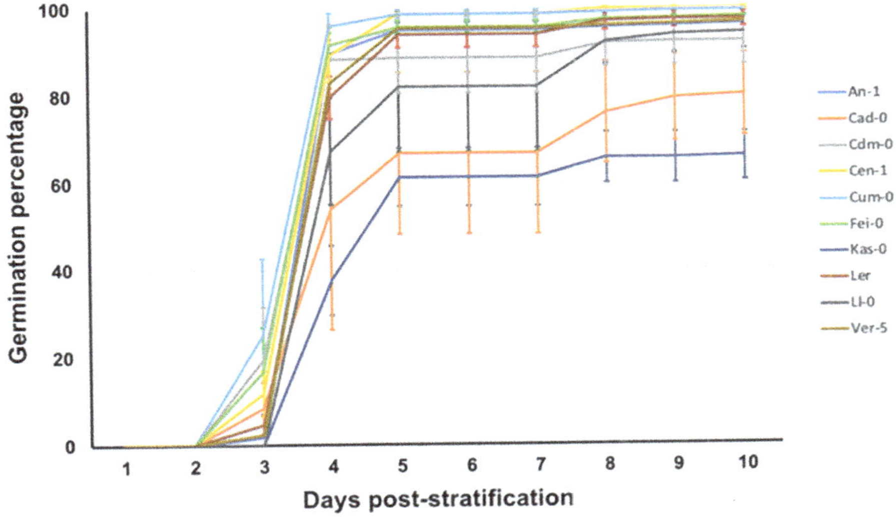

Fig. 3 Percent germination between 1 and 10 days post-stratification of seeds from 10 Arabidopsis genotypes. Genotypes are An-1 (Antwerp, Belgium), Cen-1 (Centenera, Spain), Fei-0 (Santa Maria da Feira, Portugal), L*er* (Landsberg, Poland), and Ver-5 (Verín, Spain), which are short-lived; and Cad-0 (Candelario, Spain), Cdm-0 (Caldas de Miravete, Spain), Cum-0 (Cumbres Mayores, Spain), Kas-0 (Kashmir, India), and Ll-0 (Llagostera, Spain), which are long-lived. Data are mean ± standard error of three replicates

and by column. The pooling of RNA extracts is carried out in a two-dimensional fashion, such that each RNA extract is present in exactly one row pool and exactly one column pool. Then, virus detection is performed as described in Subheading 3.3. This strategy works optimally when the analyzed trait occurs at low frequency, as is the case of *Cucumber mosaic virus* (CMV) seed transmission in Arabidopsis for most plant genotypes [21]. Nevertheless, there is always a certain percentage of indetermination. In that case, the RNA extracts in the problematic rows/columns are analyzed individually.

7. Although including duplicates increases the time and resources devoted to the detection, it also increases the accuracy of the detection as it allows the validation of positive and negative samples. It also allows the detection of doubtful samples where only one of the two replicates is positive and the other is not. This may be due to the following: (i) Amplification does not always happen because the amount of viral RNA in the sample is near the detection limit, or (ii) the positive duplicate is contaminated during the amplification process. To resolve this indetermination, detection in these samples can be repeated with twice more total plant RNA than in the initial detection.

8. For RT-qPCRs, specific primers should be designed to amplify short fragments (50–200 nt). As an example, primers to detect

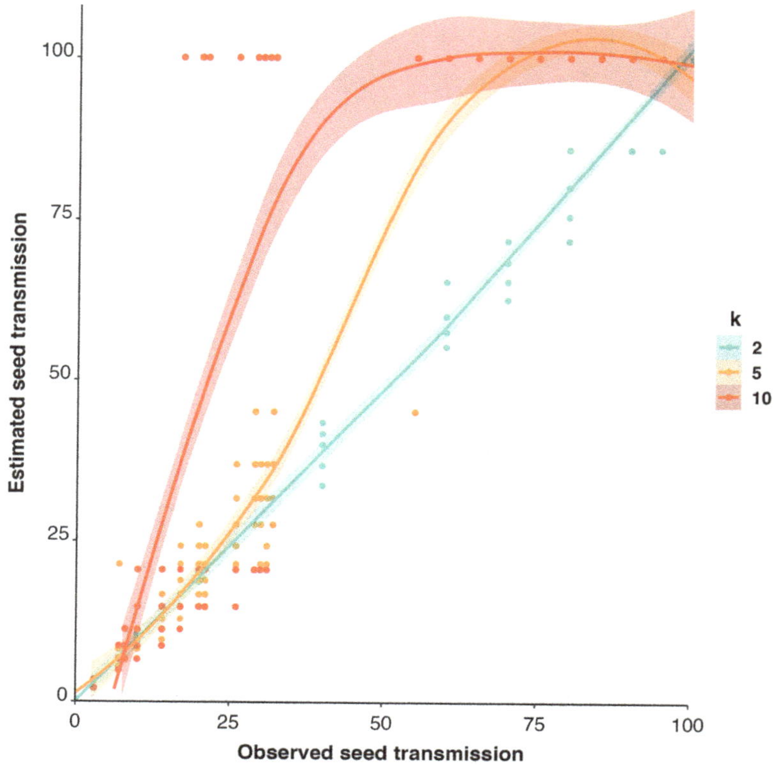

Fig. 4 Comparison of observed virus seed transmission rate with estimations derived using the equation $p = 1 - (1 - Y/N)^{1/k}$, where p is the probability of virus transmission by a single seed, y is the number of positive samples (positive pools), n is the total number of samples assayed (number of pools), and k is the number of seedlings per sample (pool). Estimations were done for three different pool sizes: $k = 2$ (blue), $k = 5$ (yellow), and $k = 10$ (red). Estimations were done for observed transmission rates ranging between 0% and 100%, with 10 pooling replicates per observed seed transmission rate value. Each dot represents one of these replicates

Turnip mosaic virus (TuMV) in seedlings consisted of an 18 nt forward primer (5′-TGTTCGGCTTGGATGGAA-3′) and a 21 nt reverse primer (5′- TTAACGTCCTCGGTCGTATGC-3′) (nucleotide positions 9515–9586 as in Acc. N. AB194802) that amplify a 70 nt fragment of the virus coat protein gene. On the other hand, primers to detect CMV in seedlings consisted of a 20 nt forward primer (5′--GAACAGTTTCCACTGATGCT-3′) and a 20 nt reverse primer (5′- ATAGGACAATCATACGTCGG-3′) (nucleotide positions 409–563 as in Acc. N. D10538) that amplify a 154 nt fragment of the movement protein gene.

9. Reverse transcription is typically carried at 50 °C for 10′, followed by 5″ of denaturation at 95 °C. For the qPCR, 40 cycles

of amplification are run, consisting of 10″ of denaturation at 95 °C and 10″ (for TuMV) or 20″ (for CMV) of amplification at 60 °C for each cycle. Fluorescence quantification is done at this point. The products are denatured again at 95 °C for 30″ and cooled down to 60 °C for 1′. Melting curves are calculated by gradually increasing the temperature until reaching 95 °C. Extending the qPCR to 40 cycles ensures a maximum detection rate, as one should bear in mind that the virus load in germinated seedlings is very low. For the same reason, it is not infrequent that a certain proportion of positive samples amplify late during the qPCR (cycles 30–35). We have validated these late positives by detecting the virus in the same samples via standard PCR with the same primers and further Sanger sequencing, such that these late amplicons are not false positives.

10. Absolute viral RNA accumulation (measured as nanograms of viral RNA per microgram of total plant RNA) can be quantified by utilizing internal standards. Internal standards consist of two-fold dilution series of genomic purified viral RNA, ranging from 50 ng to 0.003 ng. Plasmids in which the fragment to be amplified is cloned can also be used to build internal standards. Keep in mind that the dilution series must be done using RNA extracts from no-infected seedlings rather than water to homogenize the background. Because virus load in seedlings is low, if desired the internal standard series can be shortened by starting the dilutions at 12.5 ng. Ct values obtained for each sample can be interpolated in the Ct to RNA quantity regression constructed with the internal standards.

References

1. Sastry KS (2013) Seed-borne plant virus diseases. Springer, New Delhi
2. Kumar PL, Cuervo M, Kreuze JF et al (2021) Phytosanitary interventions for safe global germplasm exchange and the prevention of transboundary pest spread: the role of CGIAR germplasm health units. Plants 10:328
3. Verhoeven A, Kloth KJ, Kupczok A et al (2023) Arabidopsis latent virus 1, a comovirus widely spread in *Arabidopsis thaliana* collections. New Phytol 237:1146–1153
4. Chalam VC (2020) Elimination of plant viruses by certification and quarantine for ensuring biosecurity. In: Awasthi LP (ed) Applied plant virology. Advances, detection, and antiviral strategies, 1st edn. Academic, Cambridge, pp 749–762
5. Pagán I (2019) Movement between plants: vertical transmission. In: Palukaitis P, García-Arenal F (eds) Cucumber mosaic virus. APS Press, St. Paul, pp 185–198
6. Pagán I (2022) Transmission through seeds: the unknown life of plant viruses. PLoS Pathog 18:e1010707
7. Salem NM, Ehlers JD, Roberts PA et al (2010) Biological and molecular diagnosis of seed-borne viruses in cowpea germplasm of geographically diverse sub-Saharan origins. Plant Pathol 59:773–784
8. Pina MAS, Gómez-Aix C, Méndez-López E et al (2021) Imaging techniques to study plant virus replication and vertical transmission. Viruses 13:358
9. Torre C, Agüero J, Gómez-Aix C et al (2020) Comparison of DAS-ELISA and qRT-PCR for the detection of cucurbit viruses in seeds. Ann Appl Biol 176:158–169

10. Kaur SI, Kashyap PL, Kang SS et al (2020) Detection and diagnosis of seed-borne viruses and virus-like pathogens. In: Kumar R, Gupta A (eds) Seed-borne diseases of agricultural crops: detection, diagnosis & management. Springer, Singapore, pp 169–199
11. Ellis MD, Hoak JM, Ellis BW et al (2020) Quantitative real-time PCR analysis of individual flue-cured tobacco seeds and seedlings reveals seed transmission of *Tobacco mosaic virus*. Phytopathology 110:194–205
12. Davino S, Caruso AG, Bertacca S et al (2020) *Tomato brown rugose fruit virus*: seed transmission rate and efficacy of different seed disinfection treatments. Plants 9:1615
13. Bester R, Cook G, Breytenbach JH et al (2021) Towards the validation of high-throughput sequencing (HTS) for routine plant virus diagnostics: measurement of variation linked to HTS detection of citrus viruses and viroids. Virol J 18:1–19
14. Vishunavat K, Prabakar K, Anand T (2023) Seed health: testing and management. In: Dadlani M, Yadava DK (eds) Seed science and technology. Springer, Singapore, pp 335–364
15. International Seed Testing Association (ISTA) (2023) International rules for seed testing. https://www.seedtest.org/en/useful-links-footer/ista-rules.html. Accessed on 29 April 2023
16. Pagán I, Alonso-Blanco C, García-Arenal F (2007) The relationship of within-host multiplication and virulence in a plant-virus system. PLoS One 2:e786
17. Montes N, Alonso-Blanco C, García-Arenal F (2019) *Cucumber mosaic virus* infection as a potential selective pressure on *Arabidopsis thaliana* populations. PLoS Pathog 15: e1007810
18. Cobos A, Montes N, López-Herranz M et al (2019) Within-host multiplication and speed of colonization as infection traits associated with plant virus vertical transmission. J Virol 93:e01078–e01019
19. Gibbs AJ, Gower JC (1960) The use of a multi-transfer method in plant virus transmission studies: some statistical points arising in the analysis of results. Ann Appl Biol 48:75–83
20. Zuzarte PC, Denroche RE, Fehringer G et al (2014) A two-dimensional pooling strategy for rare variant detection on next- generation sequencing platforms. PLoS One 9:e93455
21. Montes N, Cobos A, Gil-Valle M et al (2021) *Arabidopsis thaliana* genes associated with *Cucumber mosaic virus* virulence and their link to virus seed transmission. Microorganisms 9:692

Chapter 14

Protein–Protein Interaction via Two-Hybrid Assay in Yeast

Christiane Eliza Motta Duarte and Nivea Costa Euclydes

Abstract

The yeast two-hybrid assay enables detecting interactions between proteins, which makes this tool of particular interest for plant–virus interaction studies. Basically, the reporter gene expression (*HIS3*) is activated by the binding of a transcription factor GAL4, which, like eukaryotic transcription factors, is modular in nature and consists of two structurally independent domains: DNA-binding (DB) and activation (AD) domains. The two proteins under investigation are expressed separately, one fused to the BD domain and the other to the AD domain. In the yeast strain AH109, activation of the reporter gene occurs only in cells that contain interacting proteins, reconstituting the transcription factor GAL4 which then binds to the responsive promoter and results in yeast colony growth.

Key words Protein interaction, *Saccharomyces cerevisiae*, Reporter gene, DNA-binding domain, Activation domain

1 Introduction

The yeast two-hybrid (Y2H) system is a widespread technique due to the relative ease and speed of detecting interactions between two proteins. Additionally, it offers numerous advantages compared to traditional biochemical methods. Y2H avoids costly protein purification or antibody development, leading to a substantial time reduction required to identify new protein partners [1]. This technique was developed by Fields and Song (1989) to identify binary protein–protein interactions [2] and has been considered the most widely utilized approach to identify protein–protein interactions [3]. Y2H is based on the modular nature of eukaryotic transcription factors, which are made up of two distinct domains: the DNA-binding domain (BD) that interacts with DNA and the activation domain (AD), which recruits the transcriptional machinery to initiate transcription on target promoters [2]. The system is composed of a plasmid encoding a fusion protein to the BD domain, a second plasmid containing another fusion protein to the AD domain, and a genetically modified yeast containing a

Elizabeth P. B. Fontes and Kristiina Mäkinen (eds.), *Plant-Virus Interactions*, Methods in Molecular Biology, vol. 2724, https://doi.org/10.1007/978-1-0716-3485-1_14,

reporter gene activated by the GAL4 transaction factor [4]. The separated BD and AD domains are not fully functional but can reconstitute the transcription factor activity through noncovalent interactions between two fusion proteins coexpressed in yeast cells. The expression of a reporter gene is activated if the domains are kept in proximity, allowing the yeast cell to grow on a specific selective medium [5] (See Fig. 1a, b). For the AH109 strain, the reporter gene is *HIS3*. This strain, a derivative of PJ69-2A, also has the *ADE2* and *MEL1* reporter genes. The three reporter genes are under the control of distinct GAL4 upstream activating sequences (UASs) and TATA boxes, thereby minimizing the occurrence of false positives [6].

The yeast two-hybrid technology can identify new protein interactions, confirm suspected interactions, and define interaction domains [7]. The most extensively used experimental approaches for identifying virus–host protein interactions include yeast two-hybrid and affinity purification–mass spectrometry, which together account for more than 90% of the information accessible in public databases [8].

Y2H has also been applied to identify plant–virus protein interactions [9]. Several studies have used Y2H to detect interactions between viral and host proteins. Here are few examples: The coat protein (CP) of the turnip crinkle virus interacts with a member of the NAC family of proteins that participate in plant development and defense processes [10]. The silencing suppressor protein HcPro interacts with the calmodulin-like protein rgs-CaM [11]. Y2H screening of the tobacco mosaic virus replicase protein identified interaction with an ATPase protein and a subunit of the photosystem II complex [12]. The interaction between the AL2 protein of tomato golden mosaic virus, also called transcriptional activator protein (TrAP), with SNF1-kinase [13], and the interaction between a tomato NAC protein and the viral protein replication enhancer protein (Ren) from the tomato leaf curl virus [14] were also detected by Y2H. A novel receptor-like protein kinase NSP-interacting kinase (NIK) that interacts with a geminivirus nuclear shuttle protein (NSP) was discovered by yeast two-hybrid screening [15, 16]. Y2H has also found the following cellular partners of NSP: NSP-associated kinase (NsAK) [17], NSP-interacting GTPase (NIG) [18], and NSP-interacting syntaxin domain-containing protein (NISP) [19]. The movement protein (MP) from the cabbage leaf curl virus (CabLCV) and squash leaf curl virus has been shown to interact with synaptotagmin A (SYTA), an endosomal plant protein [20], and MP from the abutilon mosaic virus has been discovered as a possible binding partner for Arabidopsis thaliana's chaperone heat shock cognate 70 kDa protein (cpHSC70-1) [21].

In addition to detecting new interactions by two-hybrid library screenings, the two-hybrid protein–protein interaction (PPI) assay

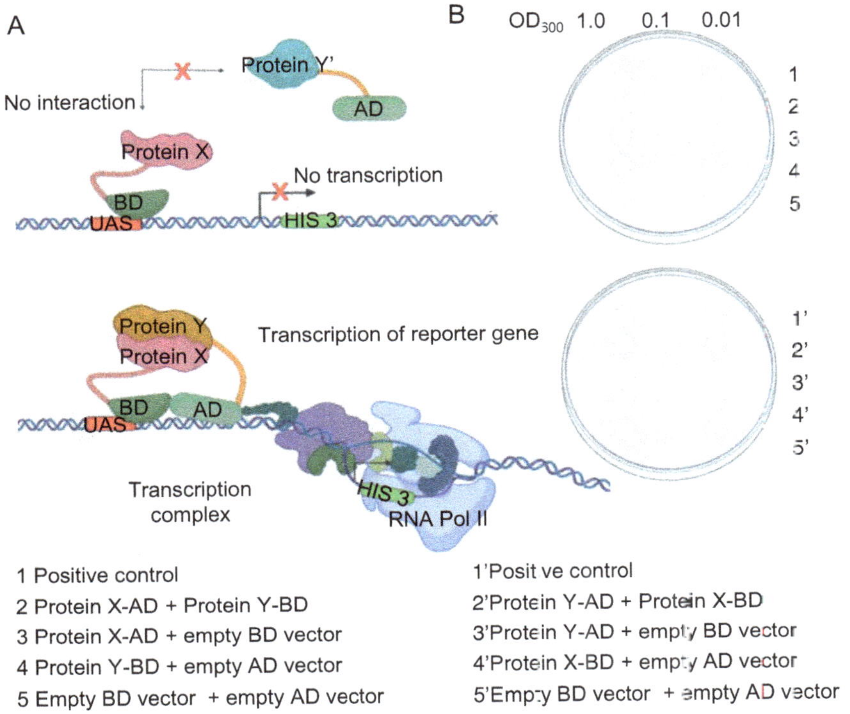

Fig. 1 One fragment of the GAL4 transfactor, which is modular, is fused to a protein X, and the other fragment is fused to a protein Y. If fusion proteins X and Y do not interact when co-expressed in yeast, there is no reconstitution of the transfactor, and the yeast that is auxotroph for histidine does not grow on triple-dropout selective medium. On the other hand, if the interaction occurs, the transfactor is reconstituted, recovering its activity, which activates the expression of the *HIS3* reporter gene resulting in yeast colony growth (**a**). Illustrative scheme of combinations: the X and Y proteins must be fused to the BD and AD domains. If X is BD-fused, it must be combined with the empty AD vector and AD-fused Y protein. The reverse combination will have protein X AD-fusion, Y fused to BD, and negative controls. Negative control strains must contain the empty vector for the BD domain and the target vector expressing a protein fused to AD and vice versa, in addition to the combination of both the empty AD and BD vectors. A positive control can be a combination of proteins whose interaction has already been documented. In all combinations, the first drop from left to right has OD600 = 1, and the subsequent two drops are culture dilutions at 10 and 100X. This same scheme should be reproduced on SD-Leu-Trp-His plates with and without 3-AT and in one SD-Leu-Trp plate that will be a positive control for the presence of both plasmids. These negative combinations should be done on the same day as the transactivation assay (**b**). The figure was created with BioRender.com

has often been used to confirm interactions identified via other in vitro PPI assays. A PPI network between the MP and NSP from the begomoviral CabLCV and Arabidopsis host proteins has been revealed using an in situ synthesized nucleic acid programmable protein array (NAPPA) [19]. The Y2H was used to confirm that NSP, but not MP, interacts specifically with NISP. We also show here by two-hybrid that the host protein At4G14560, a key regulator of auxin-modulated gene expression, interacts with both MP

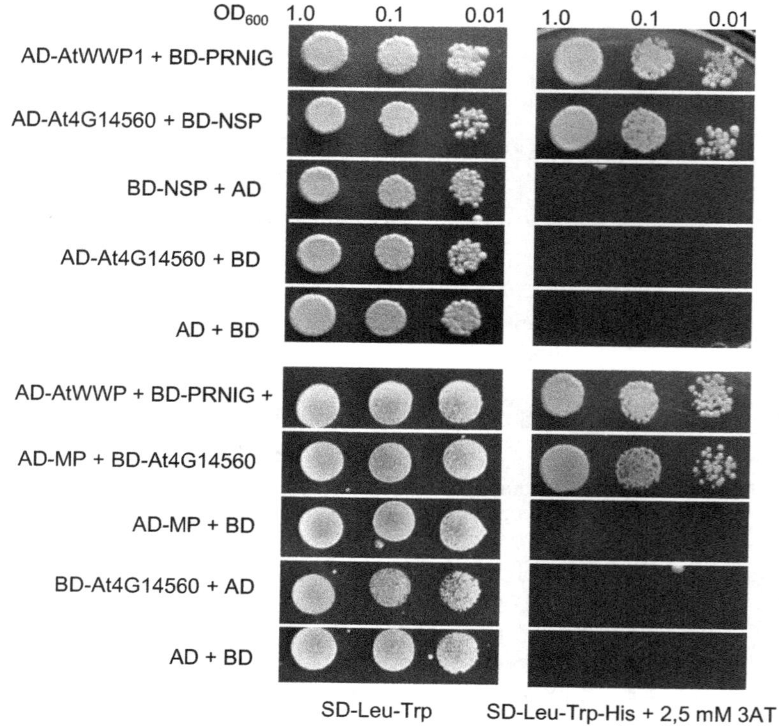

Fig. 2 Yeast cells co-transformed with the indicated constructs were evaluated for their histidine prototrophy by being plated on a selective medium deficient in leucine, tryptophan, and histidine, supplemented with 2.5 mM 3 AT. At4G14560, a key regulator of auxin-modulated gene expression, interacts with the viral proteins MP and NSP. Co-expression of AD-At4G14560 + BD-NSP and AD-MP + BD-At4G14560 promoted histidine prototrophy on a medium lacking histidine and supplemented with 3AT

and NSP (Fig. 2), confirming another relevant result from the protein microarray.

Moreover, Y2H has uncovered other significant interactions, such as protein–protein interactions that occur between host proteins and are relevant to the viral infection process: (i) LIMYB (L10-interacting Myb domain-containing protein) and ribosomal protein L10 (RPL10) [16]; (ii) WW domain-containing protein (AtWWP1) and NIG [22]; and (iii) NIK1 with brassinosteroid insensitive 1-associated kinase 1 (BAK1) and flagellin sensing 2 (FLS2) [23]. The receptor NIK1 inversely modulates antiviral and antibacterial immunity in plants [23]. Upon viral infection, the NIK1-mediated antiviral signaling pathway mediates phosphorylation of the ribosomal protein RPL10, which then translocates into the nucleus and interacts with LIMYB to suppress the expression of genes involved in translation, which culminates with global translational suppression [16]. Without pathogen invasion, autoimmunity is prevented by NIK1 interaction with BAK1 or FLS2 [23]. Another defense mechanism against viral infection is mediated by WWP1, which confers resistance to begomoviruses by sequestering NIG into nuclear bodies. NIG facilitates the nuclear export of the NSP-DNA complex. AtWWP1 interacts with NIG and confines it within nucleus, preventing its association with NSP [22]. The association of LIMYB and RPL10, WWP1 and NIG, and NIK1 with BAK1 or FLS2 was identified by two-hybrid assay. These examples reinforce the applicability of Y2H in screening protein–protein interactions, including the detection of host proteins that play an important role throughout viral infection.

The chapter presented here describes the main procedures for using the GAL4-based Y2H system and is subdivided into 4 sections: first and second transformations (which alternatively can be performed simultaneously), transactivation assay, and self-activation testing, besides a protocol for yeast protein extraction for fusion protein expression confirmation.

Another illustrative example of the Y2H assay is shown in Fig. 3. Methyl-CpG-binding domain-containing protein 2 (MBD2) from Arabidopsis was fused to the AD domain, and the viral coat protein AV1, from the CabLCV, was fused to the BD domain. The *HIS3* reporter gene was neither activated in yeast cells co-transformed with these constructions nor with the reciprocal empty vector combinations. The interaction between the host proteins AtWWP1 and the Pro-rich domain of NSP-interacting guanosine triphosphatase (PRNIG) was also evaluated by Y2H [22]. AH109 is unable to synthesize histidine; however, if there is an interaction between the tested proteins, histidine prototrophy is restored and the cells grow in selective medium deficient in histidine. Co-expression of BD-PRNIG fusion and AD-WWP1 fusion promoted histidine prototrophy in the presence of 3-amino-1,2,4-triazole (*3AT*), a competitive inhibitor of the enzyme HIS3 (imidazoleglycerol-phosphate dehydratase), which acts by reducing background growth, allowing a more straightforward distinction between background and positive colonies [7] (Fig. 3a). The interaction between NIG and WWP1, initially identified via yeast two-hybrid screen with a cDNA library from Arabidopsis thaliana aerial tissues, was confirmed through Y2H analysis. This interaction may interfere with the NIG pro-viral function, as WWP1 is able to reallocate NIG to nuclear bodies [22].

Accumulation of AD-AtWWP1, AD-MBD2, BD-PRNIG, and BD-AV1 proteins in yeast was evaluated by Western blot, using specific antibodies for the BD and AD domains (Fig. 3b). Western blots are especially important when positive interactions are not detected, such as AD-AtMBD2 + BD-AV1. Western blot validation demonstrates that the fusion proteins are stably expressed in yeast and confirms that the failure to display His prototrophy is due to the lack of protein interactions.

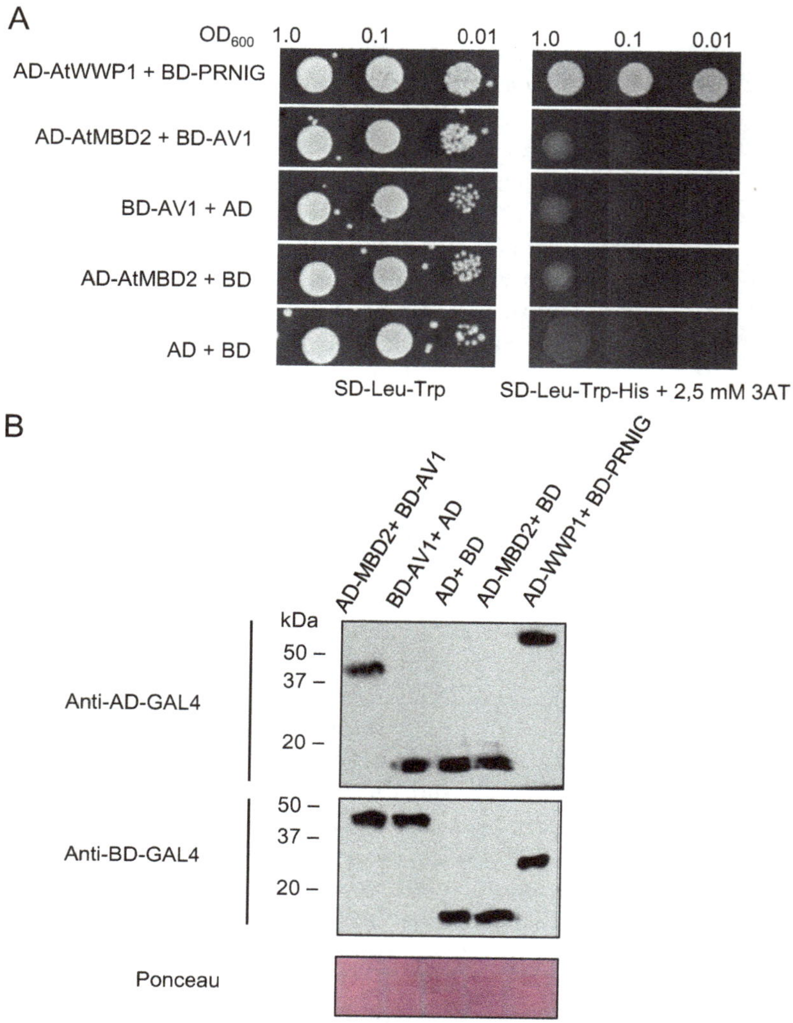

Fig. 3 Yeasts co-transformed with the indicated constructs were evaluated for their histidine prototrophy by being plated on a selective medium deficient in leucine, tryptophan, and histidine, supplemented with 2.5 mM 3 AT. Only the combination AD-AtWWP1 + BD-PRNIG showed growth. (**b**) Accumulation of the AD-AtWWP1, AD-MBD2, BD-PRNIG, and BD-AV1 proteins in yeast. Protein extracts from yeast carrying the constructs shown in (**a**) were fractionated by SDS–PAGE and immunoblotted using anti-AD-GAL4 (A-2) and anti-BD-GAL4 (RK5C1) (Santa Cruz Biotechnology®)

2 Materials

2.1 Yeast Strain and Vectors

1. Yeast strain AH109 MATα with genotype (trp1-901, leu2-3, 112, ura3-52, his3-200, gal4Δ, gal80Δ, LYS2::GAL1UAS-GAL1TATA-HIS3, GAL2UAS-GAL2TATA-ADE2, URA3:: MEL1UASMEL1TATA-lacZ) [6]. This strain is auxotrophic for leucine, tryptophan, histidine, and adenine.

2. Yeast expression of empty vectors: pDEST22 (AD domain) and pDEST32 (BD domain), both domains fused at the N-terminus. (*See* **Note 1**).
3. Appropriate plasmid DNA in solution: two-hybrid expression vectors containing the target proteins in frame with the BD and AD domains. (*See* **Note 2**).

2.2 Yeast Transformation

Medium and solutions: All solutions must be prepared with Milli-Q water.

1. Adenine 0.2%: adenine 0.2 g. Add water to a volume of 100 mL. Autoclave. Store at 4 °C for 1 month.
2. Glucose 40%: α-D-(+) anhydrous glucose 20 g. Add water to a volume of 50 ml. Autoclave. (*See* **Note 3**).
3. YPDA medium (100 mL): 1 g of yeast extract, 2 g of peptone, and 1.5 mL of 0.2% adenine. Make the volume up to 95 mL of water in a 100 mL measuring cylinder and mix thoroughly. Autoclave. If solid medium, include 1.6% agar (w/v) before autoclaving. Before first use, add 5 mL of autoclaved 40% glucose. Store the liquid medium at room temperature, while the solid medium, after pouring the plates, is stored at 4 °C for 2–3 weeks.
4. 10X Total amino acid solution (500 mL): adenine 0.1 g, arginine 0.1 g, isoleucine 0.15 g, lysine 0.15 g, methionine 0.1 g, phenylalanine 0.25 g, threonine 0.15 g, tyrosine 0.15 g, valine 0.75 g, and uracil 0.1 g. Weigh all amino acids very carefully so that contamination does not occur while handling the flasks and transfer them to a Becker. Make the volume up to 500 mL in a 500 mL measuring cylinder and mix thoroughly. Autoclave. Store at 4 °C for 1 month.
5. 10X Leucine solution: leucine 0.1 g. Distilled water q.s.p. 100 mL. Autoclave. Store at 4 °C for 1 month.
6. 10X Tryptophan solution: tryptophan 0.02 g. Distilled water q.s.p. 100 mL. Autoclave. Store at 4 °C for 1 month.
7. 10X Histidine solution: histidine 0.02 g. Distilled water q.s.p. 100 mL. Autoclave. Store at 4 °C for 1 month.
8. SD-Leu selective medium (synthetic dropout medium with all amino acids except leucine) 100 mL. Prepare liquid and solid media: 0.17 g Yeast nitrogen base w/o amino acids, 10 mL of 10X total amino acid solution, 10 mL of 10X tryptophan solution, 10 mL of 10X histidine solution. Add 80% of the final volume of water. Check the pH and, if necessary, adjust it to 5.8 with 1.0 N NaOH. Make the volume up to 95 mL and autoclave. If solid medium, include 1.6% agar (w/v) before autoclaving. Before first use, add 5 mL of autoclaved 40% glucose. Store the broth at 4 °C for 2–3 weeks.

9. SD-Trp selective media (synthetic dropout medium with all amino acids except tryptophan) 100 mL (enough for nearly six 90 mm plates). Prepare liquid and solid media: 0.17 g yeast nitrogen base w/o amino acids, 10 mL of 10X total amino acid solution, 10 ml of 10X leucine solution, and 10 mL of 10X histidine solution. Add 80% of the final volume of water. Check the pH. If necessary, adjust it to 5.8 with 1.0 N NaOH. Make the volume up to 95 mL and autoclave. Before first use, add 5 mL of autoclaved 40% glucose. If solid medium, include 1.6% agar (w/v) before autoclaving. Store the broth at 4 °C for 2–3 weeks.
10. SD-Leu-Trp selective media (synthetic dropout medium with all amino acids except leucine and tryptophan) 100 mL (enough for nearly six 90 mm plates). Prepare liquid and solid media: 0.17 g yeast nitrogen base w/o amino acids, 10 mL of 10X total amino acid solution, and 10 mL of 10X histidine solution. Add 80% of the final volume of water. Check the pH. If necessary, adjust it to 5.8 with 1.0 N NaOH. Make the volume up to 95 mL and autoclave. If solid medium, include 1.6% agar (w/v) before autoclaving. Before first use, add 5 mL of autoclaved 40% glucose. Store the broth at 4 °C for 2–3 weeks.
11. SD-Leu-Trp-His selective medium (synthetic dropout medium with all amino acids except leucine, tryptophan, and histidine) 200 mL (enough for nearly twelve 90 mm plates). Solid medium only: 0.34 g yeast nitrogen base w/o amino acids, 20 mL of 10X total amino acid solution. Add 80% of the final volume of water. Check the pH and adjust it to 5.8 with 1.0 N NaOH if necessary. Make the volume up to 190 mL, including 1.6% agar (w/v) and autoclave. Before first use, add 10 mL of 40% glucose. Prepare plates with different concentrations of 3-AT (0 mM, 2.5 mM, 5.0 mM, and 10 mM 3AT). These plates should be prepared at most 2–3 days prior to the transactivation assay.
12. 1 M Lithium acetate (50 mL): lithium acetate dihydrate 5.1 g. Adjust pH to 7.5 with acetic acid. Distilled water q.s.p. 50 ml. Autoclave and store at room temperature.
13. PEG (polyethylene glycol average mol wt 3350) 50% (w/v) (50 mL): PEG 3350 25 g. Distilled water q.s.p. 50 mL (~25 mL of water). Autoclave and store at room temperature. (*See* **Note 4**).
14. 3-amino-1,2,4-triazole (3-AT) 2 M solution (30 mL): 5.04 g of 3-AT. Distilled water q.s.p. 30 mL. Filter-sterilize and store at −20 °C in a light-protected environment. The 3AT solution is thermolabile. Add to the solid medium when it is at ~55 °C.

15. Salmon Sperm DNA (ssDNA): 2 mg/mL. The ssDNA is acquired commercially at a concentration of 10 mg/mL. From stock, prepare 2 mg/mL solution in sterile distilled water. Store ssDNA at −20 °C. Boil for 5 min before yeast transformation, and then cool for 3–4 min on ice. (*See* **Note 5**).

2.3 Protein Expression Verification

1. TCA 12.5%: Trichloroacetic acid (TCA) 1.25 g, distilled water q.s.p.10 mL. Weigh the TCA and solubilize it in water carefully, as TCA is very corrosive.
2. 1% SDS/0.1 M NaOH: 0.1 g Sodium dodecyl sulfate (SDS), 0.04 g Sodium hydroxide (NaOH). Weigh all reagents and solubilize them in distilled water q.s.p.10 mL.
3. 6X Sample buffer (Laemmli buffer) 10 mL: SDS 1.2 g, bromophenol blue 6 mg, glycerol 4.7 mL, 0.5 M Tris–HCl pH 6.8 1.2 mL, distilled water 2.1 mL. Mix all the reagents. To aid in solubilization, heat to 40 °C before adding 0.5 mL of mercaptoethanol (5% final concentration). Alternatively, dithiothreitol (DTT) can be used at a final concentration of 350 mM (54 mg/mL). Aliquot and keep frozen at −20 °C.
4. Antibodies anti-BD and anti-AD domain. (*See* **Note 17**).
5. SDS-PAGE running buffer, transfer buffer, and blocking solution for Western blot.

2.4 Equipment

1. Microcentrifuge with rotor for 1.5-mL tubes.
2. Centrifuge with rotor for 15-mL tubes.
3. Incubator at 30 °C for growth of plate cultures.
4. Incubator/shaker at 30 °C, for aerating liquid cultures.
5. Water bath or dry bath for treatment at 42 and 30 degrees.
6. Laminar flow hood.
7. Spectrophotometer.
8. Cuvettes.
9. Vortex mixer.
10. pH meter.
11. 1.5-mL polypropylene microcentrifuge tubes.
12. 15-mL conic tubes.
13. 100 x 15 mm sterile disposable Petri plates.
14. Glass spreader.
15. Electrophoresis and transfer apparatus for verification of protein expression.

3 Methods

3.1 First Transformation

1. Scrape frozen glycerol stock the AH109 yeast with a sterile toothpick or bacteriological needle, and incubate in 3–5 mL YPDA medium. To disperse cell clumps, vigorously vortex the cultures for 0.5–1 min. From this pre-inoculum, obtain fresh cultures as they result in better transformation efficiency.
2. Centrifuge the conic tubes at 3000 *g* for 5 min at room temperature.
3. Remove the supernatant and resuspend the pellet in 1 mL of sterile distilled water. Transfer to 1.5-mL microtubes.
4. Centrifuge again at 3000 *g* for 5 min at room temperature.
5. Discard the supernatant (by gently inverting the tube) and resuspend the pellet in 500 μL of 100 mM lithium acetate (freshly made from the 1 M stock solution).
6. Centrifuge at 13,000 *g* for 1 min at room temperature. Discard the supernatant, eliminating as much liquid as possible using a micropipette. Cells are now ready for the transformation step.
7. Add the following components to the yeast pellet in that order:

   240 μL PEG 3350 50% (viscous solution, aspirate slowly with the micropipette)

   36 μL of 1 M lithium acetate

   25 μL of salmon sperm (2 mg/mL), previously boiled for 5 min and immediately cooled on ice.

   500–1000 ng of plasmid DNA with the construct of interest cloned into pDEST 32 or pDEST22 (only one plasmid type per transformation).

   (*See* **Note 6**).
8. Use a micropipette to resuspend the cells thoroughly and then vortex the tubes. Incubate the tubes for 30 min at 30 °C in water or dry bath.
9. Incubate tubes for an additional 25 min at 42 °C in a water bath or dry bath.
10. Centrifuge the tubes at 13000 *g* for 1 min at room temperature. Remove the supernatant using a micropipette.
11. Resuspend the pellet in 1 mL of distilled water and centrifuge again at 13000 *g* for 1 min at room temperature.
12. Remove the supernatant, leaving about 100 μL of liquid this time. Resuspend the pellet in this volume of remaining liquid with a micropipette.
13. Spread the cell suspension over the appropriate selective medium with the aid of a glass spreader (remember to flame

the glass spreader and cool it by touching it to the agar along the edge near the rim before plating the cells). (*See* **Note 7**).

14. Incubate at 30 °C for 3–5 days. Wait until colonies reach 1–2 mm in diameter. (*See* **Notes 8** and **9**).

3.2 Second Transformation

1. The transformation protocol will be the same, except that the yeast must now be inoculated into selective media and not YPDA. Collect several colonies from the plate obtained in **step 14** (approximately 10), and incubate them in a liquid selective medium corresponding to the vector used in the first transformation (if pDEST32, SD-Leu medium; if pDEST22, SD-Trp medium). To disperse cell clumps, vigorously vortex the cultures for 0.5–1 min. Incubate for 16 to 24 h at 30 °C with a 220 rpm rotation.
2. Follow **steps 2–12** from Subheading 3.1, but remember that the cell must now be transformed with the construct in the vector opposite to that of the first transformation (e.g., if a cloned protein in pDEST32 was used in the first transformation, the second transformation must be with the other protein under analysis cloned in pDEST22, and vice versa). Include negative and positive controls. Including many controls can add confidence to the detected interactions. (*See* schemes of all combinations in Fig. 1b).
3. After the transformation, spread the cell suspension with a glass spreader over the SD-Leu-Trp selective medium since the selection will be for the presence of both vectors simultaneously.
4. Incubate at 30 °C for 3–5 days. Wait until colonies reach 1–2 mm in diameter. (*See* **Note 10**).

3.3 Transactivation Assay and Self-Activation Testing

1. Collect several colonies from **step 4**, Subheading 3.2 (approximately 10–15), and incubate them in liquid selective SD-Leu-Trp medium overnight at 30 °C with 220 rpm rotation.
2. In the flow hood, remove an aliquot of the culture and read the absorbance at 600 nm in a spectrophotometer. To prepare the drops, the culture must have reached an OD600 = 1.0. An OD600 = 1 equals 1–2 × 10^7 cells per mL.
3. Adjust the concentrations of all inoculums to the same absorbance reading. If necessary, make dilutions in water or culture medium until all are with OD600 = 1.0.
4. After adjustment to OD600 = 1, make serial dilutions of 10 and 100X in sterile distilled water for each combination (positive control, interaction, negative controls). (*See* interaction scheme: hypothetical proteins X and Y in Fig. 1b). Remember that this entire process needs to be done in the laminar flow hood.

5. Drops are made on triple-dropout selective medium (SD-Leu-Trp-His) supplemented with different concentrations of 3-AT (suggested: 0, 2.5, 5, and 10 mM). The required concentration of 3-AT will depend on the auto-activation test (see **steps** 7 and **8**). (*See* **Notes 11** and **12**).
6. Leave the plates open in the stream until the water evaporates. Seal with plastic wrap and incubate for 3–5 days at 30 °C. Photograph the plates, and prepare the figure with the results. (*See* **Note 13**).
7. For self-activation testing, use transformants with an empty vector plus a target protein fused to the complementary vector, e.g., target protein fused to the BD domain, combined with an empty vector containing an AD domain. To measure basal *HIS3* reporter gene activity, use SD-Leu-Trp-His supplemented with different concentrations of 3-AT: 0, 1, 2.5, 5, and 10 mM.
8. Make the drops in the same as in **steps 4–6**. Check for the lowest concentration of 3-AT at which background colony growth is completely blocked. (*See* **Notes 14–16**).

3.4 Verification of Protein Expression [24]. (See Note 17)

1. Cultivate yeast cells in 5 mL of appropriate selective medium until they reach an OD600 ~ 3 (takes about 3 to 4 days of incubation).
2. Transfer the culture to 2-mL microtubes and pellet the cells by centrifuging for 1 min at 14000 *g* at room temperature. Discard the supernatant. Continue centrifuging the culture in the same tube until the entire volume has been centrifuged.
3. Wash the pellet with 900 μL of water, vortex, and centrifuge again 1 min at 14000 *g* at room temperature, and then discard the supernatant.
4. Suspend the pellet in 400 μL of 12.5% TCA. Freeze samples for at least 30 min at −80 °C (we usually leave them overnight). Extraction can be stopped at this point, and samples can be stored at −80 °C for later use.
5. Thaw the sample and sediment the cells by centrifuging them for 5 min at 14000 g at room temperature. Discard the supernatant.
6. Wash the pellet twice with 500 μL of ice-cold 80% acetone by centrifuge or 5 min at 14000 *g* at room temperature.
7. Carefully remove the supernatant and let the acetone evaporate completely until the pellet is dry.
8. Solubilize the pellet in 75 μL of 1% SDS/0.1 M NaOH (at this stage, the pellet will have a whitish and crumbly appearance).

9. Add 17 μL of 6X sample buffer (Laemmli buffer). Boil the samples for 5 min at 100 °C.
10. Proceed with the separation of proteins by loading them on a standard SDS polyacrylamide gel and transfer the separated proteins to a blotting membrane. Use specific antibodies against BD and AD domains for probing.

4 Notes

1. Other commercial vectors are also available, such as pGBKT7g and pGADT7g, which produce amino-terminal BD and AD fusions, respectively. pGBKCg and pGADCg, on the other hand, fuse the BD or AD at the carboxyl terminus of the target protein. pDEST, pGBKT7g, and pGADT7g vectors are Gateway compatible.
2. The genes of interest must first be amplified using PCR and then cloned into yeast expression vectors using either classic cloning or recombination methods (e.g., the Gateway system).
3. The amount of water that must be added to make the glucose solution is quite minimal. Slowly pour in the water, being careful not to exceed the final capacity of 50 mL. It might take some time for glucose to dissolve. Continue to stir with the magnetic bar until the solubilization is complete. Because it causes caramelization of the media when autoclaved combined with the other components, the glucose is frequently prepared separately (however, it is not essential to do so) and added to the other components after autoclaving.
4. PEG is particularly tough to dissolve. Continue stirring with the magnetic bar until the solubilization is complete. Store at room temperature with the vial sealed with parafilm to prevent evaporation.
5. Avoid freeze and thaw cycles. It is best to make small aliquots from stock and thaw just enough for use.
6. Because acetate is toxic to cells at high concentrations, adding PEG before 1 M lithium acetate is crucial. The PEG will form a kind of shield around the cells.
7. The selective medium will be SD-Trp if the first transformation is with the pDEST22 vector and SD-Leu if the initial transformation is with the pDEST32 vector. The *TRP1* gene is carried via the pDEST22 vector, which allows cells to be selected for transformation by growing them on tryptophan-free media. The *LEU2* gene in pDEST32 permits yeast growth in media without leucine to select it.

8. Transformation timing: **step 1** takes 12–16 h; **steps 2–13** take 1 h to 1 h 30 min, depending on the number of transformations performed; and **Step 14** takes 3–5 days. It takes at least 3 days for the transformed yeast to develop colonies appropriate for isolation.
9. Yeast, transformed or untransformed, can be kept on a solid medium at 4 °C for 2–4 weeks before being restreaked. Alternatively, yeast strains can be stored indefinitely at −70 °C in YPD medium with 25% glycerol. If it is necessary to store for a period longer than 1 year, the temperature must be kept below −55 °C. To keep the plasmid under selection pressure, transformed yeast strains should be kept in the proper selective medium. Remove a tiny frozen aliquot from the stock and streak on a suitable SD plate for 2–3 days at 30 °C to recover strains.
10. Instead of performing two separate transformations, it is possible to carry out a single transformation of the yeast AH109 with the two plasmids (pDEST32 and pDEST22 simultaneously or co-transformation). In this case, the selective growth medium after the transformation will be SD-Leu-Trp. Still, the transformation methodology will be the same as in **steps 2–12**, Subheading 3.1, except that 500 ng of each vector will be employed in co-transformation. However, the efficiency of this type of transformation is around 40% compared to the sequential transformation of each plasmid individually due to the lower probability that a given yeast cell will take up both plasmids. Nevertheless, this can be a viable alternative when one of the proteins is toxic to the yeast and ends up interfering with growth, thereby making sequential transformation more difficult.
11. Test the interaction in media containing different concentrations of 3-AT that can vary from 1 mM to 50 mM depending on the strength of the interaction between the proteins evaluated. Higher concentrations of 3-AT are not advised as they will stop all growth. 3-AT is a competitive inhibitor of the enzyme HIS3 (imidazoleglycerol-phosphate dehydratase), which acts by reducing background growth, allowing a more straightforward distinction between background and positive colonies. The optimal concentration of this reagent should be empirically determined for each assay. Remember: 3-AT is thermolabile and will be destroyed if added to a medium hotter than 55 °C.
12. To facilitate the pipetting process, the bottom of the plate where the drops will be made can be marked with a microtube pen with dots at ~1 to 1.5 cm between them. Start pipetting on the plates from right to left, from the lowest to the highest

culture concentration (this way, it is possible to use the same tip for the same combination). Make the drops with the aid of a micropipette and a maximum of 10 μL of each cell suspension. Be very careful when pipetting not to contaminate the plates or make sudden movements so that the drops are not scattered in the medium.

13. Different concentrations of 3-AT are used in the two-hybrid test, but only the lowest concentration that inhibits background growth is shown in the figure with the final results. Photograph the plates after the lowest dilution of the positive control to indicate growth. If you wait too long, the difference in growth between the dilutions will become insignificant. The final figure should also include a plate with the drops made in SD-Leu-Trp medium. This part of the assay will act as a positive control for the presence of both plasmids in the cells.

14. Autoactivation: Growth of negative controls when the protein fused to the BD domain is transformed together with the empty AD plasmid. This finding can be explained if the tested protein has a transcriptional activation domain, which is especially likely if it is a transcription factor [5, 25, 26]. In such cases, the two-hybrid assay can only be performed with that protein fused to the AD domain or with truncated versions of the protein in which the intrinsic AD domain is inactivated. It is also possible to test cloning in a different vector. For example, suppose autoactivation occurs in a vector with the tag at the N-terminus. In that case, the protein under analysis may show no self-activation at all when expressed as carboxy-terminal fusions.

15. It is worth mentioning that some types of interactions are not detectable by the double-hybrid system. If one of the following occurs, it can interfere with the proteins' ability to interact: (i) Fusion proteins are not stably expressed in the host cell because protein fusions that are either toxic or unstable in yeast; (ii) fusion to GAL4 domains obstructs the site of interaction; (iii) the fusion protein is incorrectly folded; (iv) the fusion protein cannot be localized to the yeast nucleus; or (v) posttranslational event-dependent interactions do not occur in yeast. For problem 1, checking fusion protein expression levels by western blot is essential. For 2, other combinations of carboxy-terminal and amino-terminal fusion vectors can be tested and, for 3, alternatively, can be used partitioned domains of the protein, which do not contain a membrane localization signal.

16. In addition to failure to detect interactions, the two-hybrid technique can produce false positives. False positives can occur, for example, when the interactions are detected in

yeast but have no biological relevance. A typical example is as follows: *In vivo* proteins are not located in the same cell compartment or are expressed in different tissues and stages of development [27]. Additionally, false positives may also result from the self-activation of the reporter gene (*See* **Note 14**). It is essential to validate the interactions detected by Y2H with alternative techniques such as co-immunoprecipitation, pull-down, and bimolecular fluorescence complementation (BiFC).

17. The verification of the fusion protein expression by Western blotting is strongly recommended, especially when no interaction between the proteins is detected. For this, total yeast protein extracts are prepared from yeast liquid culture. Glass beads are not required for the methodology provided here. Use antibodies against the BD and AD domains (these are commercially available from several companies: Takara, Santa Cruz Biotechnology, and Invitrogen, for example). Electrophoretic separation, transfer, and membrane blocking can be performed according to standard laboratory protocols.

Acknowledgments

This work was supported by a Technical and Financial Cooperation Agreement for the joint execution of the structuring research project n° 145/2021, signed between the Universidade do Estado de Minas Gerais and the Fundação Arthur Bernardes. The authors also thank the Institutional Research Support Program (PAPq)—UEMG and Research Productivity Scholarship Program—PQ/UEMG.

References

1. Serebriiskii IG, Golemis EA (2001) Two-hybrid system and false positives. Approaches to detection and elimination. Methods Mol Biol (Clifton, N.J.) 177:123–134. https://doi.org/10.1385/1-59259-210-4:123
2. Fields S, Song O (1989) A novel genetic system to detect protein-protein interactions. Nature 340:245–246. https://doi.org/10.1038/340245a0
3. Shivhare D, Musialak-Lange M, Julca I, Gluza P, Mutwil M (2021) Removing auto-activators from yeast-two-hybrid assays by conditional negative selection. Sci Rep 11:5477. https://doi.org/10.1038/s41598-021-84608-9
4. Chien CT, Bartel PL, Sternglanz R, Fields S (1991) The two-hybrid system: a method to identify and clone genes for proteins that interact with a protein of interest. Proc Natl Acad Sci U S A 88:9578–9582. https://doi.org/10.1073/pnas.88.21.9578
5. Ma J, Ptashne M (1987) Deletion analysis of GAL4 defines two transcriptional activating segments. Cell 48(5):847–853. https://doi.org/10.1016/0092-8674(87)90081-x
6. James P, Haliaday J, Craig EA (1996) Genomic libraries and a host strain designed for highly efficient two-hybrid selection in yeast. Genetics 144:1425–1436. https://doi.org/10.1093/genetics/144.4.1425

7. Clontech (2007) Matchmaker™ GAL4 two hybrid system 3 and libraries user mannual. Clontech, Palo Alto
8. Brito AF, Pinney JW (2017) Protein-protein interactions in virus-host systems. Front Microbiol 8:1557. https://doi.org/10.3389/fmicb.2017.01557
9. MacFarlane SA, Uhrig JF (2008) Yeast two-hybrid assay to identify host-virus interactions. Methods Mol Biol (Clifton, N.J.) 451: 649–672. https://doi.org/10.1007/978-1-59745-102-4_44
10. Ren T, Qu F, Morris TJ (2000) HRT gene function requires interaction between a NAC protein and viral capsid protein to confer resistance to Turnip crinkle virus. Plant Cell 12: 1917–1925. https://doi.org/10.1105/tpc.12.10.1917
11. Anandalakshmi R, Marathe R, Ge X, Herr JM Jr, Mau C, Mallory A, Pruss G, Bowman L, Vance VB (2000) A calmodulin-related protein that suppresses posttranscriptional gene silencing in plants. Science 290:142–144. https://doi.org/10.1126/science.290.5489.142
12. Abbink TE, Peart JR, Mos TN, Baulcombe DC, Bol JF, Linthorst HJ (2002) Silencing of a gene encoding a protein component of the oxygen-evolving complex of photosystem II enhances virus replication in plants. Virology 295:307–319. https://doi.org/10.1006/viro.2002.1332
13. Hao L, Wang H, Sunter G, Bisaro DM (2003) Geminivirus AL2 and L2 proteins interact with and inactivate SNF1 kinase. Plant Cell 15: 1034–1048. https://doi.org/10.1105/tpc.009530
14. Selth LA, Dogra SC, Rasheed MS, Healy H, Randles JW, Rezaian MA (2005) A NAC domain protein interacts with tomato leaf curl virus replication accessory protein and enhances viral replication. Plant Cell 17:311–325. https://doi.org/10.1105/tpc.104.027235
15. Mariano AC, Andrade MO, Santos AA, Carolino SM, Oliveira ML, Baracat-Pereira MC, Brommonshenkel SH, Fontes EP (2004) Identification of a novel receptor-like protein kinase that interacts with a geminivirus nuclear shuttle protein. Virology 318:24–31. https://doi.org/10.1016/j.virol.2003.09.038
16. Zorzatto C, Machado JP, Lopes KV, Nascimento KJ, Pereira WA, Brustolini OJ, Reis PA, Calil IP, Deguchi M, Sachetto-Martins G, Gouveia BC, Loriato VA, Silva MA, Silva FF, Santos AA, Chory J, Fontes EP (2015) NIK1-mediated translation suppression functions as a plant antiviral immunity mechanism. Nature 520:679–682. https://doi.org/10.1038/nature14171
17. Florentino LH, Santos AA, Fontenelle MR, Pinheiro GL, Zerbini FM, Baracat-Pereira MC, Fontes EP (2006) A PERK-like receptor kinase interacts with the geminivirus nuclear shuttle protein and potentiates viral infection. J Virol 80:6648–6656. https://doi.org/10.1128/JVI.00173-06
18. Carvalho CM, Fontenelle MR, Florentino LH, Santos AA, Zerbini FM, Fontes EP (2008) A novel nucleocytoplasmic traffic GTPase identified as a functional target of the bipartite geminivirus nuclear shuttle protein. Plant J 55:869–880. https://doi.org/10.1111/j.1365-313X.2008.03556.x
19. Gouveia-Mageste BC, Martins LGC, Dal-Bianco M, Machado JPB, da Silva JCF, Kim AY, Yazaki J, Dos Santos AA, Ecker JR, Fontes EPB (2021) A plant-specific syntaxin-6 protein contributes to the intracytoplasmic route for the begomovirus CabLCV. Plant Physiol 187:158–173. https://doi.org/10.1093/plphys/kiab252
20. Lewis JD, Lazarowitz SG (2010) Arabidopsis synaptotagmin SYTA regulates endocytosis and virus movement protein cell-to-cell transport. Proc Natl Acad Sci U S A 107:2491–2496. https://doi.org/10.1073/pnas.0909080107
21. Krenz B, Windeisen V, Wege C, Jeske H, Kleinow T (2010) A plastid-targeted heat shock cognate 70kDa protein interacts with the Abutilon mosaic virus movement protein. Virology 401:6–17. https://doi.org/10.1016/j.virol.2010.02.011
22. Calil IP, Quadros IPS, Araújo TC, Duarte CEM, Gouveia-Mageste BC, Silva JCF, Brustolini OJB, Teixeira RM, Oliveira CN, Milagres RWMM, Martins GS, Chory J, Reis PAB, Machado JPB, Fontes EPB (2018) A WW domain-containing protein forms immune nuclear bodies against begomoviruses. Mol Plant 11:1449–1465. https://doi.org/10.1016/j.molp.2018.09.009
23. Li B, Ferreira MA, Huang M, Camargos LF, Yu X, Teixeira RM, Carpinetti PA, Mendes GC, Gouveia-Mageste BC, Liu C, Pontes CSL, Brustolini OJB, Martins LGC, Melo BP, Duarte CEM, Shan L, He P, Fontes EPB (2019) The receptor-like kinase NIK1 targets FLS2/BAK1 immune complex and inversely modulates antiviral and antibacterial immunity. Nat Commun 10:4996. https://doi.org/10.1038/s41467-019-12847-6
24. Baerends RJ, Faber KN, Kram AM, Kiel JA, van der Klei IJ, Veenhuis M (2000) A stretch of positively charged amino acids at the N terminus of Hansenula polymorpha Pex3p is

involved in incorporation of the protein into the peroxisomal membrane. J Biol Chem 275: 9986–9995. https://doi.org/10.1074/jbc.275.14.9986

25. Ruden DM, Ma J, Li Y, Wood K, Ptashne M (1991) Generating yeast transcriptional activators containing no yeast protein sequences. Nature 350:250–252. https://doi.org/10.1038/350250a0

26. Ruden DM (1992) Activating regions of yeast transcription factors must have both acidic and hydrophobic amino acids. Chromosoma 101: 342–348. https://doi.org/10.1007/BF00346013

27. Russell RB, Aloy P (2008) Targeting and tinkering with interaction networks. Nat Chem Biol 4:666–673. https://doi.org/10.1038/nchembio.119

Chapter 15

Bimolecular Fluorescence Complementation (BiFC) in Host–Virus Interactions

Fredy Davi Albuquerque Silva, João Paulo Batista Machado, and Pedro Augusto Braga dos Reis

Abstract

Bimolecular fluorescence complementation (BiFC) is an assay widely used for studying protein–protein interactions and determining the subcellular localization of proteins. This technique involves fusing the proteins of interest to separate structural domains of a fluorescent protein, followed by transient expression in cells. The interaction between the proteins of interest in vivo allows the reconstitution of the fluorescence that can be visualized by fluorescence microscopy. BiFC has been particularly useful in investigating the interactions between viral and host proteins. Here, we describe the steps involved in preparing expression cassettes that allow the expression of proteins of interest fused to nonfluorescent fragments of yellow fluorescent protein (YFP), *Agrobacterium* transformations, and agroinfiltration of *Nicotiana benthamiana* leaves to facilitate virus protein–host protein interactions. Finally, high-resolution images can be obtained by analyzing the leaves under a confocal microscope.

Key words BiFC, Viral protein–host protein interactions, Agroinfiltration, Confocal microscopy

1 Introduction

Protein–protein interaction assays are essential for uncovering the dynamics of protein function and network in the cellular machinery. Hence, they are extensively used to identify and characterize proteins in their biological context in high-throughput or low-throughput assays. Yeast two-hybrid (Y2H), co-immunoprecipitation (CoIP), and bimolecular fluorescence complementation (BiFC) are among the most used in the molecular biology field. BiFC involves the expression of the proteins of interest fused to separate structural domains without the fluorescence activity of a fluorescent protein, such as a yellow fluorescent protein (YFP) (Fig. 1), in living cells. Once the interaction occurs between the

---

Authors João Paulo Batista Machado and Pedro Augusto Braga dos Reis have equally contributed to this chapter.

Elizabeth P. B. Fontes and Kristiina Mäkinen (eds.), *Plant-Virus Interactions*, Methods in Molecular Biology, vol. 2724, https://doi.org/10.1007/978-1-0716-3485-1_15,

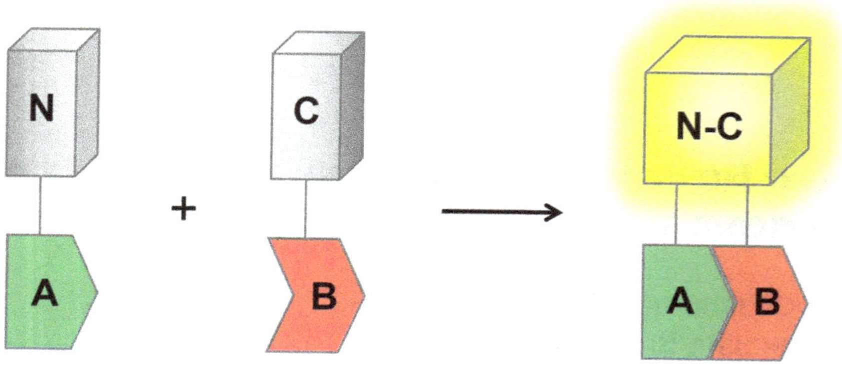

Fig. 1 Schematic representation of the BiFC assay. Two nonfluorescent fragments (N and C) of the yellow fluorescent protein (YFP) are fused to putative interaction partners (A and B). The association of the interaction partners allows the formation of a bimolecular fluorescent complex

two target proteins, the two molecular fragments of the fluorescent protein (carboxy- and amino-terminal) independently fused to the interest proteins can reconstitute the fluorescence signal. BiFC has been used in a variety of biological systems, which have allowed the identification of several protein–protein interactions [1]. Moreover, it has been used to investigate direct viral protein–host protein interactions and their subcellular localization [2].

Viruses are obligate intracellular parasites that rely entirely on the host cells' machinery to synthesize viral components such as nucleic acids, membranes, and proteins [3]. A movement protein (MP) is a specific virus-encoded protein that can recruit plant host factors to assist local and systemic viral spread throughout plant cells [4]. Therefore, identifying virus–host protein interaction allows a better understanding of virus infection in the plant. The BiFC approach has been utilized to study several protein–protein interactions in plants, such as interactions between: (1) intracellular MP, viral nuclear shuttle protein (NSP) from cabbage leaf curl virus (CabLCV), and NSP-interacting kinase 1 of *Arabidopsis thaliana* [5]; (2) the *A. thaliana* NSP-interacting syntaxin domain-containing protein (NISP) and NSP from CabLCV in vesicle-like compartments [6]; (3) plant partners related to viral interactions such as *A. thaliana* WWP1 protein and NSP-interacting GTPase (NIG) [7]; and (4) signaling processes like co-receptor interaction complex NIK1 and flagellin sensing 2/brassinosteroid insensitive 1-associated kinase 1 (FLS2/BAK1) [8]. The BiFC method allows visualization of viral protein–host protein interactions and its subcellular localization analysis without the proteins changing their chemical traits [9]. Here, we describe the BiFC approach for assessing viral protein–host protein interactions and visualization and how to perform this method while avoiding false positives during the analysis. Moreover, we described the BiFC assay performed to evaluate the interaction between two Arabidopsis proteins AtWWP1 and NIG, which are involved in virus infection responses. AtWWP1 relocates NIG from the cytoplasm to the nucleus, where it is confined to AtWWP1-formed nuclear bodies, suggesting that the NIG-AtWWP1 interaction may interfere with the NIG proviral function associated with its cytosolic localization [7].

2 Materials

2.1 General Equipment

1. Spectrophotometer.
2. Centrifuge for use at 4 °C and room temperature.
3. Laminar flow hood.
4. Electroporation system (e.g., Bio-Rad Gene Pulser).
5. Petri dish incubator for use at 28 °C and 37 °C.
6. Orbital shakers for liquid cultures at 28 °C and 37 °C, and 200 rpm.
7. Plant growth chambers.
8. Laser scanning confocal microscope (e.g., Zeiss LSM 510 META used to obtain the images shown in Fig. 2). The equipment must include a laser with an excitation wavelength capable of exciting the YFP molecule optimally, which is in the range of 490–515 nm, such as an argon laser at 514 nm. Also, a filter capable of capturing the light emitted by the YFP (maximum emission intensity in the 520–560 nm range), such as the 505–550 nm band pass filter.
9. Glass microscope slides and coverslips.
10. Image processing software (e.g., Zeiss LSM 510 software and Adobe Photoshop).
11. Plastics (Petri dishes, microtubes, pipette tips, falcon tubes).

2.2 Cloning into BiFC Vectors

1. The coding region of interest genes properly isolated and inserted into donor plasmids. The coding region of the *NIG* and *AtWWP1* genes used in the BiFC assay shown in Fig. 2 was isolated by PCR (polymerase chain reaction) using gene-specific primers and subsequently inserted into the pDONR™207 Gateway® entry vector (Thermo Scientific) via a BP recombination reaction, resulting in NIG-NS-pDONR207 and AtWWP1-NS-pDONR207 clones.
2. Plant BiFC vectors: sets of plant-compatible BiFC vectors, which allow the evaluation of protein–protein interactions using either the N- or C-terminal fusion of nonfluorescent fragments of YFP (*see* **Note 1**). In the BiFC assay illustrated in Fig. 2, we used the vectors SPYNE and SPYCE [10], which

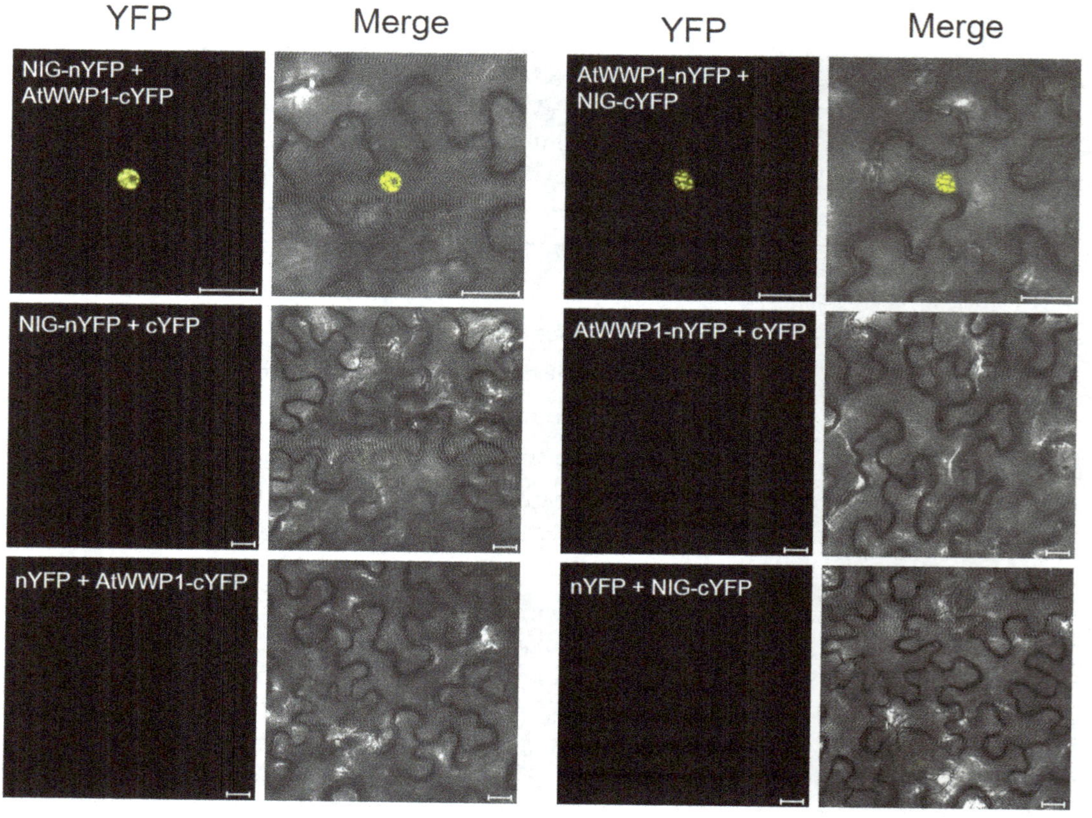

Fig. 2 Confocal microscopy image of a BiFC experiment. The confocal microscopy images of the interaction between NIG and AtWWP1 proteins from Arabidopsis were obtained after 3 days of agroinfiltration. The figure shows the different combinations of NIG and AtWWP1 proteins fused to C-terminal (cYFP) or N-terminal (nYFP) portions of YFP. Through the image resolution is possible to visualize the formation of nuclear speckles in the interaction of either NIG-nYFP + AtWPP1-cYFP or AtWPP1-nYFP + NIG-cYFP. The combination of NIG-nYFP, NIG-cYFP, AtWWP1-nYFP, and AtWWP1-cYFP with the empty vectors nYFP and cYFP showed no fluorescence signal. Scale bars, 20 μm

contain the N-terminus (nYFP) and C-terminal (cYFP) region of YFP, respectively, fused in frame at the C-terminal position of the proteins of interest. Thus, the genes of interest should be cloned into SPYNE and SPYCE without the stop codon.

3. *Escherichia coli* cells (e.g., DH5α strain) for cloning procedures.
4. LB (Luria-Bertani) medium: Weigh 10 g of tryptone, 10 g of sodium chloride (NaCl), and 5 g of yeast extract. Combine the chemicals, add deionized water to a volume of 900 mL, and stir the solution until the solutes dissolve. Adjust the pH to 7.0 with 5 N NaOH. Adjust the volume of the solution to 1 L with deionized water. Add 15 g agar for a solid medium. Sterilize by autoclaving at 121 °C for 30 min at 15 psi of pressure on the liquid cycle.

5. Antibiotics: Prepare stock solutions of antibiotics to be used in the selection of the plasmids following specific recommendations. Sterilize stock solutions of antibiotics dissolved in water by filtration through a 0.22 μm filter and store at −20 °C (*see* **Note 2**).

2.3 Transformation of Agrobacterium Cells

1. *Agrobacterium tumefaciens* cells: For the transient transformation protocol, we have adopted the strain GV3101, which carries genes for resistance to gentamycin. Additionally, to improve the protein expression, an *Agrobacterium* strain encoding a viral protein suppressor of RNA silencing in transformed tobacco leaves, such as the potyviral helper component proteinase (HCPro) and P19 protein of tomato bushy stunt virus, can be included in the agroinfiltrations.
2. LB medium: Prepare as described in the previous section.
3. GYT medium: 10% (v/v) glycerol, 0,125% (w/v) yeast extract, and 0,25% (w/v) tryptone. Dissolve 0.0125 g of yeast extract and 0.025 g of tryptone in 5 mL of deionized water. Add 1 mL of glycerol. Using a graduated cylinder, make up the volume to 10 mL with deionized water. Sterilize the medium by passing it through a 0.22 μm filter. Store in small batches at 4 °C.
4. Antibiotics: To be used to select the *Agrobacterium* strain carrying the constructs of interest. Prepare stock solutions as described in the previous section.
5. Electroporation device and cuvettes fitted with electrodes spaced 0.1–0.2 cm apart.

2.4 Agroinfiltration

All prepared solutions should use ultrapure Milli-Q ion-free water, resistivity 18.2 MΩ-cm, low total organic carbon (TOC), and analytical grade reagents. Previously, prepare and store all reagents at 4 °C or follow specific recommendations. Please carefully follow all waste disposal regulations.

1. Magnesium chloride (100×): 1 M magnesium chloride ($MgCl_2$). Weigh 9521 g and transfer to a suitable container. Add pure water to a volume of 50 mL and mix. Transfer the solution to a graduated cylinder and add pure water to bring the final volume to 100 mL (*see* **Note 3**).
2. MES (100×): 1 M 2-(N-morpholino) ethanesulfonic acid (MES). Dissolve 19,524 g of MES in 80 mL of pure water. Adjust the pH of the solution to 5.6 with 5 M KOH. Make up to 100 mL with pure water (*see* **Note 3**).
3. Acetosyringone (1000×): 200 mM 3′,5′-dimethoxy-4′--hydroxyacetophenone (acetosyringone). Weigh 0,392 g acetosyringone and transfer it to a suitable container. Add 99,5% dimethyl sulfoxide (DMSO) to a volume of 5 mL (*see* **Note 4**). Make up to 10 mL with DMSO. Store at −20 °C (*see* **Note 5**).

4. Sterile 1 mL insulin syringe.
5. *Nicotiana benthamiana* or *Nicotiana tabacum* plants grown in separated pots with 4- to 6-week-old.

3 Methods

3.1 Cloning into BiFC Vectors

Depending on the cloning strategy to be used, the coding region of the genes of interest can be transferred to the BiFC vectors through cleavage and ligation or performing an LR recombination reaction between an attL-containing entry clone and an attR-containing destination vector to generate an expression clone (Gateway). The SPYNE and SPYCE vectors used in the BiFC assay, shown in Fig. 2, contain cloning sites compatible with the Gateway system. Therefore, we performed an LR recombination reaction to transfer the coding region of NIG and AtWWP1 without the stop codon from entry clones (NIG-NS-pDONR207 and AtWWP1-NS-pDONR207) to the vectors SPYNE and SPYCE, resulting in NIG-nYFP, AtWWP1-nYFP, NIG-cYFP, and AtWWP1-cYFP expressing clones.

1. Transform *Escherichia coli* (e.g., DH5α strain) with recombinant expression plasmids following your preferred protocol.
2. Spread *E. coli* cells on LB agar plates supplemented with appropriate antibiotics and incubate them at 37 °C overnight. The SPYNE and SPYCE vectors contain the kanamycin resistance gene. Therefore, NIG-nYFP, AtWWP1-nYFP, NIG-cYFP, and AtWWP1-cYFP expression vectors were selected in LB medium containing kanamycin 100 mg/L.
3. Confirm the colonies obtained by PCR using specific primers.
4. Grow the individual colonies in 5 mL of LB liquid medium supplemented with the appropriate antibiotics at 37 °C under rotation of 200 rpm on a horizontal shaker.
5. Extract the plasmid DNA using a miniprep kit following the manufacturers' information.
6. Sequence the recombinant plasmids to confirm their correct sequence.

3.2 Transformation of Agrobacterium Cells by Electroporation

3.2.1 Preparation of Cells

1. Inoculate a single colony of *Agrobacterium tumefaciens* GV3101 from a fresh agar plate into a flask containing 50 mL of LB medium supplemented with gentamycin 50 mg/L. Incubate the culture overnight at 28 °C with vigorous aeration (220 rpm in a rotary shaker) to an OD_{600} of 0.4–0.6.
2. Harvest the cells by centrifugation at 3000 × *g* for 10 min at 4 °C. Decant the supernatant and resuspend the cell pellet in 50 mL of pure ice-cold water previously sterilized by autoclaving. Repeat washing the cells in water 4–5 times.

3. Harvest the cells by centrifugation at 3000 × *g* for 10 min at 4 °C. Decant the supernatant (*see* **Note 6**) and resuspend the pellet in 2 mL of ice-cold GYT medium (*see* **Note 7**).
4. Dispense 40 μL of the cell suspension aliquots into a sterile, ice-cold 0.5 mL microtube, drop into a liquid nitrogen bath, and transfer to a − 80 °C freezer until required.

3.2.2 Electroporation

1. Remove an appropriate number of aliquots of *Agrobacterium* cells from the −80 °C freezer. Thaw the frozen cells on ice and add 0.5–1 μL (1–10 ng/μL) of plasmid DNA to 40 μL of recipient cells (*see* **Note 8**).
2. Pipette the DNA/cell mixture into a cold electroporation cuvette (*see* **Note 9**).
3. Carry out the electroporation by applying a single pulse of 25 μF capacitance, 2.4 kV, and 200 Ω resistance (e.g., Bio-Rad Gene Pulser) (*see* **Note 10**).
4. After applying the pulse, remove the electroporation cuvette and add 1 mL of LB medium at room temperature.
5. Transfer the cells to a 1,5-mL microtube and incubate the cultures with gentle rotation for 2 h at 28 °C.
6. Recover the cells by short centrifugation, resuspend them in 0.1 mL LB medium, and spread on LB agar medium containing the appropriate antibiotic. *A. tumefaciens* GV3101 carrying the NIG-nYFP, AtWWP1-nYFP, NIG-cYFP, and AtWWP1-cYFP expression vectors were selected in LB medium containing gentamycin 50 mg/L and kanamycin 100 mg/L.
7. Invert the plates and incubate them at 28 °C. Let colonies grow for 2–3 days.

3.3 Agroinfiltration of *Nicotiana benthamiana* or *Nicotiana tabacum*

1. Grow the previously transformed *Agrobacterium* cultures for 16 h at 28 °C under 200 rpm shaking in 5 mL of LB medium plus appropriate antibiotics (*see* **Note 11**).
2. Prepare the infiltration buffer (10 mM $MgCl_2$, 10 mM MES, pH 5.6 adjusted with KOH) by mixture 1% (v/v) magnesium chloride (100×) and 1% (v/v) MES (100×) in deionized water (*see* **Note 12**).
3. Centrifuge the *Agrobacterium* cultures containing the vectors of interest at 3000 × *g* for 5 min at room temperature and discard the supernatant.
4. Resuspend the cell pellet in 3 mL infiltration buffer (*see* **Note 13**).
5. Harvest the cells by centrifugation at 3000 × *g* for 5 min at room temperature. Decant the supernatant and resuspend the cell pellet in 2 mL of infiltration buffer (*see* **Note 14**).

6. Add 200 μM acetosyringone and keep the cultures at room temperature for a minimum of 2 h and a maximum of 4 h in the dark (*see* **Note 15**).
7. Measure the OD_{600} of each agrobacterial culture transformed with interested constructs using 100 μL of the cell suspension diluted in 900 μL of water.
8. Dilute each agrobacterial suspension to OD_{600} of 1.0 in infiltration buffer supplemented with 200 μM acetosyringone.
9. Prepare working suspension by mixing different combinations of the *A. tumefaciens* carrying the appropriated BiFC constructs (Fig. 3) and the *A. tumefaciens* encoding a viral protein suppressor of RNA silencing (e.g., HC-Pro and P19) at a 1:1:1 ratio. Prepare also negative controls by combining each protein partner fused to nYFP or cYFP with the cognate empty vector.
10. Infiltrate agrobacterial suspensions on the abaxial side of the leaves of *N. benthamiana* or *N. tabacum* using a sterile 1 mL insulin syringe without the needle using gentle pressure (Fig. 4b) (*see* **Notes 16** and **17**).
11. Keep the plants incubated in a growth chamber under long-day conditions (16 h light/8 h dark) (*see* **Note 18**).

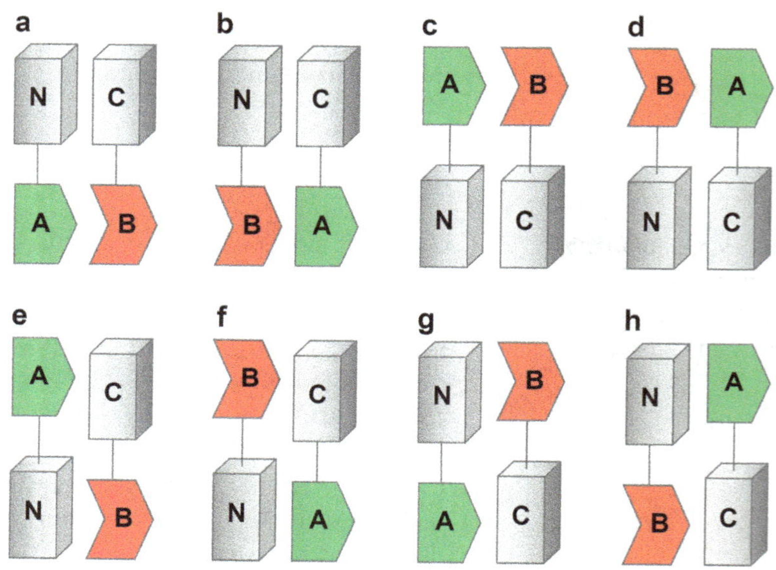

Fig. 3 Multiple combinations of fusion proteins should be tested for BiFC assay. The nonfluorescent fragments of YFP (N and C, which correspond to the N-terminal and C-terminal portion of YFP, respectively) can be fused in frame at the N- or C-terminal position of the proteins under study, resulting in eight possible distinct combinations (a through h)

Fig. 4 *Nicotiana benthamiana* agroinfiltration process. (**a**) Five-week-old *N. benthamiana* plant used for the agroinfiltration. Arrows indicate the leaves at the optimal developmental stage for infiltration. (**b**) Infiltration of the *Agrobacterium* suspension on the abaxial side of the selected leaves of *N. benthamiana*

3.4 Fluorescence Detection by a Confocal Microscope and Image Acquisition

Observe the fluorescence in the epidermal cell layer of the abaxial side expressing the fusion proteins 2–3 days post-infiltration (*see* **Note 19**) under confocal microscopy (Fig. 2) (*see* **Note 20**). YFP is a fluorescent protein with an optimal excitation wavelength in the range of 490–515 nm (excitation peak at 513 nm) and maximal emission intensity in the range of 520–560 nm (emission peak at 530 nm). Therefore, the YFP fluorescence might be visualized using an appropriate laser and filter set. In this protocol, the images shown in Fig. 2 were obtained using a Zeiss inverted LSM 510 META laser scanning microscope equipped with an argon laser as an excitation source and a 40× or 63× oil immersion lens. The fluorescence detection and image acquisition procedures are described below:

1. With a scalpel, make a 0.5–1.0 cm^2 cut in the infiltrated area leaf.
2. Transfer the specimen to a glass slide, add a drop of distilled water (mounting medium) onto it (*see* **Note 21**) and cover it with the coverslip (*see* **Note 22**).
3. Turn on only the laser(s) needed to excite the fluorophores in your sample. In the case of YFP, we use an argon laser (*see* **Note 23**). Set the output intensity to an appropriate level (40% is a good starting point).
4. Select single-track mode and choose a filter configuration for the particular fluorophores you are using. To obtain the images shown in Fig. 2, YFP was excited with a 514 nm wavelength of an argon laser (utilizable laser intensity to 20%). The emission was collected using a 521–553 nm filter.

5. Select the appropriate scan average from the available values 2, 4, 8, and 16 to obtain the final image (*see* **Note 24**). The images in Fig. 2 were obtained from 8 scans in average line mode.
6. Set the pinhole diameter (*see* **Note 25**).
7. Start scanning the sample. The preliminary image, whether too dim or too bright, usually needs to be optimized.
8. To optimize the image, go to the image window, and then select 'Palette' from the toolbar. Choose "ranger indicator" from the color palette list. The range indicator palette is used to optimize the gain and offset setting of images before scanning. Restart scanning the sample and set the Detector Gain up to no more than a small percentage of your stained areas have red pixels. Adjust the Amplifier Offset until the background is a mixture of blue and black pixels.
9. If necessary, the zoom and scan field rotation can be set. The images in Fig. 2 were captured with an objective of 40×. Those images showing protein–protein interaction were obtained with a scan zoom of 2.5.
10. Acquire and save the image.
11. Perform image processing using Zeiss LSM 510 software and Adobe Photoshop.

4 Notes

1. Different sets of vectors are available for N-terminal or C-terminal YFP fusions. They contain the N-terminal and C-terminal portion of YFP fused in frame at the N- or C-terminal position of the proteins under study. Because fusion can interfere with the interaction domains of proteins of interest, deciding whether the N-terminal or C-terminal YFP fusion is more suitable for the study is essential.
2. Antibiotics dissolved in ethanol do not need to be sterilized. Store solutions in containers protected from light.
3. Sterilize the solution by filtration through a disposable filter (0.22 μm pore size). Divide into 10 mL aliquots and store at 4 °C.
4. Acetosyringone is diluted in 99,5% dimethyl sulfoxide (DMSO), ethanol, or dimethyl formamide (DMF).
5. Prepare acetosyringone stocks in small batches, store them at −20 °C, and discard them after a month.
6. Carefully decant the supernatant and use an automatic pipette to remove any remaining drops of water.

7. This is best done by gentle swirling rather than pipetting or vortexing.
8. Ideally, the DNA to be electroporated should be resuspended in water or TE buffer (Tris–HCl pH 8.0, 10 mM EDTA pH 8.0) at a concentration of 1–10 ng/μL. The presence of ions in the electroporation cuvette increases the conductivity of the solution and causes the electrical current to arc or skip through the solution of cells and DNA. If arcing occurs in the presence of DNA but not in its absence, remove ions from the DNA preparation through dilution or purification by spin-column chromatography.
9. Tap the solution to ensure that the suspension of bacteria and DNA sits at the bottom of the cuvette.
10. Before placing the cuvette in the electroporation device, dry the condensation and moisture from outside the cuvette.
11. Always use fresh *Agrobacterium* cultures for all infiltrations.
12. To 100 mL of buffer, add 1 mL of 1 M magnesium chloride (100×) and 1 mL of 1 M MES (100×) in 98 mL of deionized water.
13. Infiltration buffer must be prepared fresh on the day of assay.
14. The cells can be resuspended in a smaller volume if they do not recover large quantities of cells.
15. It is recommended to keep the cells in the agroinfiltration buffer for 2–4 h to activate virulence gene expression.
16. Select individual leaves of *N. benthamiana* or *N. tabacum* with 4- to 6-week-old. Do not pick too old or very young leaves to infiltrate (Fig. 4a). Be careful not to damage the leaf. Dry any amount of suspension leaked during infiltration and mark the infiltrated area with a pen because the infiltrated area will not be distinguishable from the non-infiltrated area.
17. Perform a mock infiltration with infiltration buffer to assess background fluorescence or protein level.
18. It is recommended to keep the plants in a high-light environment after infiltration. This will optimize the expression of the constructs.
19. The transient expression tends to peak around 60–72 h post-infiltration and rapidly declines after that.
20. To complement the interpretation of the images obtained, it is suggested to analyze the expression level of the fusion proteins in the infiltrated leaf area employing western blotting to make sure that all the proteins were expressed in the experiment, even the control conditions.

21. If a large amount of water has been added and the coverslip has floated, or if the water has bled out to the edges of the slide, you can use the edge of the paper towel and place it next to the edge of the coverslip and just let the paper towel extract some of the excess water.
22. Avoid the formation of air bubbles because they will make the visualization of the specimen more difficult. Air bubbles have a different refractive index than water, making the bubble appear darker and thicker under the microscope lens. To minimize the formation of air bubbles, try placing the coverslip on edge and slowly lowering it onto the slide. Another alternative is to place the specimen in a vacuum chamber and pump it. However, the presence of air bubbles does not mean that you cannot fully observe the sample. If air bubbles do not obstruct or impair the observation of the sample, just work around the existing air bubbles.
23. To turn on the argon laser, click standby to warm it up. After the initial warm-up period, the on button will appear, allowing the laser to be fully activated.
24. Higher the number of scan averages selected in the mean average method, the better will be the image quality; however, the scanning time will be longer. When using the mean method, the image information is generated by adding up all scans, pixel by pixel, and then calculating the mean value. The averaging scan number necessary to obtain a good image depends on the signal-to-noise ratio of your sample image. Low-contrast images require a higher averaging scan number, but be aware that this could cause photobleaching. Most samples will require an average scan number of 4 or 8.
25. The smaller the pinhole, the thinner the optical slice will be, and the less light will be collected. A pinhole set to 1 Airy unit will provide optimal resolution. However, the pinhole must be adjusted according to the sample's fluorescence, using higher Airy values for darker samples.

Acknowledgments

The authors acknowledge the microscopy facility at Universidade Federal de Viçosa. This work was supported by the Council for Advanced Professional Training (CAPES), National Council for Scientific and Technological Development (CNPq), The Minas Gerais Research Funding Foundation (FAPEMIG- APQ-01357-22), and Financial Agency for Studies and Projects (FINEP).

References

1. Kodama Y, Hu CD (2012) Bimolecular fluorescence complementation (BiFC): a 5-year update and future perspectives. BioTechniques 53(5):285–298. https://doi.org/10.2144/000113943
2. Citovsky V, Lee LY, Vyas S, Glick E, Chen MH, Vainstein A, Gafni Y, Gelvin SB, Tzfira T (2006) Subcellular localization of interacting proteins by bimolecular fluorescence complementation in planta. J Mol Biol 362(5): 1120–1131. https://doi.org/10.1016/j.jmb.2006.08.017
3. Hyodo K, Okuno T (2020) Hijacking of host cellular components as proviral factors by plant-infecting viruses. Adv Virus Res 107: 37–86. https://doi.org/10.1016/bs.aivir.2020.04.002
4. Silva FDA, Raimundo GS, Fontes EPB (2022) Begomovirus–host protein-protein interactions in intracellular virus movement. In: Gaur RK, Sharma P, Czosnek H (eds) Geminivirus: Detection, diagnosis and management. Academic, pp 347–356. https://doi.org/10.1016/B978-0-323-90587-9.00023-7
5. Brustolini OJB, Machado JPB, Condori-Apfata JA, Coco D, Deguchi M, Loriato VAP, Pereira WA, Alfenas-Zerbini P, Zerbini FM, Inoue-Nagata AK, Santos AA, Chory J, Silva FF, Fontes EPB (2015) Sustained NIK-mediated antiviral signalling confers broad-spectrum tolerance to begomoviruses in cultivated plants. Plant Biotechnol J 13(9): 1300–1311. https://doi.org/10.1111/pbi.12349
6. Gouveia-Mageste BC, Martins LGC, Dal-Bianco M, Machado JPB, da Silva JCF, Kim AY, Yazaki J, dos Santos AA, Ecker JR, Fontes EPB (2021) A plant-specific syntaxin-6 protein contributes to the intracytoplasmic route for the begomovirus CabLCV. Plant Physiol 187(1):158–173. https://doi.org/10.1093/plphys/kiab252
7. Calil IP, Quadros IPS, Araújo TC, Duarte CEM, Gouveia-Mageste BC, Silva JCF, Brustolini OJB, Teixeira RM, Oliveira CN, Milagres RWMM, Martins GS, Chory J, Reis PAB, Machado JPB, Fontes EPB (2018) A WW domain-containing protein forms immune nuclear bodies against begomoviruses. Mol Plant 11(12):1449–1465. https://doi.org/10.1016/j.molp.2018.09.009
8. Li B, Ferreira MA, Huang M, Camargos LF, Yu X, Teixeira RM, Carpinetti PA, Mendes GC, Gouveia-Mageste BC, Liu C, Pontes CSL, Brustolini OJB, Martins LGC, Melo BP, Duarte CEM, Shan L, He P, Fontes EPB (2019) The receptor-like kinase NIK1 targets FLS2/BAK1 immune complex and inversely modulates antiviral and antibacterial immunity. Nat Commun 10(1):4996. https://doi.org/10.1038/s41467-019-12847-6
9. Stefano G, Renna L, Brandizzi F (2015) BiFC for protein-protein interactions and protein topology: discussing an integrative approach for an old technique. Methods Mol Biol 1242:173–182. https://doi.org/10.1007/978-1-4939-1902-4_15
10. Schütze K, Harter K, Chaban C (2009) Bimolecular fluorescence complementation (BiFC) to study protein-protein interactions in living plant cells. In: Pfannschmidt T (ed) Plant signal transduction. Methods in molecular biology, vol 479. Humana Press, Totowa, NJ. https://doi.org/10.1007/978-1-59745-289-2_12

Chapter 16

Co-immunoprecipitation Assays to Detect Protein–Protein Interactions

Mengling Huang, Xiao Yu, and Bo Li

Abstract

Proteins usually do not function as monomers but rather perform their functions by interacting with themselves or other proteins. Co-immunoprecipitation is an essential assay for detecting protein interactions in vivo. In this chapter, we describe how to use co-immunoprecipitation to detect protein interactions in Arabidopsis protoplasts, seedlings, and *Nicotiana benthamiana* leaves. When using co-immunoprecipitation assays to detect protein interactions, it is necessary to pay attention to the design of the experimental and control groups.

Key words Protein–protein interaction in vivo, Co-immunoprecipitation, Arabidopsis protoplasts, Arabidopsis seedlings, *Nicotiana benthamiana* leaves

1 Introduction

Co-immunoprecipitation (Co-IP) is a classical method to study protein–protein interaction based on the specificity of the interaction between antibodies and antigens [1]. It is an effective method for determining the interactions of multiple proteins in vivo. Here, we will mainly describe the specific experimental steps of Co-IP and considerations. In brief, Co-IP consists of the following steps, including sample preparation, sample lysis, target protein capture, protein complex purification, and result analysis by Western blot (Fig. 1). To explore protein interactions in plant cells, samples could be obtained from various materials, including Arabidopsis protoplasts [2], Arabidopsis seedlings [3], and *Nicotiana benthamiana* leaves [4].

Co-IP has been successfully and widely applied to detect protein interactions in various aspects of plant research. The specific interaction between the secreted peptides SMALL PHYTOCYTOKINES REGULATING DEFENSE AND WATER LOSS (SCREWs) and the cognate cell surface-localized receptor kinase

Elizabeth P. B. Fontes and Kristiina Mäkinen (eds.), *Plant-Virus Interactions*, Methods in Molecular Biology, vol. 2724, https://doi.org/10.1007/978-1-0716-3485-1_16,

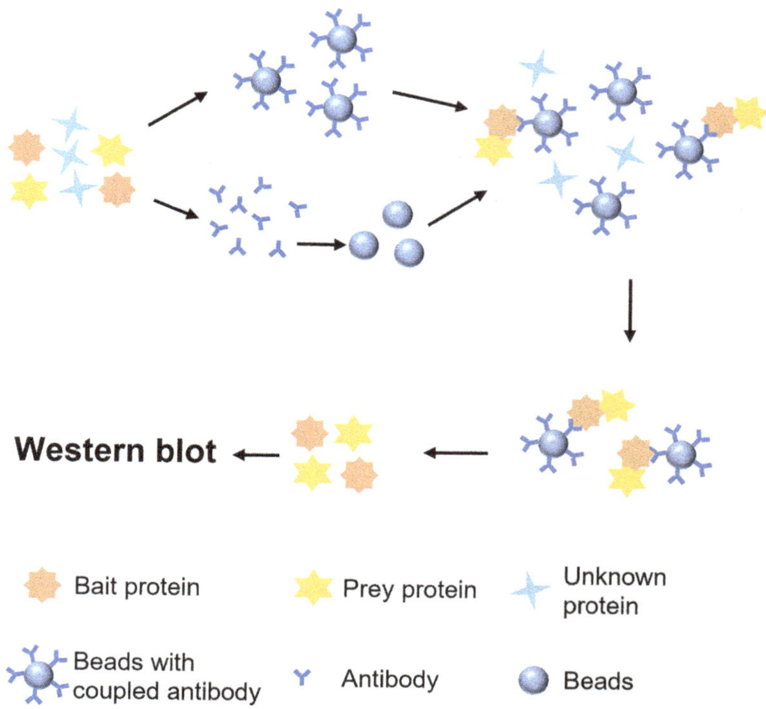

Fig. 1 Flowchart of the co-immunoprecipitation assay procedure. Cell lysate-containing antigen-fused protein samples are incubated with beads coupled with specific antibodies or antibody and affinity beads sequentially. Protein-bound beads are collected by centrifugation and washed by washing buffer to get rid of the unbound proteins. The purified protein complex is further analyzed by Western blot

PLANT SCREW UNRESPONSIVE RECEPTOR (NUT) was demonstrated by Co-IP [5]. The dynamic interaction among the components in the receptor complex was shown, such as the leucine-rich repeat receptor kinases flagellin-sensitive 2 (FLS2)/BRI1-associated receptor kinase 1 (BAK1)/*Botrytis*-induced kinase 1 (BIK1) interaction in a stimulus-dependent manner [6, 7]. The *Ralstonia solanacearum* type III effector (T3E) PopP2 and *Pseudomonas syringae* pv. *tomato* T3E AvrRps4 interact with the nucleotide-binding leucine-rich repeat (NB-LRR) receptor resistance (R) protein RRS1 (resistance to *Ralstonia solanacearum* 1) [8]. A recent study has shown that the *R. solanacearum* type III effector RipAB disrupts salicylate signaling by targeting Arabidopsis TGACG SEQUENCE-SPECIFIC BINDING PROTEIN (TGA) transcription factors, and the interaction between RipAB and TGA transcription factors was demonstrated by Co-IP [2]. Co-IP has also been used to demonstrate interactions between other virulence factors and targets; for example, the C4 protein from the geminivirus tomato yellow leaf curl virus (TYLCV) interacts with calcium-sensing receptor (CAS) in chloroplasts to suppress downstream defense responses [9]. Several independently evolved virulence factors from unrelated plant RNA viruses hijack and repress the jasmonic acid (JA) pathway by disturbing the mediator subunit (OsMED25) and basic helix–loop–helix (bHLH) transcription factors (OsMYC3) complex and cooperating with Jasmonate ZIM domain (JAZ) to synergistically attenuate JA signaling. Co-IP was used to analyze the interaction between viral proteins and host proteins [10]. Therefore, Co-IP is a commonly used and effective assay for detecting protein interactions.

2 Materials

2.1 Sample Preparation

1. Swing bucket centrifuge.
2. Centrifuge tubes.
3. 6-well culture plates.
4. The vectors used in this chapter: *HBT-35S:ASR3-HA*; *HBT-35S:TGA4-HA*; *HBT-35S:GFP*; *HBT-35S:RipAB-GFP*.
5. Half-strength Murashige and Skoog medium (½ MS): One liter contains 2.23 g MS salt, 0.5 g MES, and 5 g sucrose; adjust pH to 5.8.
6. Petri dishes (90 mm in diameter).
7. Liquid nitrogen.
8. *Agrobacterium* strain GV3101 or other suitable strain.
9. UV spectrophotometer.
10. *Agrobacterium* infiltration (see *Two-Hybrid Systems:* Methods and Protocols, Methods in Molecular Biology Chapter 19 for *Agrobacterium* infiltration) [12].
11. Hole punching machine (5 mm in diameter).

2.2 Cell Lysate and Input Sample Preparation

1. Co-IP buffer: 50 mM Tris–HCl (pH 7.5), 150 mM NaCl, 5 mM EDTA, 10% glycerol, 0.5% triton X100, 1 mM DTT, 2 mM NaF, 2 mM Na_3VO_3, 1% proteinase inhibitor cocktail (Roche). Prepare fresh and precool it to 4 °C before use.
2. 2 × SDS protein loading buffer: 25 mM Tris–HCl (pH 6.8), 1% sodium dodecyl sulfate, 5% glycerol, 0.01% bromophenol blue, 5% β-mercaptoethanol, and 4 mM DTT. Store at −20 °C.
3. Centrifuge tubes.
4. Vortex.
5. Refrigerated centrifuge.
6. Metal bath.
7. Handheld grinder.

2.3 Incubation with Suitable Beads and Antibody Capture

1. Select the proper beads. Specific antibody-coupled beads are available for commonly used tags, such as anti-HA magnetic beads (Bimake), ANTI-FLAG® M2 Affinity Gel (Sigma), MYC-Trap_A (Chromo Tek), and GFP-trap_A (Chromo Tek). If the antibody-coupled beads are not available, Pierce Protein A/G Agarose (Thermos Fisher) can be used, which specifically binds to the conserved region of an antibody. The affinity beads are divided into two types (*see* **Note 1**).
2. Swing bucket centrifuge.
3. Centrifuge tubes.
4. Co-IP washing buffer: 50 mM Tris–HCl (pH 7.5), 150 mM NaCl, 5 mM EDTA, 10% glycerol, and 0.1 ~ 1% triton X100. Store at room temperature and precool it to 4 °C before use.
5. Mixer.

2.4 Washing

1. Co-IP washing buffer.
2. Vacuum pump.
3. Swing bucket centrifuge.
4. Magnetic stand.

2.5 Elution

1. 2 × SDS protein loading buffer.
2. Metal bath.

2.6 Analyze the Samples by Western Blot

1. SDS-PAGE and Western blotting equipment.
2. Blotting PVDF membrane (Bio-Rad).
3. Substrate for detection, such as SuperSignal West Pico PLUS Chemiluminescent Substrate, SuperSignal West Femto Maximum Sensitivity Substrate (Thermos Fisher).
4. Antibodies: anti-HA-HRP (Roche), anti-GFP (Roche), anti-mouse HRP (Biosharp), anti-FLAG-HRP (Sigma), anti-MYC-HRP (Roche).
5. Image J program.

3 Methods

3.1 Preparing Sample

Co-IP assays are generally performed using plant materials from the following sources.

1. Arabidopsis protoplasts

Isolate Arabidopsis protoplasts (see *Plant Signal Transduction:* Methods and Protocols, Methods in Molecular Biology Chapter 8 for isolation protocol) [13]; transfect 500 μL protoplasts with 25 μL bait gene vector (1.6 μg/μL) and 25 μL prey gene vector

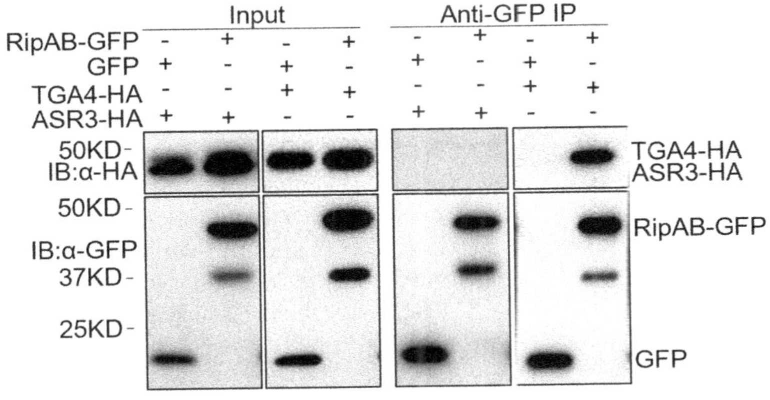

Fig. 2 Co-IP assays to examine the interactions between RipAB and TGA4 in vivo. RipAB-GFP was co-expressed with TGA4-HA in Arabidopsis protoplasts. Total protein extracts were immunoprecipitated with GFP-trap beads and immunoblotted using an anti-HA or anti-GFP antibody

(1.6 μg/μL) or control vector (1.6 μg/μL); incubate the transfected protoplasts in the 6-well culture plates under low light for 8–12 h, you can set different treatments if you need; harvest the protoplasts in a 1.5 mL centrifuge tube and freeze them in −80 °C.

For example, to detect the interaction between RipAB and TGA4, a vector can be constructed with the *TGA4* gene fused with an HA tag as the prey and the *RipAB* gene with an GFP tag as the bait, the experimental setup is as follows: negative control group: ASR3-HA (unrelated protein-HA) + empty vector-GFP, ASR3-HA + RipAB-GFP or TGA4-HA + empty vector-GFP; experimental group: TGA4-HA + RipAB-GFP (Fig. 2).

2. Arabidopsis seedlings

Seedlings are grown on ½ MS plates for 10 ~ 12 days and transferred to ddH_2O for overnight recovery (perform the proper treatment if you need); collect five seedlings in a 1.5-mL centrifuge tube, and twenty seedlings are harvested per sample.

Considering that the antibodies for endogenous proteins sometimes do not work efficiently, complementary transgenic plants carrying target proteins fused with proper tags can be constructed. For example, to verify the interaction between ENHANCED DISEASE SUSCEPTIBILITY (EDS1) and PBS1-LIKE 19 (PBL19) *in planta*, transgenic plants expressing *pEDS1:EDS1-FLAG/pPBL19:PBL19*^{C3A}*-GFP-HA* in the *eds1 pbl19* background was generated, and *pPBL19:PBL19*^{C3A}*-GFP-HA/pbl19* transgenic plants were used as a control [11]. Here, EDS1-FLAG was the prey protein, and PBL19-GFP-HA was the bait protein.

3. *N. benthamiana* leaves

Agrobacterium tumefaciens strain GV3101 contains the binary vector for the bait gene and prey gene, respectively; before infiltration, the bacterial suspension is adjusted to OD600 of 1.5; mix the bacterial cultures containing the bait gene vector and the prey gene vector at a 1:1 ratio and infiltrate into the leaves of three-week-old soil-grown *N. benthamiana*; after infiltration for 1 ~ 3 days, collect five leaf disks (5 mm in diameter) in a 1.5-mL centrifuge tube and around 25 leaf disks are harvested per sample.

The experimental setup refers to that in Arabidopsis protoplasts.

3.2 Cell Lysate and Input Sample Preparation

3.2.1 Samples from Arabidopsis Protoplasts

1. Add 200 μL Co-IP buffer in the samples and vortex to resuspend the protoplasts.
2. Take 20 μL supernatant into a new 1.5-mL centrifuge tube with 20 μL 2 × SDS protein loading buffer. Heat the samples in a metal bath at 95 °C for 5 min. These samples will be used for the input (*see* **Note 2**).
3. Add another 300 μL Co-IP buffer into the tubes prepared from step "1." Vortex vigorously and incubate on ice for 10 ~ 15 min (*see* **Note 3**). You can precool the centrifuge to 4 °C in this step.
4. Centrifuge at 4 °C at 12,000 rpm for 10 ~ 15 min (the pellet should be white), and collect the supernatant into a new centrifuge tube.

3.2.2 Samples from Arabidopsis Seedlings and N. benthamiana Leaves

1. Ground the seedlings or leaves with the motor driven homogenizing pestle in liquid nitrogen. Add 150 μL Co-IP buffer per centrifuge tube and keep on ice for 10 ~ 15 min (*see* **Note 4**).
2. Centrifuge the samples at 4 °C at 12,000 rpm for 10 ~ 15 min.
3. Combine the supernatant from the same sample into one tube and centrifuge again to remove the residue. Transfer the supernatant to a new tube.
4. Take 20 μL supernatant into a new 1.5-mL microfuge tube with 20 μL 2 × SDS protein loading buffer. Heat the samples in a metal bath at 95 °C for 5 min. These samples will be used for the input.

3.3 Incubation with Suitable Beads and Antibody Capture

1. Preclean can remove the host proteins that bind nonspecifically to the protein A/G agarose from the lysate and help reduce the background signal. To preclean, add 5 μL protein A/G agarose into the supernatant from steps "4" in 3.2.1 or steps "3" in 3.2.2, incubate at 4 °C with gentle shaking for 30 min, spin with the swing bucket centrifuge at 1000 rpm for 1 min, and take the supernatant to a new tube for subsequent steps.

If the Beads Are Coupled with Antibodies

2. Wash the beads twice with Co-IP washing buffer. Use 5 ~ 10 μL beads for each sample. Resuspend the beads after washing with 20 μL Co-IP buffer for each sample. Make sure beads are mixed evenly before being added to different tubes (*see* **Note 5**).
3. Add the beads to the supernatant from step "1" in 3.3, making sure to allocate an equal volume of beads to each sample.
4. Incubate at 4 °C with gentle shaking for 2 ~ 3 h (*see* **Note 6**).

If the Beads Are Not Coupled with Antibodies

5. Transfer the supernatant from step "1" in 3.3 into a new tube with proper antibody (usually 1 μL) and incubate at 4 °C with gentle shaking for 1 ~ 2 h.
6. Wash protein A/G agarose beads twice with Co-IP buffer, and use 5 μL beads per sample.
7. Add the beads into the tubes prepared from step "5" and incubate at 4 °C with gentle shaking for 1 ~ 2 h (*see* **Note 6**).

3.4 Washing

1. After step "4" or step "7" in 3.3, spin with the swing bucket centrifuge at 1000 rpm for 1 min (*see* **Note 7**). Discard supernatant with vacuum (*see* **Note 8**).
2. Wash the beads 3 ~ 5 times with Co-IP washing buffer (*see* **Note 9**). Try to eliminate the supernatant completely without touching the beads (*see* **Note 10**).

3.5 Elution

1. Add 30 μL 2 × SDS protein loading buffer, heat for 10 min at 95 °C, and then spin for 2 min. These will be the Co-IP samples.

3.6 Sample Analysis by Western Blot

1. Run input and Co-IP samples including all controls in SDS-PAGE gels for Western blot. 10 μL of input samples are used to detect the prey and bait proteins, respectively. A 5 μL Co-IP sample is used to detect the prey proteins, and a 15 μL Co-IP sample is used to detect the bait proteins.
2. Use appropriate antibodies to detect the prey and bait proteins to determine whether they interact with each other or to evaluate the strength of their interaction.
3. The band intensity can be calculated using Image J if you need to quantify the band intensities and compare the affinity among different conditions.

4 Notes

1. The comparison of the properties of agarose resin and magnetic beads: agarose resin—porous structure, strong binding ability, high background signal, and relatively inexpensive.

   Magnetic beads: solids, relatively weak binding ability, low background signal, and relatively expensive.
2. The input sample preparation is important because these samples are used to prove the presence of the target proteins that you want to test for interaction.
3. Ensure that the samples are sufficiently lysed so that the proteins are completely released in the buffer.
4. Arabidopsis seedlings and *N. benthamiana* leaves should be well grounded, and the supernatant should be free of plant tissues; if not, centrifuge again and carefully transfer the supernatant to a new tube.
5. Assign same amount of beads to each sample to ensure the homogeneity of the experiment.
6. Incubating the beads and proteins for too long at 4 °C will not enhance the binding; instead, you will get a much higher background or nonspecific precipitation.
7. If you use magnetic beads, you can use a magnetic stand to separate the beads from the liquid.
8. With the help of a vacuum, the supernatant is sucked out quickly and thoroughly.
9. The washing buffer should be kept on ice to ensure that the buffer is used in an ice-cold state.
10. Try to remove the supernatant completely to avoid the nonspecific binding of proteins.

Acknowledgements

This work was supported by National Natural Science Foundation of China (Grant no. 31970125) and National Key R&D Program of China (Grant no. 2022YFA1304400).

References

1. Lin JS, Lai EM (2017) Protein-protein interactions: co-immunoprecipitation. Methods Mol Biol 1615:211–219
2. Qi P, Huang M, Hu X, Zhang Y, Wang Y, Li P, Chen S, Zhang D, Cao S, Zhu W et al (2022) A Ralstonia solanacearum effector targets TGA transcription factors to subvert salicylic acid signaling. Plant Cell 34:1666–1683
3. Li B, Jiang S, Yu X, Cheng C, Chen S, Cheng Y, Yuan JS, Jiang D, He P, Shan L (2015) Phosphorylation of trihelix transcriptional repressor ASR3 by MAP KINASE4

negatively regulates Arabidopsis immunity. Plant Cell 27:839–856

4. Wang Y, Tang M, Zhang Y, Huang M, Wei L, Lin Y, Xie J, Cheng J, Fu Y, Jiang D et al (2023) Coordinated regulation of plant defense and autoimmunity by paired trihelix transcription factors ASR3/AITF1 in Arabidopsis. New Phytol 237:914–929
5. Liu Z, Hou S, Rodrigues O, Wang P, Luo D, Munemasa S, Lei J, Liu J, Ortiz-Morea FA, Wang X et al (2022) Phytocytokine signalling reopens stomata in plant immunity and water loss. Nature 605:332–339
6. Chinchilla D, Zipfel C, Robatzek S, Kemmerling B, Nurnberger T, Jones JD, Felix G, Boller T (2007) A flagellin-induced complex of the receptor FLS2 and BAK1 initiates plant defence. Nature 448:497–500
7. Zhang J, Li W, Xiang T, Liu Z, Laluk K, Ding X, Zou Y, Gao M, Zhang X, Chen S et al (2010) Receptor-like cytoplasmic kinases integrate signaling from multiple plant immune receptors and are targeted by a *Pseudomonas syringae* effector. Cell Host Microbe 7:290–301
8. Sarris PF, Duxbury Z, Huh SU, Ma Y, Segonzac C, Sklenar J, Derbyshire P, Cevik V, Rallapalli G, Saucet SB et al (2015) A plant immune receptor detects pathogen effectors that target WRKY transcription factors. Cell 161:1089–1100
9. Medina-Puche L, Tan H, Dogra V, Wu M, Rosas-Diaz T, Wang L, Ding X, Zhang D, Fu X, Kim C, Lozano-Duran R (2020) A defense pathway linking plasma membrane and chloroplasts and co-opted by pathogens. Cell 182:1109–1124 e1125
10. Li L, Zhang H, Chen C, Huang H, Tan X, Wei Z, Li J, Yan F, Zhang C, Chen J, Sun Z (2021) A class of independently evolved transcriptional repressors in plant RNA viruses facilitates viral infection and vector feeding. Proc Natl Acad Sci U S A 118:118
11. Li Y, Xue J, Wang FZ, Huang X, Gong BQ, Tao Y, Shen W, Tao K, Yao N, Xiao S et al (2022) Plasma membrane-nucleo-cytoplasmic coordination of a receptor-like cytoplasmic kinase promotes EDS1-dependent plant immunity. Nat Plants 8:802–816
12. Munoz A, Castellano MM (2018) Coimmunoprecipitation of interacting proteins in plants. Methods Mol Biol 1794:279–287
13. Schapire AL, Lois LM (2016) A simplified and rapid method for the isolation and transfection of arabidopsis leaf mesophyll protoplasts for large-scale applications. Methods Mol Biol 1363:79–88

Chapter 17

Split-Luciferase Complementation Imaging Assay in Virus–Plant Interactions

Huang Tan, Xi Zhang, and Rosa Lozano-Duran

Abstract

Protein–protein interactions constitute the interface between a virus and the cell it infects and are crucial determinants of the outcome of the viral infection. Multiple techniques have been developed to study how viral and host proteins interact in plants; among them, the split-luciferase complementation imaging assay stands out due to its capacity to detect protein–protein interactions in vivo, in the context of the infection, if desired, in an easy, fast, quantitative, and inexpensive manner. In this chapter, we use the interaction between the V2 protein from the geminivirus tomato yellow leaf curl virus (TYLCV) and *Nicotiana benthamiana* Argonaute 4 (AGO4) as an example to present how to perform this simple yet powerful assay using transient *Agrobacterium tumefaciens*-mediated transformation of *N. benthamiana* leaves to test the protein–protein interactions of choice.

Key words Viral protein, Protein–protein interaction, Split-luciferase complementation assay, In vivo, *Nicotiana benthamiana*

1 Introduction

Viruses are intracellular parasites that require the host molecular machinery to accomplish their own vital functions. The interaction between virus-encoded and cellular proteins constitutes the interface between the pathogen and host. Therefore, their detection and analysis are essential to fully understand how the infection takes place.

In plants, multiple techniques have been used to test virus–host protein–protein interactions; among them, the split-luciferase complementation assay [1] stands out since it offers several advantages over other alternatives. To begin with, this approach can use transient *Agrobacterium tumefaciens*-mediated transformation of *Nicotiana benthamiana*, a popular model species in plant virology, and is fast, simple, and inexpensive. In addition, it is quick, sensitive, and quantitative and likely detects direct interactions. Finally,

Elizabeth P. B. Fontes and Kristiina Mäkinen (eds.), *Plant-Virus Interactions*, Methods in Molecular Biology, vol. 2724, https://doi.org/10.1007/978-1-0716-3485-1_17,

this technique is applied in the cellular context, where the proteins of interest are in their natural subcellular localization, and the experiments can be performed both in infected and uninfected cells.

In this assay, the proteins of interest are fused to one half of the luciferase enzyme each (N- or C-terminal fragments, NLuc or CLuc) and co-expressed in plant cells—in the proposed protocol here, following the transient transformation of *N. benthamiana* leaves mediated by *A. tumefaciens*. The luciferase protein is only reconstituted when the two fusion proteins interact, driven by the proteins of interest, bringing the NLuc and CLuc fragments into close spatial proximity and hence restoring enzymatic activity. When the substrate luciferin is provided, this activity results in light emission, which is detectable by a luminometer or a charge-coupled device (CCD) imaging system dynamically and quantitatively.

In the first description of the split-luciferase complementation imaging protocol, the translational fusion of the proteins of interest to NLuc or CLuc was accomplished through traditional molecular cloning following restriction and ligation [1]; in order to simplify the cloning step, Gateway-compatible versions of the vectors were later generated [2, 3]. This approach has been successfully used to identify interactions between plant and viral proteins or among viral proteins in plant cells [e.g., 4–6].

In this chapter, we use the interaction between the V2 protein from the geminivirus tomato yellow leaf curl virus (TYLCV) and the Argonaute 4 (AGO4) protein from its experimental host *N. benthamiana* [4] as a case example of how to perform this simple yet powerful assay to uncover protein–protein interactions in plant virology.

2 Materials

2.1 General Equipment

1. Standard laboratory plasticware (Falcon tubes, microcentrifuge tubes, Petri dishes, pipette tips) and micropipettes.
2. Laminar flow hood. Alternatively, an alcohol burner can be used to maintain sterility for bacterial work.
3. Temperature-adjustable Petri dish incubators (37 °C, 28 °C).
4. Temperature-adjustable shakers for liquid cultures (37 °C and 28 °C with speed at 200 rpm).
5. Centrifuge.
6. Electronic balance.

2.2 Plant Cultivation

1. Plant material: *N. benthamiana*.
2. Plastic growth pots.
3. Plastic trays.
4. Transparent plastic lids or plastic film.
5. Soil mixture: 3 parts of potting substrate and 1 part of vermiculite.

2.3 Cloning

1. Solid (agar plate) and liquid Luria Broth (LB) preparation: 10 g NaCl, 10 g tryptone, and 5 g yeast extract in 900 mL distilled water (dH_2O). The final volume of 1000 mL is adjusted using dH_2O until the components are dissolved. For the solid LB, 15 g of agar is added. Autoclave at 121 °C for 15 min and cool to 55 °C before adding the appropriate antibiotics. Pour plates directly after.
2. Antibiotics preparation: 50 mg/mL kanamycin in dH_2O; 50 mg/mL rifampicin in dimethylsulfoxide (DMSO); 50 mg/mL gentamycin in dH_2O. The antibiotic solutions are sterilized by filtration through 0.22 μm membrane filters, aliquoted, and stored at −20 °C until use.
3. *Escherichia coli* strain DH5α (for maintenance and amplification of general vectors and plasmids in cloning procedures) and DB3.1 (for maintenance and amplification of vectors containing the *ccdB* gene) and corresponding competent cells for plasmids transformation.
4. *A. tumefaciens* strain GV3101 (containing pMP90, Rif^R $Gent^R$) and corresponding competent cells. The GV3010 strain is widely used for efficient transient expression in *N. benthamiana* leaves.
5. Gateway entry vector (e.g., pDONR™ 207, Thermo Scientific).
6. Gateway-compatible split-luciferase binary vectors for *in planta* expression (pGWB-NLuc (Addgene: 174050) and pGWB-CLuc (Addgene: 174051)) are used in this protocol ([2, 3, 7]; see **Notes 1** and **2**).
7. Primer design for BP cloning (forward primer: 5′-GGGGA CAAGTTTGTACAAAAAAGCAGGCTTC + gene-specific sequence-3′; reverse primer: 5′-GGGGACCACTTTGTACAA GAAAGCTGGGTC + gene-specific sequence (with and without stop codon)-3′) (*see* **Note 3**).
8. Gateway™ BP Clonase™ II Enzyme mix (Thermo Scientific).
9. Gateway™ LR Clonase™ II Enzyme mix (Thermo Scientific).
10. Proofreading DNA polymerase.

11. Standard gel extraction kit.
12. Standard plasmid miniprep kit. Alternatively, the alkaline lysis protocol for plasmid extraction can be used [8].
13. Thermocycler.
14. DNA gel electrophoresis equipment.
15. Temperature-adjustable water bath (42 °C, 37 °C).

2.4 Agrobacterium-Mediated Transient Transformation

1. $MgCl_2$ stock solution (1 M): 20.33 g $MgCl_2{\cdot}6H_2O$ in 80 mL dH_2O. Adjust the final volume to 100 mL with dH_2O. The stock solution is sterilized by autoclaving (121 °C for 15 min) and stored at room temperature.
2. 2-(N-morpholino)ethanesulfonic acid (MES)-KOH stock solution (0.5 M, pH 5.6): 48.81 g MES free acid in 400 mL dH_2O. The pH is adjusted to 5.6 using 1 M KOH. Add dH_2O to the final volume of 500 mL. The stock solution is sterilized by filtration through 0.22 μm membrane filters and kept at 4 °C for long-term storage (*see* **Note 4**).
3. Acetosyringone (AS) stock solution (150 mM): 0.2943 g AS powder in 10 mL DMSO. The stock solution is sterilized by filtration through 0.22 μm membrane filters, aliquoted, and stored at −20 °C until use (*see* **Note 5**).
4. Infiltration buffer: 10 mM $MgCl_2$, 10 mM MES-KOH, and 150 μM AS. Infiltration buffer should be prepared each time freshly.
5. pH meter.
6. Vortexer.
7. Spectrophotometer.
8. Plastic spectrophotometer cuvettes (1 mL).
9. 1-mL syringe (needleless).
10. Paper towel.

2.5 Luciferase Assay

1. D-luciferin stock solution (10 mM, 10X): 16 mg sodium D-Luciferin in 5 mL dH_2O. Aliquot 500 μL into sterilized 1.5 mL tubes. Store the solution at −20 °C or preferably at −80 °C (*see* **Note 6**).
2. D-luciferin working solution (1 mM): Quickly thaw the stock solution of D-luciferin and dilute 1:10 in dH_2O. D-luciferin working solution should be prepared each time freshly.
3. 1-mL syringe (needleless).
4. Paper towel.
5. CCD imaging system (e.g., NightShade LB985) (*see* **Note 7**).

3 Methods

3.1 Plant Cultivation

1. Sow *N. benthamiana* seeds in a 3:1 mixture of potting substrate and vermiculite. Cover the pots with plastic film, and then put them in a controlled growth room or growth chamber in long-day conditions (16 h light and 8 h dark).
2. Check regularly and water the plants if needed to keep the soil humid.
3. 10–14 days after sowing, transfer seedlings into individual plastic pots and cover them again with transparent plastic lids or plastic film for 2–3 days to minimize stress caused by transplanting (*see* **Note 8**).
4. Water plants regularly and add fertilizer if needed. Four-week-old plants are used for the split-luciferase complementation assay (*see* **Note 9**).

3.2 Molecular Cloning

In this protocol, the Gateway-compatible binary vectors (pGWB-NLuc and pGWB-CLuc, [2, 3]) are used (*see* **Notes 1** and **2**). The Gateway cloning procedure is outlined below.

1. PCR-amplify the coding sequence of genes of interest (GOI) using gene-specific primers (see Subheading 2.3).
2. Run the PCR products in a 1% agarose gel by electrophoresis, cut the bands with the expected size, and purify them using a gel extraction kit according to the manufacturer's instructions.
3. Clone the purified gene fragments into a Gateway entry vector (e.g., pDONR™ 207, Thermo Scientific) via a BP reaction following the manufacturer's instructions.
4. Transform each BP reaction into *E. coli* (e.g., DH5α strain) competent cells by transformation following your preferred protocol. Plate them on selective LB plates supplemented with appropriate antibiotics and incubate them in a 37 °C oven overnight.
5. Check individual colonies by PCR using specific primers. Grow a confirmed colony in 5 mL LB liquid medium containing appropriate antibiotics and incubate the culture in a 37 °C shaker (200 rpm) overnight.
6. Purify plasmid DNA using a miniprep kit following the manufacturer's instructions or the classic homemade alkaline lysis protocol [8]. Confirm the cloned sequence by DNA sequencing.
7. Perform an LR reaction to subclone the GOI into the split-luciferase Gateway-compatible binary vectors (Fig. 1a, b, *see* **Note 2**).

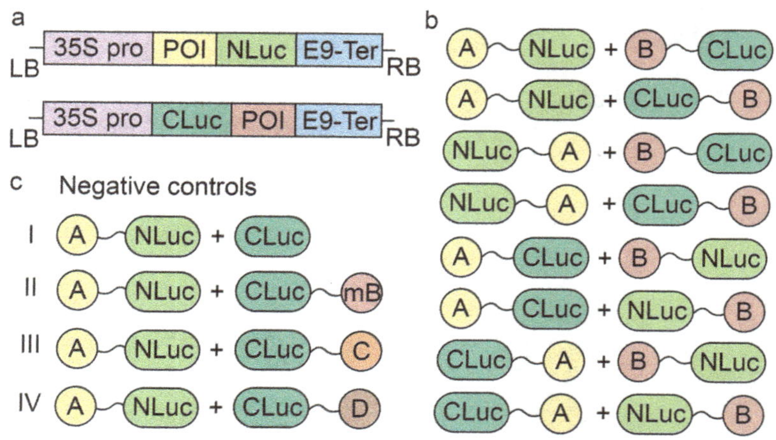

Fig. 1 Schematic representation of fusion orientations and negative controls used in split-luciferase complementation assay. (**a**) Cassettes to express the fusion proteins of interest (POI) in Gateway-compatible vectors: pGWB-NLuc and pGWB-CLuc. NLuc: N-terminal part of luciferase; CLuc: C-terminal part of luciferase; 35S pro: 35S promoter, which is a constitutive promoter derived from cauliflower mosaic virus (CaMV); E9-Ter: E9-terminator, containing terminator and polyadenylation signal from the pea rbcS-E9 gene; LB: left border; RB: right border. (**b**) Two fragments of luciferase (NLuc and CLuc) can be fused to either the N- or C-terminus of the POI. Therefore, eight pairwise combinations can be tested for each potential protein–protein interaction. (**c**) A negative control must be included in the assay. Four types of negative controls are listed here: replacing one POI (e.g., protein B) by (I) an empty vector expressing luciferase fragment only (CLuc); (II) a mutated version of the protein B (mB); (III) a protein that is known not to interact (protein C); and (IV) an unrelated protein with similar subcellular localization (protein D)

8. Proceed as indicated in **steps 4, 5**, and **6**. After this, constructs to express the fusion proteins for split-luciferase complementation imaging analysis are successfully generated.

3.3 Agrobacterium-Mediated Transient Transformation

1. Transform the plasmids generated in Subheading 3.2 into *A. tumefaciens* (e.g., GV3101 strain) competent cells by transformation following your preferred protocol. Select the colonies containing the corresponding binary vectors by colony PCR using specific primers.
2. Grow a single positive colony in 5 mL LB liquid medium containing appropriate antibiotics and incubate the culture at a 28 °C shaker (200 rpm) for 16–24 h until the OD_{600} = 0.5–1.5 (*see* **Note 10**). We recommend making a glycerol stock (in 25% glycerol) for each *A. tumefaciens* clone and storing it at −80 °C. Before using, streak out *A. tumefaciens* clones harboring the desired binary vectors on LB solid plates supplemented with appropriate antibiotics and incubate them in a 28 °C oven for 2 days.

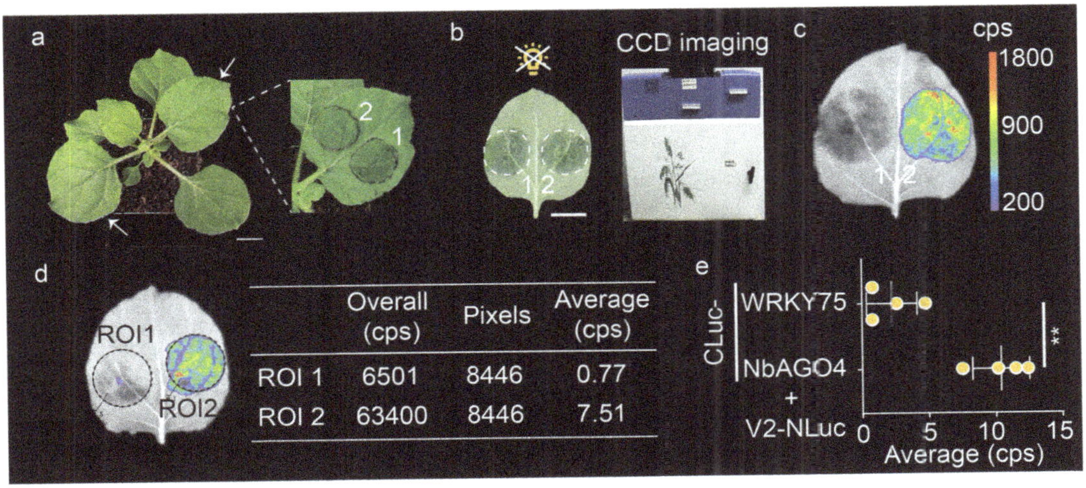

Fig. 2 Detection of a host–virus protein–protein interaction by split-luciferase imaging assay. A case study: the V2 protein encoded by tomato yellow leaf curl virus (TYLCV) interacts with the Argonaute 4 (AGO4) protein from *N. benthamiana*. (**a**) The third or fourth youngest leaves from 4-week-old *N. benthamiana* plants, indicated by white arrows, are selected for *Agrobacterium*-mediated transient transformation. "1," as a negative control, refers to the co-expression of V2-NLuc and SlWRKY75-CLuc. "2," as an experimental sample, refers to the co-expression of V2-NLuc and NbAGO4-CLuc. Experimental samples and negative controls should be infiltrated in the same leaf. (**b**) Infiltrate luciferin into the *Agrobacterium*-inoculated zone at 36–48 hours post-infiltration (hpi). Following a 5-min dark treatment, capture the luminescence signal in a CCD imaging apparatus (e.g., NightShade LB985). (**c**) Representative images of the luminescence signal. The color bar indicates the detected counts per second (cps) of luminescence emission in leaves. (**d**) Define the regions of interest (ROIs) and export the overall cps and the number of pixels from the NightShade LB985 controlling software Indigo™. The average value (cps) of the relative luciferase activity = overall cps/number of pixels. (**e**) Representative graph showing the average values (cps) of relative luciferase activity of ROIs in 1 and 2 (see (a)). Error bars represent the standard deviation from four biological replicates. Asterisks indicate a statistically significant difference between the experiment sample and the negative control according to a two-tailed t-test. **: $p < 0.01$

3. Harvest the *A. tumefaciens* cells by centrifugation at 4000 × *g* for 10 min and discard the supernatant completely.
4. Resuspend the pellets in the infiltration buffer (see Subsection 2.4) and measure the optical density (OD_{600}) using a spectrophotometer.
5. Adjust the final concentration of bacteria to $OD_{600} = 0.1–0.5$ (*see* **Note 11**).
6. Incubate the infiltration solution in the dark and at room temperature for 2–4 h (*see* **Note 12**).
7. Before co-infiltration, mix equal volumes (1:1) of two *A. tumefaciens* clones carrying binary vectors to express the NLuc- and CLuc-fused protein, respectively. Both positive and negative controls are required (Fig. 1c, see **Notes 13–15**).

8. Infiltrate fully expanded leaves (usually the third or fourth youngest) from 4-week-old *N. benthamiana* plants with a 1-mL needleless syringe (*see* **Notes 16** and **17**). Infiltration is performed by pressing the syringe on the abaxial side of the leaf and applying pressure with a fingertip on the other side. Infiltrate the experimental groups and negative controls side by side in the same leaf to avoid leaf-to-leaf variation (Fig. 2a, *see* **Note 18**).
9. Remove the excess infiltration solution from the leaf surface using tissue paper. Label the infiltration area using a permanent marker pen.
10. Put the plants back into the growth room or growth chamber. The luciferase activity will be measured at 36–48 h post-inoculation (hpi) (*see* **Note 19**).

3.4 Luminescence Detection by a CCD Imaging System and Data Analysis

1. At 36–48 hpi, infiltrate 1 mM D-luciferin solution (diluted in H_2O) into the marked infiltration zone.
2. Cut off the infiltrated leaves and place them in the dark for 5–10 min before imaging to remove the background caused by the chlorophyll luminescence.
3. Put the leaves upside down inside the CCD imaging apparatus (e.g., NightShade LB985). Use the NightShade LB985 controlling software Indigo™ to set the acquisition parameters (e.g., exposure time) and to capture the luminescence signal (Fig. 2b, c, *See* **Note 20**).
4. Define the regions of interest (ROI) for all samples and controls. Export the values of the relative luciferase activity (counts per second (cps)) and the size of ROIs (number of pixels). Calculate the average value (cps) of the relative luciferase activity (total luminescence value/the number of pixels) in the ROI (*see* **Note 7**).
5. Use the average values (cps) from different leaves for data representation and statistical analyses (Fig. 2d, e).

4 Notes

1. Alternatively, the original vectors pCAMBIA1300-NLuc/CLuc [1] can be used to generate constructs that express the fusion proteins of interest (POIs) via traditional cloning. In addition, if Golden Gate or Green Gate cloning is preferred, the modules containing NLuc or CLuc can be generated.
2. In pGWB-NLuc, the NLuc is fused to the C-terminal part of the POI, while in pGWB-CLuc, the CLuc is fused to the N-terminal of the POI. Fusion orientation might have a significant impact on the results of the split-luciferase assay. Ideally,

without prior knowledge, we recommend that both constructs be generated for each POI so that all possible pairwise combinations can be tested (Fig. 1b).

3. The genes of interest should be cloned into pGWB-NLuc and pGWB-CLuc with and without stop codon, respectively.
4. Autoclaving is not recommended for MES buffer.
5. Acetosyringone is a natural phenolic product and induces the expression of virulence genes of *Agrobacterium*, which are required for T-DNA transfer. It is known to improve transformation efficiency in both dicot and monocot plants.
6. The sodium or potassium salt of D-Luciferin is more convenient for experiments than the free acid form since it readily dissolves in water. The D-Luciferin solution should be protected from light. To avoid repeated freezing and thawing, we recommend aliquoting the D-Luciferin solution in small volumes. In our hands, 10 mM D-Luciferin stocks are still suitable for use after 6 months of storage at −20 °C.
7. The luciferase activity can also be quantitatively analyzed using a luminometer.
8. We use 7 × 5 cm round pots, placing 1 *N. benthamiana* plant per pot.
9. To ensure reliable and reproducible results, only healthy *N. benthamiana* plants, which are critical for the efficient transient expression of the POIs, should be used. Five-week-old plants can also be used for transient expression assays. Using plants older than 5-week-old should be avoided.
10. Typically, 5 mL of overnight cultured *A. tumefaciens* cells should be enough to infiltrate at least three fully expanded leaves. Freshly prepared *Agrobacterium* culture/plates should be used each time.
11. We use a final OD_{600} = 0.5 in most cases. However, we recommend optimizing the final OD_{600} experimentally to ensure that accumulation of the POI is appropriate, but there is no tissue damage caused by high bacterial load.
12. Prior to infiltration, incubation of the *Agrobacterium* suspensions in the dark is needed to fully induce the *Agrobacterium vir* genes by acetosyringone, which is essential for efficient transformation.
13. To indicate that the assay works, infiltrating a positive control is essential. We usually infiltrate the negative controls and the samples to be tested in one leaf and the positive control in another leaf.

14. To exclude non-specific and artifactual signals, appropriate negative controls should be incorporated into the split-luciferase assay. In many examples available in the literature, empty vectors (either expressing NLuc or CLuc only) serve as negative interaction control. However, for examining the interaction between protein A and B, the ideal negative control would entail replacing protein B with a mutated version known not to interact with protein A or an unrelated protein with the same subcellular localization as protein A.
15. The interaction between plant and viral proteins can also be examined in the context of the viral infection. To this end, an infectious viral clone of interest should be co-infiltrated with clones harboring binary vectors expressing both NLuc- and CLuc-fused proteins.
16. Choose healthy leaves of similar size for infiltration. If the leaves are difficult to infiltrate, pricking gently on the abaxial side of the leaf with a needle will help.
17. To avoid cross-contamination, wipe the wet surface of the infiltration zone with tissue paper and clean your gloves by using 75% ethanol before moving to the next inoculum.
18. We infiltrate at least four independent leaves from different plants for each biological replicate.
19. The POI accumulation normally reaches a high level at 36–48 h post-infiltration (hpi). However, it is advisable to optimize the time for your POI experimentally.
20. We strongly recommend collecting the leaf samples from the infiltrated area for protein extraction and checking the accumulation of all tested proteins and controls by western blotting. In our laboratory, we use anti-luciferase antibodies (rabbit polyclonal anti-luciferase, Sigma), which recognize both NLuc and CLuc.

Acknowledgments

The authors would like to thank Liping Wang for critically reading this chapter.

References

1. Chen HM, Zou Y, Shang YL et al (2008) Firefly luciferase complementation imaging assay for protein–protein interactions in plants. Plant Physiol 146(2):323–324
2. Wang YZ, Li YS, Rosas–Diaz T et al (2019) The IMMUNE–ASSOCIATED NUCLEOTIDE–BINDING 9 protein is a regulator of basal immunity in *Arabidopsis thaliana*. Mol Plant-Microbe Interact 32(1): 65–75
3. Yu G, Xian L, Xue H et al (2020) A bacterial effector protein prevents MAPK–mediated phosphorylation of SGT1 to suppress plant immunity. PLoS Pathog 16(9):e1008933

4. Wang LP, Ding Y. He L et al (2020) A virus–encoded protein suppresses methylation of the viral genome through its interaction with AGO4 in the Cajal body. elife 9:e55542
5. Wang LP, Tan H, Medina–Puche L et al (2022) Combinatorial interactions between viral proteins expand the potential functional landscape of the tomato yellow leaf curl virus proteome. PLoS Pathog 18(10): e1010909
6. Wang YJ, Wu YY, Gong Q et al (2019) Geminiviral V2 protein suppresses transcriptional gene silencing through interaction with AGO4. J Virol 93(6):e01675–e01618
7. Wang LP, Yu G, Macho A et al (2021) Split–luciferase complementation imaging assay to study protein–protein interactions in *Nicotiana benthamiana*. Bio Proto 11(23):e4237
8. Birnboim HC, Doly J (1979) A rapid alkaline extraction procedure for screening recombinant plasmid DNA. Nucleic Acids Res 7(6): 1513–1523

Chapter 18

The Split-Luciferase Complementation Assay to Detect and Quantify Protein–Protein Interactions in Planta

Eduardo Bassi Simoni and Célio Cabral Oliveira

Abstract

Protein–protein interactions play a critical role in plant viral infection and defense responses against pathogens. This protocol provides a detailed and reliable methodology for investigating protein–protein interactions using a luciferase-based complementation assay that includes easy luminescence-based normalization within a single plate. The protocol includes step-by-step procedures, reagent lists, and considerations for data interpretation, ensuring robust and reproducible results. By following this protocol, researchers can advance on understanding of the crucial role of protein–protein interactions in plant viral infection and defense responses to other pathogen attacks.

Key words Luciferase, HiBiT, protein–protein interactions, complementation

1 Introduction

Protein–protein interactions are crucial in plant viral infection, as well as in the plant's defense response to viral attacks. Plants have evolved pattern recognition receptors (PRRs) that recognize pathogen-associated molecular patterns (PAMPs), triggering defense responses. Plant immunity can also be inhibited by viral proteins. For example, the *Begomovirus* nuclear shuttle protein (NSP) interacts with the *Arabidopsis* receptor-like kinase (RLK) NSP-interacting kinase 1 (NIK1) to suppress the activation of antiviral responses [1, 2].

Various methodologies are used to detect and measure interactions between two or more proteins. Co-immunoprecipitation (Co-IP), yeast two-hybrid (Y2H), and bimolecular fluorescence complementation (BiFC) are examples of methodologies that use antibodies [3], domains of interaction with reporter genes [4], and reconstitution of fluorescent proteins [5], respectively, to detect protein interactions. However, these methods have some limitations in specific contexts or for certain types of data, such as the

Elizabeth P. B. Fontes and Kristiina Mäkinen (eds.), *Plant-Virus Interactions*, Methods in Molecular Biology, vol. 2724, https://doi.org/10.1007/978-1-0716-3485-1_18,

need for antibodies with high specificity, potential false positives, differences in study organisms, microscopy limitations, and the inability to quantify interactions. Therefore, the split-luciferase assay has emerged as an attractive alternative methodology to overcome some these limitations.

The split-luciferase assay is a protein-protein interaction assay that relies on the reconstitution of a luciferase enzyme. The proteins of interest are fused to the N-terminal (nLUC) and C-terminal (cLUC) fragments of a luciferase enzyme. If these proteins interact, the nLUC and cLUC fragments are brought into proximity, allowing them to reconstitute a functional luciferase enzyme that emits light when luciferin, the substrate, is added. For studies in yeast, the split-luciferase assay can be performed in either the cytoplasm or nucleus of the yeast cell [6]. For studies in mammalian cells, the split-luciferase assay can be performed in various cell lines, including HEK293, COS-7, and HeLa cells [7]. The split-luciferase assay in plants is commonly conducted in *Nicotiana benthamiana* (*N. benthamiana*) leaves due to their high expression capacity and ease of infiltration [8–12].

In addition to being highly sensitive, the assay is also quantitative and can be conducted in real time using protein quantification techniques. For example, the HiBiT tag is an 11 amino acid peptide sequence that can be fused to a protein of interest, allowing detection and quantification of the protein. In split-luciferase assays using floated leaf disks, the HiBiT tag can be used as an internal control to normalize the variability in protein amounts and assay performance [13, 14]. However, certain points need to be carefully determined prior to the experiment, such as the selection of luciferase fragments and their proper positioning. When working with membrane proteins, the effectiveness of the split-luciferase assay may be reduced due to the limited accessibility of the split fragments across the membrane barrier. To address this issue, one fragment can be fused to the cytoplasmic domain of a membrane protein, while the other fragment is fused to a soluble protein partner. This design enables the split-luciferase fragments to come into proximity within the membrane, minimizing interference from membrane lipids.

Several studies have demonstrated the utility of split-luciferase assays in detecting protein–protein interactions in plants. For example, the interaction between glycosyltransferases in the Golgi apparatus was confirmed using split-luciferase assays in *Nicotiana* leaves [15]. Wang et al. (2020) used split-luciferase assays in *Arabidopsis* protoplasts to elucidate the heterodimerization of chitin receptor homologs [16]. In this protocol (Fig. 1), the strong interaction between the *Arabidopsis thaliana* heterotrimeric G-protein α subunit (AtGPA1) and its regulator, regulator of G signaling 1 (AtRGS1), as well as the weak-interacting mutant AtRGS1^{E320K} (Fig. 2) [17]. These interactions were already confirmed through

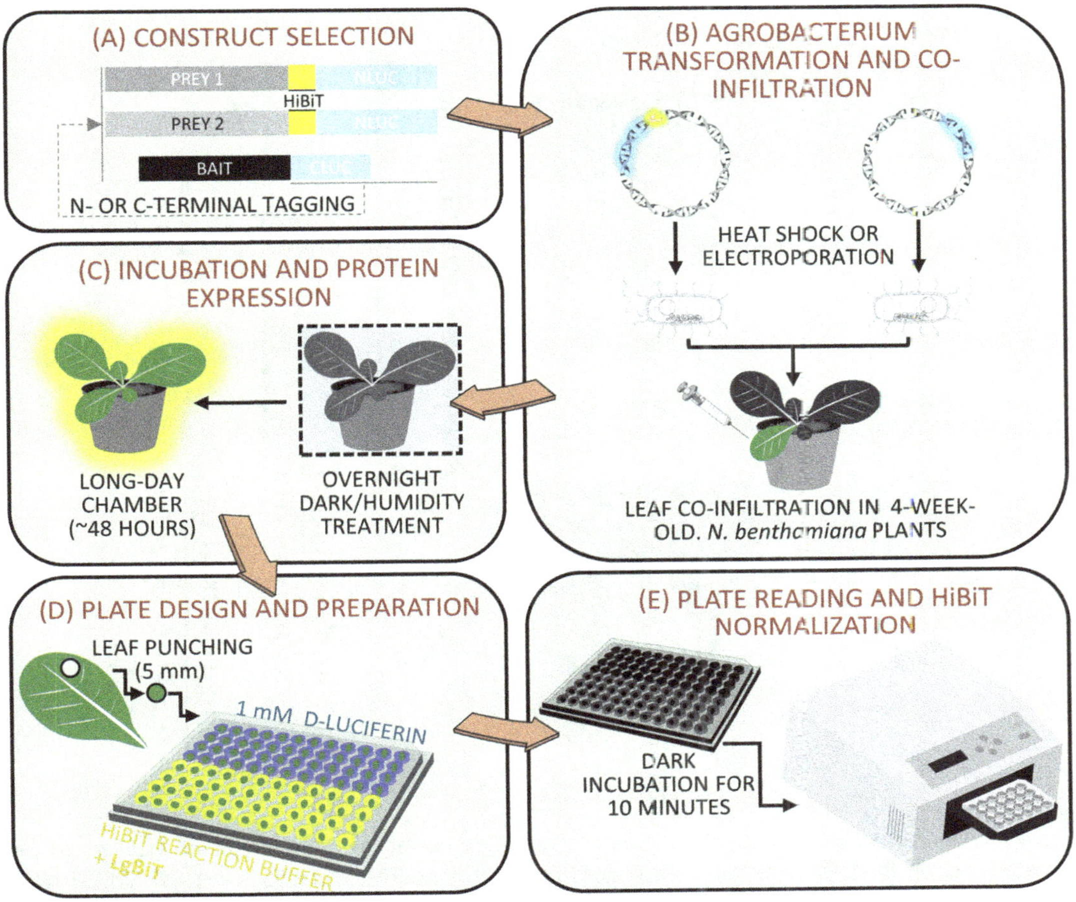

Fig. 1 Split-luciferase assay protocol in *N. benthamiana* leaf disks. (**a**) Construct selection. The coding sequences of the firefly luciferase (FLUC) N-terminal fragment (nLUC) and luciferase C-terminal fragment (cLUC) are amplified with the HiBiT small tag and cloned into respective plant expression vectors. (**b**) *Agrobacterium* transformation and co-infiltration. *Agrobacterium* strains are transformed by electroporation and clones carrying HiBiT-nLUC- and cLUC-fused constructs are transformed into *N. benthamiana* leaves via vacuum infiltration. Co-infiltration is performed by mixing equal volumes of *Agrobacterium* cultures expressing both complementary constructs. (**c**) Incubation and protein expression. Infiltrated plants are kept in the dark overnight and then cultivated for an additional 48 h under normal growth conditions. During this time, nLUC and cLUC proteins are expected to be expressed and interact to reconstitute functional luciferase. (**d**) Plate design and preparation. Leaf disks are excised from infiltrated leaves and placed on a 96-well white microplate. Proper plate design and preparation, including appropriate wells for HiBiT protein concentration control, must be ensured to obtain accurate luciferase activity measurements. (**e**) Plate reading and HiBiT normalization. Luminescence signals from the leaf disks are measured using a luminometer at a wavelength of 560 nm. HiBiT normalization is performed by dividing the luciferase activity by the concentration of the variable protein. Data are analyzed to determine protein–protein interaction strength based on normalized luciferase activity levels

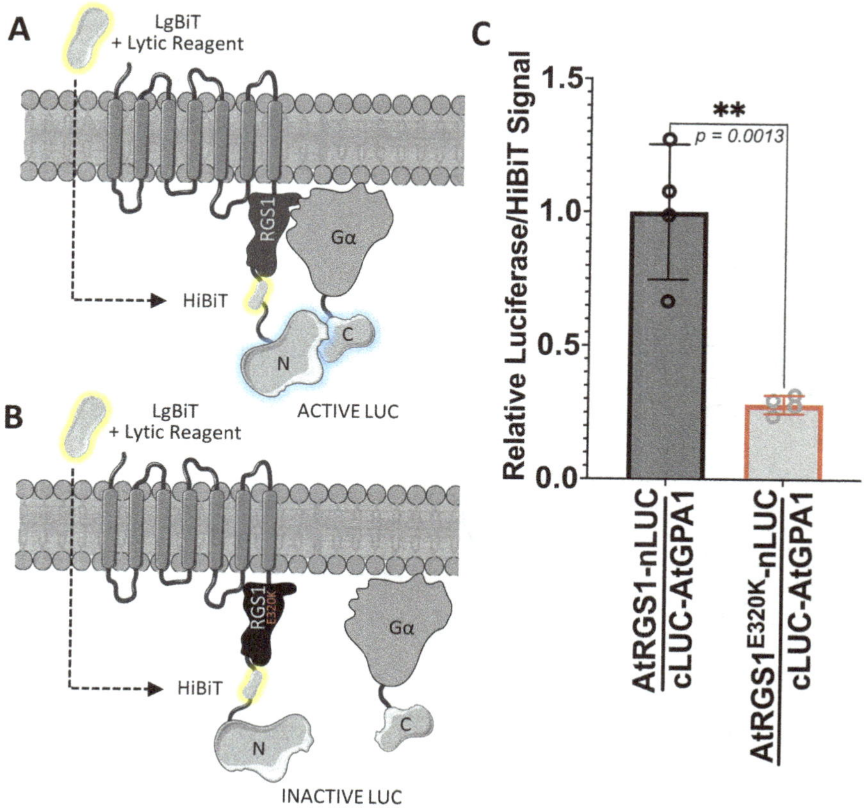

Fig. 2 Mutagenesis-induced changes in G-protein complex interactions in *Arabidopsis thaliana. In vivo* representation of the split luciferase method used to measure AtGPA1 interactions with (**a**) AtRGS1 (wild-type) or (**b**) the GAP-dead version E320K [15]. The signal from the HiBiT system components is depicted in yellow, while the signal from activated luciferase domains is shown in blue. Figures were created using the Bioicons library (https:/bioicons.com/ accessed on Monday, April 10th, 2023). (**c**) Quantification of luciferase signal normalized by the expression levels of prey proteins (HiBiT) and further normalized by the wild-type interaction. Error bars represent standard deviation, and individual data points are indicated as circles. Statistical significance is denoted by ** ($p < 0.01$), with the actual p-value (0.0013) indicated below (unpaired *t*-test)

other protein–protein interaction methodologies, such as BiFC and microscale thermophoresis (MST) [17–20].

2 Materials

2.1 Plasmid Transformation into Agrobacterium *Cells*

1. Sterile 1.5-mL microtubes and consumables.
2. GV3101 *Agrobacterium* Electrocompetent Cells (GoldBio or similar).
3. Liquid Luria-Bertani (LB) medium.

4. Plant expression plasmids that include the coding sequences for HiBiT-tagged prey proteins fused to the N-terminal domain of firefly luciferase (RGS1-HiBiT-nLUC and RGS1^{E320K}-HiBiT-nLUC), the C-terminal luciferase domain fused to Gα (cLUC-AtGPA1), and the *Tombusvirus* p19 protein-expressing construct. (*see* **Note 1**) [21].
5. LB agar plates containing gentamicin 50 μg/uL, for *Agrobacterium* GV3101 selection, and kanamycin 100 μg/uL, for selection of pCAMBIA-based plasmid containing bacteria (*see* **Note 1**).
6. Electroporator and sterile cuvettes.
7. Sterile hood.

2.2 Infiltration of N. benthamiana Leaves

1. 3-to-4 weeks old *N. benthamiana* plants.
2. Infiltration buffer: quarter-strength Murashige and Skoog (MS) liquid medium, 10 mM MES, 200 μM 5′-Dimethoxy-4′-hydroxyacetophenone (acetosyringone, fresh), pH 5.6.
3. 1-mL disposable syringes.
4. Tray with dome cover.
5. Spectrophotometer and appropriate cuvettes.
6. Falcon™ 50-mL conical centrifuge tubes.

2.3 Plate Preparation and Reading

1. Leaf striker.
2. Tweezer.
3. Luciferase reaction buffer: quarter-strength Murashige and Skoog (MS) liquid medium, 10 mM MES, 1 mM D-Luciferin Na-salt (fresh), pH 5.6.
4. Nano-Glo® HiBiT Lytic Detection System (Promega or similar).
5. Multichannel pipette.
6. Plate-reader for 550–570 nm luminescence reading.

3 Methods

3.1 Plasmid Transformation into Agrobacterium Cells

1. Prepare a sterile working area using a sterile hood.
2. Add approximately 50–200 ng of the plant expression vectors to individual sterile microtubes containing electrocompetent *Agrobacterium* cells.
3. Transfer the mixture to sterile electroporation cuvettes.
4. Electroporate cells with a 2 kV pulse or according to the recommended instrument instructions, using sterile techniques.

5. Recover cells by adding 1 mL of sterile Luria-Bertani (LB) liquid medium without any antibiotics to each cuvette and incubating at 28 °C for 2 h with constant agitation to allow resistance to establish.
6. Centrifuge cells in sterile 1.5-mL microtubes for 3 min at 3500 rpm.
7. Carefully remove the supernatant and resuspend the cell pellets in 100 μL of LB liquid medium by gently pipetting.
8. Plate cells on LB agar plates supplemented with gentamicin and kanamycin.
9. Incubate the plates at 28 °C for 48 h to allow for growth and selection.

3.2 Infiltration of N. benthamiana Leaves

1. Cultivate *Agrobacterium* clones (previously confirmed by PCR and/or digestion analysis) in approximately 10 mL of liquid LB medium, enriched with the suitable antibiotics (*see* **Note 2**).
2. Centrifuge *Agrobacterium* cells at 5000 rpm for 10 min and wash cells twice with infiltration buffer.
3. Resuspend the cell pellets in the infiltration buffer to an OD600 of 0.6.
4. Incubate *Agrobacterium* cells with gentle agitation for 2 h at room temperature.
5. Select healthy *N. benthamiana* plants with 3-to-4 weeks of optimal growth, using 4 biological replicates for each construct (*see* **Note 3**).
6. Mix equal parts of bait, prey, and p19 constructs into one infiltration solution.
7. Using a needleless syringe (*see* **Note 4**), infiltrate the abaxial side of the *N. benthamiana* leaves with the *Agrobacterium* suspensions in the final infiltration mixture (*see* **Note 5**). Carefully identify the infiltrated area with a marker pen.
8. Cover the tray containing the infiltrated plants to maintain high humidity and allow for efficient protein expression. Incubate the plants in the dark overnight.
9. Remove the cover and grow plants in optimal conditions for 24–48 h to allow for protein expression (*see* **Note 6**).

3.3 Luciferase Assay

1. Prepare the luciferase reaction buffer and keep it at room temperature.
2. Prepare the HiBiT reagent mixture as recommended by the manufacturer.
3. Strike 6 leaf disks of each leaf using the 5 mm striker, cutting close leaf disks within the leaf without catching nervure parts (*see* **Note 7**).

4. Build the plate with appropriate repetitions: 3 leaf disks are used as technical replicates for luciferase and 3 are used for HiBiT measurements (*see* **Note 8**).
5. With a multichannel pipette, carefully distribute 50 μL of the luciferase or HiBiT reaction solutions into each of the appropriate wells.
6. Cover the plate with aluminum foil and keep it under gentle agitation for 10 min at room temperature to allow enzyme activity (*see* **Note 9**).
7. Use a luminometer to measure the luciferase activity and HiBiT activity in the floated leaf disks at 550–570 nm with a 0.5-s integration time.
8. Normalize the luciferase activity measurements by dividing the luciferase activity of the proteins of interest by the HiBiT activity (*see* **Note 10**).
9. Normalize the interaction values by splitting data points by the average of positive control (RGS1-nLUC) (*see* **Note 11**).

4 Notes

1. As exemplified here, Gateway® vectors can be employed for split-luciferase assays, as they facilitate efficient cloning and expression of fusion proteins and tags, which can aid in protein quantification [22, 23]. There are a few plant expression vectors available for this assay. Here, we have used *cauliflower mosaic virus* 35S-based plasmids [10, 11]. Additionally, HCPro [24] may be used as an alternative to the p19 silencing suppressor [21].
2. *Agrobacterium* clones must be transformed with a minimum interval from transfection once it directly influences on the plant expression of exogenous constructs; culture volume will depend on the number of leaves to infiltrate.
3. *N. benthamiana* plants grow well at 25 °C in a long-day photoperiod (16 h light/8 h dark).
4. While a syringe needle can be used to make a small cut in the underside of a leaf for easier infiltration, it is not recommended to use it for the actual infiltration process. This is because doing so can cause more damage to the leaf and affect the accuracy of measurements.
5. Do not infiltrate more than two leaves per plant; different constructs may be compared in the same plant for better randomization. Leaves from the same age must be used for comparison.

6. Protein expression time may vary depending on the intrinsic characteristics of the protein or the conditions, but constant parameters are recommended during the experiment and repetitions.
7. Leaf disk strikers must be cleaned and constantly sharpened with the appropriate tools, and leaf disks must be carefully placed into the well with tweezers to avoid leaf damage.
8. It is recommended to add the positive control (RGS1-nLUC here) to all the plates if more than one comparison is being made. This will avoid plate or equipment-related errors.
9. In a split-luciferase assay, luminescence can be measured at a specific time point after initiating the assay or at the point of maximum luminescence. The decision on when to measure luminescence depends on the experimental design and study objectives. While specific time point measurement can be more efficient, as it may not require waiting for maximum luminescence. It may not capture the full extent of the protein–protein interaction and may require optimization to determine the optimal time point. On the other hand, measuring maximum luminescence can capture the full extent of the interaction but may be more time-consuming.
10. Another approach to quantify luciferase protein levels is through western blotting using a specific antibody against the protein. Commercial nLUC and cLUC antibodies may be optimized once.
11. The choice of measurement protocol generally depends on the experimental design, study objectives, and practical considerations such as time and resources available.

Acknowledgments

We acknowledge Dr. Alan M. Jones for providing the constructs and the structure for E.B.S. to conduct the experiments at The University of North Carolina in Chapel Hill. E.B.S. is primarily supported by CAPES (Coordination for the Improvement of Higher Education Personnel).

References

1. Zorzatto C, MacHado JPB, Lopes KVG et al (2015) NIK1-mediated translation suppression functions as a plant antiviral immunity mechanism. Nature 520:679–682. https://doi.org/10.1038/nature14171
2. Calil IP, Fontes EPB (2017) Plant immunity against viruses: antiviral immune receptors in focus. Ann Bot 119:711–723. https://doi.org/10.1093/AOB/MCW200
3. Muñoz A, Castellano MM (2018) Coimmunoprecipitation of interacting proteins in plants. Methods Mol Biol 1794:279–287. https://doi.org/10.1007/978-1-4939-7871-7_19/COVER
4. Fields S, Song OK (1989) A novel genetic system to detect protein–protein interactions. Nature 340(6230):245–246. https://doi.org/10.1038/340245a0

5. Walter M, Chaban C, Schütze K et al (2004) Visualization of protein interactions in living plant cells using bimolecular fluorescence complementation. Plant J 40:428–438. https://doi.org/10.1111/J.1365-313X.2004.02219.X
6. Kim HK, Cho EJ, Jo SM et al (2012) A split luciferase complementation assay for studying in vivo protein-protein interactions in filamentous ascomycetes. Curr Genet 58:179–189. https://doi.org/10.1007/S00294-012-0375-5/METRICS
7. Paulmurugant R, Gambhir SS (2003) Monitoring protein-protein interactions using split synthetic renilla luciferase protein-fragment-assisted complementation. Anal Chem 75: 1584–1589. https://doi.org/10.1021/AC020731C
8. Sheludko YV, Sindarovska YR, Gerasymenko IM et al (2007) Comparison of several Nicotiana species as hosts for high-scale agrobacterium-mediated transient expression. Biotechnol Bioeng 96:608–614. https://doi.org/10.1002/BIT.21075
9. Zhou Z, Bi G, Zhou JM (2018) Luciferase complementation assay for protein-protein interactions in plants. Curr Protoc Plant Biol 3:42–50. https://doi.org/10.1002/CPPB.20066
10. Lin ZJD, Liebrand TWH, Yadeta KA, Coaker G (2015) PBL13 is a serine/threonine protein kinase that negatively regulates Arabidopsis immune responses. Plant Physiol 169:2950–2962. https://doi.org/10.1104/PP.15.01391
11. Watkins JM, Ross-Elliott TJ, Shan X et al (2021) Differential regulation of G protein signaling in Arabidopsis through two distinct pathways that internalize AtRGS1. Sci Signal 14. https://doi.org/10.1126/SCISIGNAL.ABE4090/SUPPL_FILE/SCISIGNAL.ABE4090_SM.PDF
12. Chen H, Zou Y, Shang Y et al (2008) Firefly luciferase complementation imaging assay for protein-protein interactions in plants. Plant Physiol 146:368. https://doi.org/10.1104/PP.107.111740
13. Gbahou F, Levin S, Tikhonova IG et al (2022) Luminogenic HiBiT peptide-based NanoBRET ligand binding assays for melatonin receptors. ACS Pharmacol Transl Sci 5:668–678. https://doi.org/10.1021/ACSPTSCI.2C00096
14. Arakawa M, Morita E (2023) Protein pull-down assay using HiBiT-tag-dependent luciferase activity measurement. Bio Protoc 13. https://doi.org/10.21769/BIOPROTOC.4640
15. Lund CH, Bromley JR, Stenbæk A et al (2015) A reversible Renilla luciferase protein complementation assay for rapid identification of protein–protein interactions reveals the existence of an interaction network involved in xyloglucan biosynthesis in the plant Golgi apparatus. J Exp Bot 66:85–97. https://doi.org/10.1093/JXB/ERU401
16. Wang FZ, Zhang N, Guo YJ et al (2020) Split Nano luciferase complementation for probing protein-protein interactions in plant cells. J Integr Plant Biol 62:1065–1079. https://doi.org/10.1111/JIPB.12891/SUPPINFO
17. Johnston CA, Taylor JP, Gao Y et al (2007) GTPase acceleration as the rate-limiting step in Arabidopsis G protein-coupled sugar signaling. Proc Natl Acad Sci U S A:104. https://doi.org/10.1073/pnas.0704751104
18. Jones JC, Temple BRS, Jones AM, Dohlman HG (2011) Functional reconstitution of an atypical G protein heterotrimer and regulator of G protein signaling protein (RGS1) from Arabidopsis thaliana. J Biol Chem 286: 13143–13150. https://doi.org/10.1074/JBC.M110.190355
19. Lou F, Abramyan TM, Jia H et al (2020) An atypical heterotrimeric Gα protein has substantially reduced nucleotide binding but retains nucleotide-independent interactions with its cognate RGS protein and Gβγ dimer. J Biomol Struct Dyn 38:5204–5218. https://doi.org/10.1080/07391102.2019.1704879
20. Romain M, Thiroux B, Tardy M et al (2020) Measurement of protein-protein interactions through microscale thermophoresis (MST). Bio Protoc 10. https://doi.org/10.21769/BIOPROTOC.3574
21. Lakatos L, Szittya G, Silhavy D, Burgyán J (2004) Molecular mechanism of RNA silencing suppression mediated by p19 protein of tombusviruses. EMBO J 23:876. https://doi.org/10.1038/SJ.EMBOJ.7600096
22. Karimi M, Inzé D, Depicker A (2002) GATEWAY™ vectors for agrobacterium-mediated plant transformation. Trends Plant Sci 7:193–195. https://doi.org/10.1016/S1360-1385(02)02251-3
23. Katzen F (2007) Gateway® recombinational cloning: a biological operating system. 2:571–589. https://doi.org/10.1517/17460441.2.4.571
24. Ivanov KI, Eskelin K, Bašic M et al (2016) Molecular insights into the function of the viral RNA silencing suppressor HCPro. Plant J 85:30–45. https://doi.org/10.1111/TPJ.13088

Chapter 19

TurboID-Based Proximity Labeling: A Method to Decipher Protein–Protein Interactions in Plants

Yuanyuan Li, Yongliang Zhang, and Savithramma P. Dinesh-Kumar

Abstract

Proteins form complex networks through interaction to drive biological processes. Thus, dissecting protein–protein interactions (PPIs) is essential for interpreting cellular processes. To overcome the drawbacks of traditional approaches for analyzing PPIs, enzyme-catalyzed proximity labeling (PL) techniques based on peroxidases or biotin ligases have been developed and successfully utilized in mammalian systems. However, the use of toxic H_2O_2 in peroxidase-based PL, the requirement of long incubation time (16–24 h), and higher incubation temperature (37 °C) with biotin in BioID-based PL significantly restricted their applications in plants. TurboID-based PL, a recently developed approach, circumvents the limitations of these methods by providing rapid PL of proteins under room temperature. We recently optimized the use of TurboID-based PL in plants and demonstrated that it performs better than BioID in labeling endogenous proteins. Here, we describe a step-by-step protocol for TurboID-based PL in studying PPIs *in planta*, including *Agrobacterium*-based transient expression of proteins, biotin treatment, protein extraction, removal of free biotin, quantification, and enrichment of the biotinylated proteins by affinity purification. We describe the PL using plant viral immune receptor N, which belongs to the nucleotide-binding leucine-rich repeat (NLR) class of immune receptors, as a model. The method described could be easily adapted to study PPI networks of other proteins in *Nicotiana benthamiana* and provides valuable information for future application of TurboID-based PL in other plant species.

Key words Proximity labeling, TurboID, Biotinylation, Protein–protein interactions, NLR immune receptor

1 Introduction

Plant viruses are simple, obligate, and intracellular parasites. Due to their small genome size, plant viruses usually encode limited numbers of proteins. To successfully establish a viral infection cycle in host plants, plant viruses rely on various host machineries consisting of proteins, nucleic acids, membranes, and metabolites [1]. Therefore, dissecting molecular networks of plant–virus interactions is important for understanding the mechanism of viral infection and the defense responses against viruses by the host plant. Among

Elizabeth P. B. Fontes and Kristiina Mäkinen (eds.), *Plant-Virus Interactions*, Methods in Molecular Biology, vol. 2724,
https://doi.org/10.1007/978-1-0716-3485-1_19,

them, protein–protein interactions (PPIs) have attracted the most interest because PPIs are intrinsically involved in almost all cellular functions and biological processes.

Traditional approaches, including yeast two-hybrid (Y2H) screening and antibody-based affinity purification coupled with mass spectrometry (AP-MS), have been widely applied to identify PPIs during plant–virus interactions [2]. However, both approaches suffer from some disadvantages. For example, although Y2H screening is a high throughput method, it is a heterologous system and may have false positives due to overexpression of proteins. Those PPIs with cell and organelle type restrictions may not take place in the heterologous yeast system [3]. Furthermore, Y2H screening requires available high-quality libraries of the target plant species, constructions of which are labor-intensive and costly. The affinity purification approach allows the enrichment of stable interactions between the protein of interest and its partners in plant cells, but it suffers from capturing the transient or weak PPIs due to stringent cell lysis conditions and the subsequent washing steps typically used in AP [4]. Affinity purification is also unsuitable for insoluble proteins or targets with low abundance. These limitations, therefore, greatly hamper our understanding of the protein interaction networks in plant–virus interactions.

Enzyme-catalyzed proximity labeling (PL) is an approach developed for mapping molecular interaction networks [5]. Generally, an engineered enzyme fused to the protein of interest can produce short-lived reactive biotin species, such as radical and activated ester, to preferentially label proximal proteins within a certain radius (<20 nm) [6] (*see* Fig. 1). Because the proximate proteins are covalently tagged with biotin, the cells/tissue expressing the fusion protein can be lysed under harsh conditions, resulting in efficient disruption and solubilization of membranes and protein complexes. Subsequently, labeled proteins are enriched by streptavidin-based affinity purification and analyzed by mass spectrometry (MS) (*see* Fig. 2). The high affinity of streptavidin–biotin interaction also allows stringent washing steps to remove false positives. PL compensates for the drawbacks of the traditional approaches for identifying PPIs as it enables the capture of weak and transient interactions and insoluble protein complexes in their native cellular environment. Two major groups of enzymes commonly utilized for proximity-dependent biotinylation are peroxidases and biotin ligases. Peroxidases [such as ascorbate peroxidase 2 (APEX2), horseradish peroxidase (HRP)] generate biotin-phenoxyl radicals in the presence of hydrogen peroxide (H_2O_2) [7, 8] (*see* Fig. 1a). Wild-type biotin ligases convert biotin to the reactive biotin-adenosine-5′-monophosphoester (biotinoyl-AMP) intermediate and catalyze the direct transfer of activated biotin to a specific lysine residues of the target protein. Promiscuous biotin ligases (BioID, BioID2, BASU, TurboID, miniTurboID, and

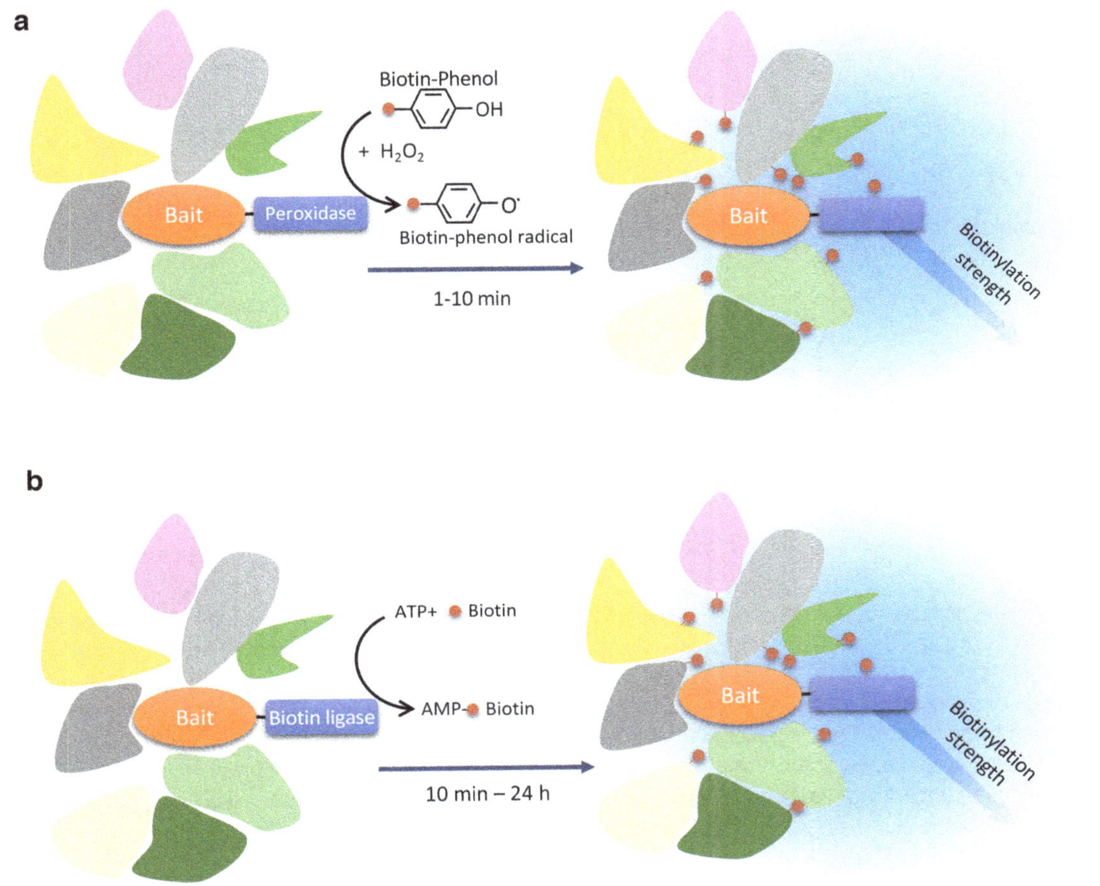

Fig. 1 Schematic of peroxidase- and biotin ligase-based proximity labeling. (**a**) Peroxidases, such as APEX or HRP, oxidize biotin–phenol into reactive radicals in the presence of H_2O_2, which labels proximal endogenous proteins. (**b**) Biotin ligases, such as BioID or TurboID, catalyze the formation of reactive biotin-AMP from biotin, which diffuses and labels proximal endogenous proteins

UltraID) modified from the wild-type enzymes have low affinity for the intermediate, allowing the generated intermediates diffuse out of the enzyme active site, and covalently biotinylate proximal proteins [9–12] (*see* Fig. 1b).

Although APEX offers rapid labeling kinetics (within minutes), the requirement of toxic H_2O_2 during labeling and high endogenous plant peroxidase activity make it unsuitable for PL studies in plants [7, 8]. In contrast, the biotin ligases provide simple and nontoxic labeling. BioID, a mutant of *Escherichia coli* biotin ligase BirA, represents the first generation of biotin ligase developed for PL. Compared to peroxidase-based PL, BioID requires 18–24 h for efficient labeling, making it difficult to capture transient PPIs. Moreover, the higher incubation temperature (37 °C) required by

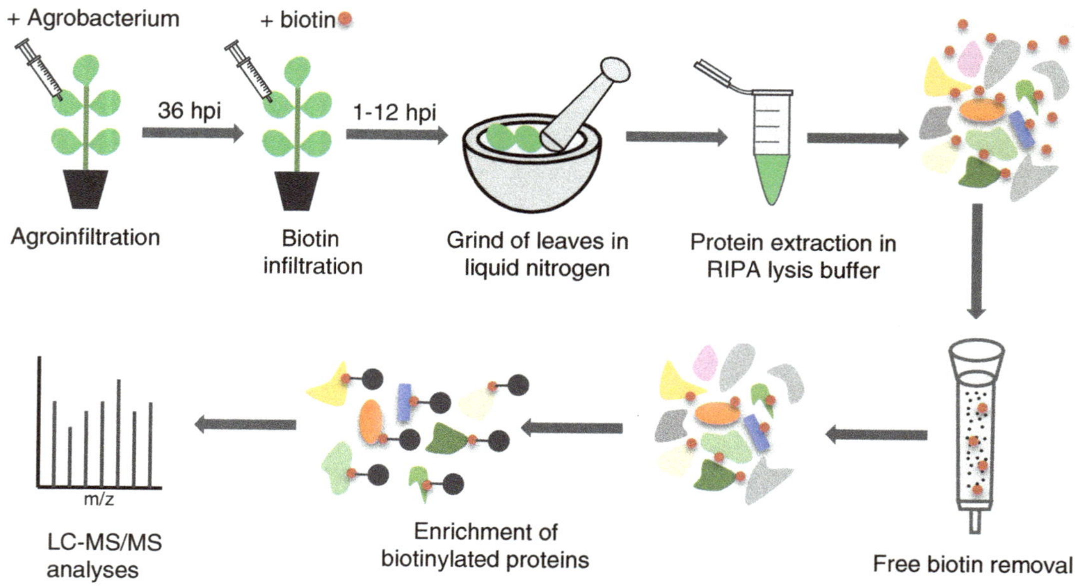

Fig. 2 Example workflow for mapping PPIs using TurboID-based proximity labeling method in *N. benthamiana*. TurboID fused to the protein of interest and a reference control are expressed in separate plants by agroinfiltration. 36 h post-infiltration, 200 μM biotin is infiltrated into the same leaves to initiate the proximity labeling of endogenous proteins. The infiltrated leaves were harvested after 1–12 h of incubation at RT and followed by grinding in liquid nitrogen. Proteins were extracted using a lysis buffer, such as RIPA lysis buffer, and a desalting column was employed to remove the free biotin in the extract. The biotinylated proteins were then affinity-purified with streptavidin-conjugated beads and identified via quantitative mass spectrometry. LC-MS/MS, liquid chromatography, and tandem mass spectrometry. (Reproduced from ref. [28] with permission from Nature Springer)

BioID is not optimal for *in vivo* PL in plants [11]. To reduce the interference with trafficking and function of the fusion protein, BioID2, a smaller biotin ligase derived from the *Aquifex aeolicus* was developed [10]. However, BioID2 requires the same labeling conditions as BioID, resulting in limited utilities of BioID/BioID2-based PL in plants [13–17]. Recently, new promiscuous biotin ligase variants of *E. coli* BirA, TurboID, and miniTurboID have been generated using yeast display-based directed evolution [9]. TurboID and miniTurboID have much higher catalytic activity than BioID or BioID2, allowing rapid nontoxic PL within 10 min under room temperature.

TurboID-based PL has been successfully conducted in a variety of cell types and organisms [18–25]. We and others recently optimized the utility of TurboID-based PL for PPIs studies in different plant systems such as *N. benthamiana*, *Arabidopsis*, and tomato [26–28]. TurboID showed a better PL performance than BioID in plants under room temperature [27, 28]. We used this method to identify interactors of plant viral immune receptor N that belong to the nucleotide-binding leucine-rich repeat (NLR) class of immune receptors [28]. N contains toll/interleukin-1 receptor (TIR) homology domain at the N-terminus [29]. In this chapter, the detailed procedures for TurboID-based investigation of PPIs *in planta* are described, using the interactome analysis of the TIR domain of N NLR immune receptor in *N. benthamiana* as an example [28, 30]. With some recently published studies [26, 27, 31, 32], it is expected that the application of this approach can be extended to other plant species and will serve as a powerful tool for PPIs studies *in planta*.

2 Materials

Prepare solutions using distilled–deionized water (ddH_2O) and analytical-grade reagents. Prepare and store reagents at room temperature (RT) unless indicated otherwise. Diligently follow all waste disposal regulations when disposing of waste materials. Use autoclaved DNase/RNase-free low-binding tubes for all work with proteins.

2.1 Agroinfiltration and Biotin Treatment

1. *N. benthamiana* seeds.
2. Soil (Sunshine® Mix #1, SunGro Horticulture), 3.5′ × 3.5′ pots, and a 23–25 °C growth chamber.
3. *Agrobacterium tumefaciens* strain (GV3101) (*see* **Note 1**).
4. TurboID plasmids (p35S:Citrine-TurboID-3xHA, Addgene #199243; pUBQ:Citrine-TurboID-3xMyc, Addgene #199244).
5. Luria-Bertani (LB) medium: 10 g/L tryptone, 5 g/L yeast extract, and 10 g/L NaCl are dissolved in ddH_2O. Adjust pH to 7.0 and autoclave to sterilize.
6. LB agar plates with appropriate antibiotics.
7. 50 mg/mL kanamycin: Dissolve in ddH_2O, filter-sterilize (0.22 μm filter) and store at −20 °C.
8. 50 mg/mL gentamicin: Dissolve in ddH_2O, filter-sterilize (0.22 μm filter) and store at −20 °C.
9. 50 mg/mL rifampicin: Dissolve in DMSO, filter-sterilize (0.22 μm filter) and store at −20 °C.
10. 1 M 2-(N-morpholine)-ethanesulfonic acid (MES): Dissolve in ddH_2O, adjust pH to 5.6 with NaOH, filter-sterilize (0.22 μm filter), and store at −20 °C.
11. 250 mM acetosyringone (AS): Dissolve in DMSO and store at −20 °C.
12. 1 M $MgCl_2$: Dissolve in ddH_2O. Autoclave to sterilize.

13. Agroinfiltration medium: 10 mM $MgCl_2$, 10 mM MES (pH 5.6), and 250 μM AS.
14. 1-mL Disposable syringe.
15. 50 mM Biotin (Sigma): Dissolve in 100 mM Na_2HPO_4. Adjust pH to 7.2 with 100 mM NaH_2PO_4. Bring to desired volume with 100 mM sodium phosphate buffer (pH 7.2). Filter-sterilize (0.22 μm filter) and store at −20 °C (*see* **Note 2**).

2.2 Protein Extraction and Desalting

1. Liquid nitrogen.
2. Mortar and pestle.
3. 2.0-ml Eppendorf tubes.
4. 15-ml or 50-ml Falcon tubes.
5. 1 M Tris–HCl buffer: Dissolve in ddH_2O. Adjust pH to 7.5. Autoclave to sterilize.
6. 5 M NaCl: Dissolve in ddH_2O. Autoclave to sterilize.
7. 0.5 M stock EDTA: Dissolve in ddH_2O. Adjust pH to 8.0. Autoclave to sterilize.
8. IGEPAL® CA-630.
9. 10% (w/v) SDS in ddH_2O.
10. 10% (w/v) Sodium deoxycholate in ddH_2O. Store in the dark.
11. 100 mM DTT: Dissolve in ddH_2O and store at −20 °C.
12. Complete EDTA-free protease inhibitor cocktail (Roche) or equivalent protease inhibitor cocktail.
13. RIPA lysis buffer: 50 mM Tris–HCl (pH 7.5), 500 mM NaCl, 1 mM EDTA (pH 8.0), 1% IGEPAL® CA-630, 0.1% SDS, 0.5% sodium deoxycholate, 1 mM DTT, 1 × complete EDTA-free protease inhibitor cocktail. Prepare fresh before use.
14. Zeba™ Spin Desalting Columns, 7 K MWCO, 10 mL (Thermo Scientific™): Store at 4 °C (*see* **Note 3**).

2.3 Quantification of Protein

1. 2 mg/mL Bovine Serum Albumin (BSA) Standard Ampules (Pierce™).
2. Coomassie brilliant blue G250.
3. Methanol.
4. 85% phosphoric acid.
5. 5× Bradford regent: 100 mg Coomassie brilliant blue G250, 47 mL of methanol, 100 mL of 85% phosphoric acid, 53 mL of ddH_2O.
6. Test tubes.
7. ELISA plate.

2.4 Enrichment of Biotinylated Proteins

1. Dynabeads™ MyOne™ Streptavidin C1 (Invitrogen™): Store at 4 °C.
2. Magnetic separation rack.
3. 1.5-mL Eppendorf tube.
4. 1 M HEPES (pH 7.5): Dissolve in ddH_2O, adjust pH to 7.5 with NaOH, filter-sterilize (0.22 μm filter), and store at 4 °C.
5. Triton X-100.
6. 8 M LiCl.
7. Wash buffer I: 2% SDS in water.
8. Wash buffer II: 50 mM HEPES (pH 7.5), 500 mM NaCl, 1 mM EDTA, 0.1% deoxycholic acid (w/v), 1% triton X-100. Prepare fresh before use.
9. Wash buffer III: 10 mM Tris–HCl 7.5, 250 mM LiCl, 1 mM EDTA, 0.1% deoxycholic acid (w/v), 1% IGEPAL® CA-630. Prepare fresh before use.
10. 50 mM Tris–HCl (pH 7.5): Prepare fresh before use and store at 4 °C.
11. 50 mM Ammonium bicarbonate: Dissolve in ddH_2O, filter-sterilize (0.22 μm filter). Prepare fresh before use and store at 4 °C.

3 Methods

3.1 Growth of Plant

1. Sow *N. benthamiana* seeds at high density (about 40 seeds) into a pot filled with wet soil and place it in a plastic growing tray covered with a transparent humidity dome. Maintain them in a growth chamber with a photoperiod of 16 h light/8 h dark (75 μmol/m^2/s) at 23–25 °C and 40–60% humidity.
2. After 1 week, carefully transfer each germinated seedling to a 3.5′ × 3.5′ pot filled with wet soil and maintain plants in the same growth chamber for 3–4 weeks.

3.2 Plasmid Construction

1. Use a standard molecular cloning method to fuse the gene of interest in-frame with TurboID vectors (Addgene plasmids #199243; #199244). Here, we made a construct expressing TurboID-fused to TIR domain of N NLR immune receptor driven by cauliflower mosaic virus 35S promoter (p35S::TIR-TurboID-3xHA) [28] (*see* **Note 4**).
2. Generate a construct expressing TurboID-fused to citrine driven by the same promoter (p35S::Citrine-TurboID-3xHA) to serve as the control for subsequent quantitative analysis (*see* **Note 5**).

3. Confirm the sequence of all the constructs by Sanger sequencing.

3.3 Agroinfiltration and Biotin Treatment

3.3.1 Agroinfiltration

1. Transform the plasmids from Subheading 3.2 into *A. tumefaciens* strains GV3101 separately [33] (*see* **Note 1**).
2. Pick and streak *Agrobacterium* colony harboring the construct onto a fresh LB agar plate supplemented with appropriate antibiotics to select the plasmid, as well as for the *Agrobacterium* (for GV3101: 50 mg/L gentamicin and 50 mg/L rifampicin) and grow at 28 °C overnight (*see* **Note 6**).
3. Inoculate *Agrobacterium* harboring the construct into 3 mL of LB medium with appropriate antibiotics (*see* **step 2**) and incubate by shaking at 28 °C in a shaker overnight until the OD_{600} reaches approximately 2.0.
4. Collect *Agrobacterium* cells by centrifugation at 3000×*g*, 10 min, and resuspend the pellet in agroinfiltration media to $OD_{600} = 1.0$ (*see* **Note 7**).
5. Use a 1 mL needleless syringe to infiltrate the suspensions into the fully mature leaves of 3- to 4-week-old *N. benthamiana* through the (abaxial) epidermis (*see* **Note 8**).
6. Maintain plants in the growth chamber (*see* Subheading 3.1, **step 1**).

3.3.2 Biotin Treatment

1. At 36 h post-infiltration (hpi) (*see* **Note 9**), infiltrate 200 μM biotin (in 10 mM $MgCl_2$ solution) into the leaves pre-infiltrated with TurboID constructs (*see* **Note 10**).
2. Maintain the plant for an additional 1–12 h in the growth chamber (*see* Subheading 3.1, **step 1**) before sample collection (*see* **Note 11**).

3.4 Protein Extraction

To minimize keratin contamination, the subsequent procedure should be processed in keratin-free condition, wear powder-free sterile gloves, and keep all reagents keratin-free when possible.

1. Cut off the infiltrated leaves from the base of the petiole, remove the leaf vein, and then flash-freeze the tissue in liquid nitrogen (*see* **Note 12**).
2. Grind the collected tissue in liquid nitrogen with a pestle and mortar into a fine powder, and store in 15-mL or 50-mL Falcon tubes at −80 °C for subsequent use. Verify expression and biotinylation of proteins by western blot (*see* **Note 13**). Typical western blot results are shown in Fig. 3.
3. Transfer 0.35 g of tissue powder into a 2.0-mL Eppendorf tube. Prepare two tubes for each sample (0.7 g/sample in total).

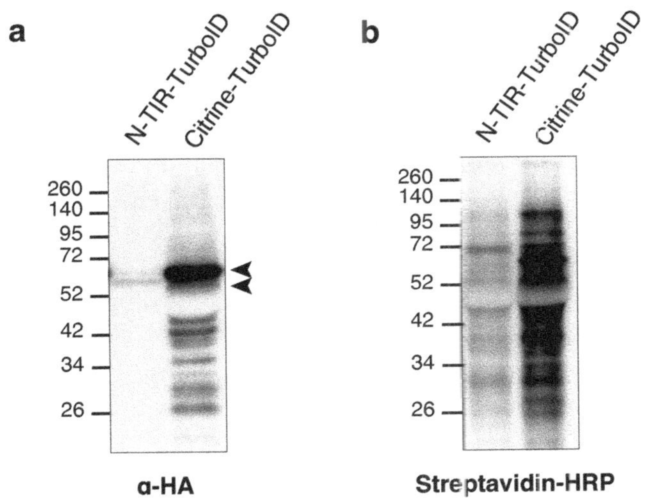

Fig. 3 Immunoblot detection of HA-tagged TurboID-fused proteins (**a**) and biotinylated proteins (**b**) from the sample collected in Subheading 3.4, **step 9**. Arrows indicate the specific band of different HA fusion proteins. (Reproduced from ref. [28] with permission from Nature Springer)

4. Add 700 μL of RIPA lysis buffer to each tube containing 0.35 g tissue powder.
5. Vortex for 10 min to mix thoroughly.
6. Leave the mixture on ice for 30 min. Mix the contents every 4–5 min by turning the tubes upside down several times.
7. Centrifuge at 16,500×*g* for 10 min at 4 °C.
8. Combine the supernatant from two tubes of the same sample and transfer it into a new 2.0-mL Eppendorf tube.
9. Take out 50 μL aliquot and store at −80 °C for subsequent immunoblot analysis of target proteins in the leaf extracts. Keep the rest samples on ice temporarily.

3.5 Removal of Free Biotin Using a Desalting Column

This usually takes 30–50 min. To save time, column equilibration can be done during the protein extraction (Subheading 3.4, **steps 6–9**). The protocol given here is for Zeba™ Spin Desalting Columns.

3.5.1 Equilibrate the Desalting Column

1. Remove the desalting column's bottom sealing material, loosen cap (do not remove cap).
2. Place the column into a collection tube (50-mL Falcon tube) and centrifuge at 1000×*g* for 2 min at 4 °C to remove the storage solution (*see* **Note 14**).
3. Discard flowthrough and replace the column back into the collection tube. Add 5 mL of RIPA lysis buffer on top of the

resin. Centrifuge at 1000×*g* for 2 min at 4 °C and discard the flowthrough.

4. Repeat **step 3** two additional times.
5. Blot the bottom of the column or plate to remove excess liquid. Transfer the column to a new collection 50-mL Falcon tube.
6. Store temporarily at 4 °C.

3.5.2 Desalting the Protein Extracts

1. Apply 1400–1500 μL of protein extract on top of the resin of the equilibrated desalting column (from Subheading 3.4, **step 9**) (*see* **Note 15**).
2. Add a stacker (another 100 μL of RIPA lysis buffer) on top of the resin as soon as the added sample has completely entered the resin (*see* **Note 16**).
3. Centrifuge at 1000×*g* for 2 min at 4 °C.
4. Keep the flowthrough (desalted protein extracts) on ice temporarily. Discard the column.

3.6 Quantification of the Desalted Protein Extracts Using a Bradford Assay (See Note 17)

1. Prepare a set of diluted BSA standards (50 μL each): 0 mg/mL, 0.2 mg/mL, 0.4 mg/mL, 0.6 mg/mL, 0.8 mg/mL, and 1 mg/mL.
2. Dilute the desalted protein extract by mixing 5 μL of the sample with 45 μL of ddH_2O.
3. Prepare 1 × Bradford regent by diluting the 5 × Bradford regent with ddH_2O.
4. Pipette 50 μL of each BSA standard and diluted protein extract into an appropriately labeled test tube.
5. Add 2.5 mL of 1 × Bradford regent into each tube and mix well.
6. Incubate at RT for 10 min.
7. Add 200 μL of solution from each tube to a well of the ELISA plate (three technical replicates per sample).
8. Measure the 562 nm absorbance using a microplate reader (*see* **Note 18**).
9. Prepare the standard curve by plotting the average Blank-corrected 562 nm measurement for each BSA standard vs. its concentration in μg/mL. Use the standard curve to calculate the concentration of each desalted protein sample (*see* **Note 19**).
10. Use 6–8 mg of desalted protein extract for subsequent biotinylated protein enrichment.

3.7 Enrichment of Biotinylated Proteins by Streptavidin Beads

3.7.1 Prewash Streptavidin-C1-Conjugated Magnetic Beads

1. Resuspend the streptavidin-C1-conjugated magnetic beads in the vial (i.e., vortex for >30 s, or tilt and rotate for 5 min).
2. For each sample, transfer 200 μL of streptavidin-C1-conjugated magnetic beads into a 2.0-mL Eppendorf tube.
3. Add 1 mL of RIPA lysis buffer to suspend the beads.
4. Place tubes on a magnetic separation rack for 3 min and gently remove the solution by pipette.
5. Remove the tube from the magnet and add 1 mL of RIPA lysis buffer to resuspend the beads.
6. Repeat **steps 4** and **5** for a total of three washes.

3.7.2 Enrichment of Biotinylated Proteins

1. Transfer each desalted protein extract to a 2.0-mL Eppendorf tube containing prewashed equilibrated streptavidin-C1-conjugated magnetic beads.
2. Incubate at 4 °C for 12–16 h (or overnight) with end-over-end rotation.
3. Separate the beads with a magnetic separation rack for 4 min at RT until the beads collect at one side of the tube, and then gently remove the supernatant by pipette (*see* **Note 20**).
4. Add 1.7 mL of wash buffer I into each tube and wash with end-over-end rotation at RT for 8 min. Repeat **step 3**.
5. Add 1.7 mL of wash buffer II into each tube and wash with end-over-end rotation at RT for 8 min. Repeat **step 3**.
6. Add 1.7 mL of wash buffer III into each tube and wash with end-over-end rotation at RT for 8 min. Repeat **step 3**.
7. Add 1.7 mL of 50 mM Tris–HCl (pH 7.5) into each tube and wash with end-over-end rotation at RT for 8 min. Repeat **step 3**.
8. Add 1 mL of 50 mM Tris–HCl (pH 7.5) into each tube, resuspend beads, and transfer to a new 1.5-mL Eppendorf tube. Wash with end-over-end rotation at RT for 8 min. Repeat **step 3**.
9. Add 1 mL of 50 mM ammonium bicarbonate buffer into each tube and wash with end-over-end rotation at 4 °C for 5 min.
10. Separate the beads with a magnetic separation rack for 4 min at 4 °C until the beads collect at one side of the tube, and then gently remove the supernatant by pipette.
11. Repeat **steps 9** and **10** five more times, for a total of six times wash with 50 mM ammonium bicarbonate buffer.
12. Resuspend beads with 1 mL of 50 mM ammonium bicarbonate buffer.

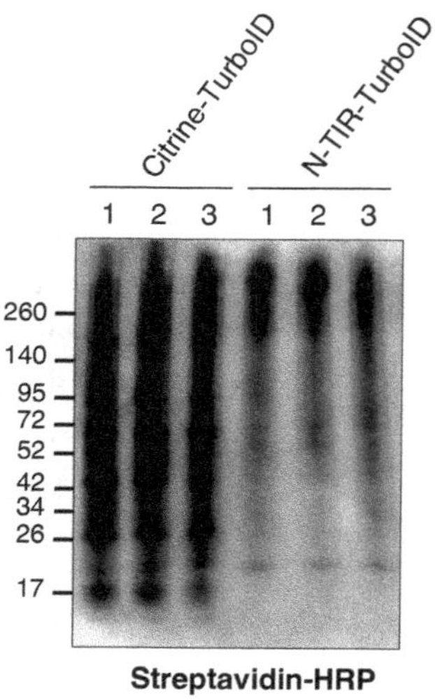

Fig. 4 Immunoblot analysis of the biotinylated proteins captured on Streptavidin beads in Subheading 3.7.2, **step 13** to confirm the enrichment of the biotinylated proteins. There are three independent replicates for each protein (reproduced from ref. [28] with permission from Nature Springer). Generally, citrine expresses at higher level, and hence, there is more cis-biotinylation in Citrine-TurboID samples

13. Take out 100 μL of beads for immunoblot analysis to confirm the enrichment of biotinylated proteins. A typical western blot is shown in Fig. 4.
14. Flash-freeze the rest of the protein samples and store them at −80 °C or send them immediately for liquid chromatography–tandem mass spectrometry (LC-MS/MS) analysis on dry ice (*see* **Note 21**).

4 Notes

1. Other *Agrobacterium* strains, such as EHA105, GV2260, and C58CI, can also be used.
2. Biotin is a carboxylic acid, and as such it is less soluble under acidic conditions, but more soluble in DMSO and dilutes alkali. Biotin stock can also be prepared in DMSO.
3. PD-10 columns (GE Healthcare) can also be used for free biotin removal (*see* ref. [26]). It will be better to compare the efficiency of these columns.
4. The TurboID fusion should not interfere with the function or localization of the protein. Previously characterized functional fusion with fluorescent protein or epitope tag can be a promising indicator of whether TurboID should be fused to

the C- and N-terminus of protein. Usually, for cytoplasmic proteins, both termini should be acceptable. However, for membrane-localized proteins, the protein topology needs to be characterized in advance. If unsure, make and test both versions.

5. It is important to include a TurboID fusion control for validating the proximal proteins to the protein of interest. The control should be expressed at a level similar to that of the TurboID-fused target protein. This can be empirically determined by adjusting the *Agrobacterium* concentration during agroinfiltration. In addition, it is important that the control protein and the target protein of interest share a similar subcellular localization pattern or at least the control protein's subcellular distribution covers or encompasses the space that the TurboID fusion touches.
6. It is more optimal to confirm the presence of the specific binary construct in the *Agrobacterium* by colony PCR.
7. Although it is optimal to incubate the inoculum in the infiltration media for 2–3 h prior to agroinfiltration at RT, based on our experience with GV3101, there is no difference between the protein expression with and without preincubation. The preincubation requirement for other *Agrobacterium* strains needs to be verified.
8. To prepare enough leaf materials for three biological replicates, we suggest infiltrating three to four plants per construct. Each plant has three entire leaves infiltrated. For each leaf, 1.5–2.0 mL of resuspended *Agrobacterium* should be sufficient.
9. The reason for picking 36 hpi for biotin infiltration is that, according to our previous studies [28], the protein of interest reaches its peak expression at this time point. It is recommended to determine the time required for optimal expression of the protein of interest.
10. A comparable biotinylation level of TurboID-fused bait protein and its proximate proteins should be achieved with the addition of 50–200 μM biotin. It is reported that increasing the biotin concentration from 50 μM to 2 mM adversely affected the capture of proximate proteins, as only the bait itself could be identified. This is probably due to high residual free biotin accumulated in the protein extract even after desalting, thus occupying the streptavidin binding sites on the beads (*see* ref. [26]).
11. The incubation time post-biotin infiltration depends on the features of the bait protein and the experimental design. Usually, longer biotin incubation time clearly increases the robustness of PL, but 1–12 h of biotin treatment should be sufficient

for the labeling of most proteins proximal to the bait protein by the TurboID fusions.

12. Harvest three to four pieces of leaves for each biological replicate from different plants. Wear gloves when handling liquid nitrogen or operating any object cooled by liquid nitrogen.
13. The protocol can be paused here. Prior to subsequent steps, it is recommended to verify protein expression and biotinylation of the target protein by immunoblot analyses.
14. After each spin, the resin should appear white and free of liquid. If a liquid is present, make sure you are using the correct centrifugation speed and time. Some models of centrifuge will time down right after pushing the "Start" button, but here requires timed runs at speed. Incomplete centrifugation may result in poor sample recovery or sample dilution.
15. A combination of the samples from each group, as described in Subheading 3.4 can result in at least 1500 μL of protein extract per sample, as the total volume after protein extraction was invariably increased to some extent relative to the original volume of added RIPA lysis buffer.
16. Adding a stacker is optional but recommended for dilute protein solutions or small sample volumes to ensure maximum sample recovery.
17. Other protein quantification methods, such as BCA, can also be used.
18. A spectrometer with an appropriate wavelength range combined with the use of cuvettes can be an alternative for measuring 562 nm absorbance.
19. Usually, the total protein concentration obtained from 0.7 g of leaves ranges from 3 to 6 mg/mL.
20. Collect beads from the lid of the tubes by brief centrifugation each time before placing them on the magnetic separation rack.
21. Typical MS results can be found in our previous publication (*see* ref. [28] Fig. 2, Supplementary Data 1, and Supplementary Data 2 in ref. [28]).

Acknowledgments

This work was supported by grants from NIH-GM132582, NSF-MCB-EAGER-2028283 and NSF-IOS-2139987 to SPD-K.

References

1. Nelson RS, Citovsky V (2005) Plant viruses. Invaders of cells and pirates of cellular pathways. Plant Physiol 138:1809–1814
2. Berggard T, Linse S, James P (2007) Methods for the detection and analysis of protein-protein interactions. Proteomics 7:2833–2842
3. Bruckner A, Polge C, Lentze N, Auerbach D, Schlattner U (2009) Yeast two-hybrid, a powerful tool for systems biology. Int J Mol Sci 10: 2763–2788
4. Dunham WH, Mullin M, Gingras AC (2012) Affinity-purification coupled to mass spectrometry: basic principles and strategies. Proteomics 12:1576–1590
5. Kim DI, Roux KJ (2016) Filling the void: proximity-based labeling of proteins in living cells. Trends Cell Biol 26:804–817
6. Qin W, Cho KF, Cavanagh PE, Ting AY (2021) Deciphering molecular interactions by proximity labeling. Nat Methods 18:133–143
7. Rhee HW, Zou P, Udeshi ND, Martell JD, Mootha VK, Carr SA et al (2013) Proteomic mapping of mitochondria in living cells via spatially restricted enzymatic tagging. Science 339:1328–1331
8. Lam SS, Martell JD, Kamer KJ, Deerinck TJ, Ellisman MH, Mootha VK et al (2015) Directed evolution of APEX2 for electron microscopy and proximity labeling. Nat Methods 12:51–54
9. Branon TC, Bosch JA, Sanchez AD, Udeshi ND, Svinkina T, Carr SA et al (2018) Efficient proximity labeling in living cells and organisms with TurboID. Nat Biotechnol 36:880–887
10. Kim DI, Jensen SC, Noble KA, Birendra KC, Roux KH, Motamedchaboki K et al (2016) An improved smaller biotin ligase for BioID proximity labeling. Mol Biol Cell 27:1188–1196
11. Roux KJ, Kim DI, Raida M, Burke B (2012) A promiscuous biotin ligase fusion protein identifies proximal and interacting proteins in mammalian cells. J Cell Biol 196:801–810
12. Kubitz L, Bitsch S, Zhao XY, Schmitt K, Deweid L, Roehrig A et al (2022) Engineering of ultraID, a compact and hyperactive enzyme for proximity-dependent biotinylation in living cells. Commun Biol 5:1–14
13. Conlan B, Stoll T, Gorman JJ, Saur I, Rathjen JP (2018) Development of a rapid in planta BioID system as a probe for plasma membrane-associated immunity proteins. Front Plant Sci 9:1882
14. Khan M, Youn JY, Gingras AC, Subramaniam R, Desveaux D (2018) In planta proximity dependent biotin identification (BioID). Sci Rep 8:1–8
15. Lin QP, Zhou ZJ, Luo WB, Fang MC, Li MR, Li HQ (2017) Screening of proximal and interacting proteins in rice protoplasts by proximity-dependent biotinylation. Front Plant Sci 8:749
16. Macharia MW, Tan WYZ, Das PP, Naqvi NI, Wong SM (2019) Proximity-dependent biotinylation screening identifies NbHYPK as a novel interacting partner of ATG8 in plants. BMC Plant Biol 19:1–11
17. Das PP, Macharia MW, Lin QS, Wong SM (2019) In planta proximity-dependent biotin identification (BioID) identifies a TMV replication co-chaperone NbSGT1 in the vicinity of 126 kDa replicase. J Proteome 204:103402
18. Larochelle M, Bergeron D, Arcand B, Bachand F (2019) Proximity-dependent biotinylation mediated by TurboID to identify protein-protein interaction networks in yeast. J Cell Sci 132:jcs232249
19. Shinoda N, Hanawa N, Chihara T, Koto A, Miura M (2019) Dronc-independent basal executioner caspase activity sustains Drosophila imaginal tissue growth. Proc Natl Acad Sci U S A 116:20539–20544
20. Takano T, Wallace JT, Baldwin KT, Purkey AM, Uezu A, Courtland JL et al (2020) Chemico-genetic discovery of astrocytic control of inhibition in vivo. Nature 588:296–302
21. Chua XEY, Aballo T, Elnemer W, Tran M, Salomon A (2021) Quantitative interactomics of Lck-TurboID in living human T cells unveils T cell receptor stimulation-induced proximal Lck interactors. J Proteome Res 20:715–726
22. Ambekar SV, Beck JR, Mair GR (2022) TurboID identification of evolutionarily divergent components of the nuclear pore complex in the Malaria model Plasmodium berghei. MBio 13: e01815–e01822
23. Hollstein LS, Schmitt K, Valerius O, Stahlhut G, Poggeler S (2022) Establishment of in vivo proximity labeling with biotin using TurboID in the filamentous fungus Sordaria macrospora. Sci Rep 12:1–11
24. Kanzler CR, Donohue M, Dowdle ME, Sheets MD (2022) TurboID functions as an efficient biotin ligase for BioID applications in Xenopus embryos. Dev Biol 492:133–138
25. Zhang YH, Shang LM, Zhang J, Liu YC, Jin CZ, Zhao YN et al (2022) An antibody-based proximity labeling map reveals mechanisms of SARS-CoV-2 inhibition of antiviral immunity. Cell Chem Biol 29 5–18

26. Arora D, Abel NB, Liu C, Van Damme P, Yperman K, Eeckhout D et al (2020) Establishment of proximity-dependent biotinylation approaches in different plant model systems. Plant Cell 32:3388–3407
27. Mair A, Xu SL, Branon TC, Ting AY, Bergmann DC (2019) Proximity labeling of protein complexes and cell-type-specific organellar proteomes in Arabidopsis enabled by TurboID. elife 8:e47864
28. Zhang YL, Song GY, Lai NK, Nagalakshmi U, Li YY, Zheng WJ et al (2019) TurboID-based proximity labeling reveals that UBR7 is a regulator of N NLR immune receptor-mediated immunity. Nat Commun 10:3252
29. Whitham S, Dinesh-Kumar S, Choi D, Hehl R, Corr C, Baker B (1994) The product of the tobacco mosaic virus resistance gene N: similarity to toll and the interleukin-1 receptor. Cell 78:1101–1115
30. Zhang Y, Li Y, Yang X, Wen Z, Nagalakshmi U, Dinesh-Kumar SP (2020) TurboID-based proximity labeling for in planta identification of protein-protein interaction networks. J Vis Exp 159:e60728
31. Teplova AD, Serebryakova MV, Galiullina RA, Chichkova NV, Vartapetian AB (2021) Identification of phytaspase interactors via the proximity-dependent biotin-based identification approach. Int J Mol Sci 22:13123
32. Xu FF, Jia M, Li X, Tang Y, Jiang KN, Bao JS et al (2021) Exportin-4 coordinates nuclear shuttling of TOPLESS family transcription corepressors to regulate plant immunity. Plant Cell 33:697–713
33. McCormac AC, Elliott MC, Chen DF (1998) A simple method for the production of highly competent cells of *Agrobacterium* for transformation via electroporation. Mol Biotechnol 9:155–159

Chapter 20

Virus-Induced Heritable Gene Editing in Plants

Ugrappa Nagalakshmi, Nathan Meier, and Savithramma P. Dinesh-Kumar

Abstract

Gene editing using clustered, regularly interspaced short palindromic repeats (CRISPR) and CRISPR-associated (Cas) nuclease is an excellent tool for assessing gene function in plants. However, delivery of CRISPR/Cas-editing components into plant cells is still a major bottleneck and requires tissue culture-based approaches and regeneration of plants. To overcome this limitation, several plant viral vectors have recently been engineered to deliver single-guide RNA (sgRNA) targets into *SpCas9*-expressing plants. Here, we describe an optimized, step-by-step protocol based on the tobacco rattle virus (TRV)-based vector system to deliver sgRNAs fused to mobile tRNA sequences for efficient heritable editing in *Nicotiana benthamiana* and *Arabidopsis thaliana* model systems. The protocol described here could be adopted to study the function of any gene of interest.

Key words Gene editing, Cas9, Guide RNA, Tobacco Rattle Virus (TRV), *Nicotiana benthamiana*, *Arabidopsis*, *Phytoene desaturase* (*PDS*), Heritable

1 Introduction

The discovery and development of clustered, regularly interspaced short palindromic repeats (CRISPR) and CRISPR-associated (Cas) nucleases for targeted genome editing over the last decade, have revolutionized gene function studies in many organisms, including plants [1, 2]. In plants, delivery of Cas nucleases and single-guide RNA (sgRNA) targeted to edit genes is generally accomplished through Agrobacterium-mediated plant transformation or direct delivery to protoplasts followed by plant regeneration. These approaches, which depend on tissue culture, are time-consuming and labor-intensive. In addition, many plants are recalcitrant to transformation or regeneration.

One approach to delivering CRISPR/Cas-editing components without transgenics is through viral vectors. Over the last four decades, engineered plant viral vectors have been a great tool in various fundamental and translational studies in plant biology [3]. Various plant viral vectors have been engineered and used to

Elizabeth P. B. Fontes and Kristiina Mäkinen (eds.), *Plant-Virus Interactions*, Methods in Molecular Biology, vol. 2724,
https://doi.org/10.1007/978-1-0716-3485-1_20,

knock down the expression of genes through virus-induced gene silencing (VIGS) and to express proteins, including those for biotechnological purposes [3–10]. Many plant viral vectors have recently been engineered to deliver gene editing reagents [11, 12]. Many vectors can efficiently deliver sgRNAs into *SpCas9*-expressing plants for gene editing without requiring tissue culture. In most cases, viral vector-mediated delivery of sgRNAs has either not been shown to induce heritable gene editing or shown to induce low-efficiency heritable editing.

An RNA virus, tobacco mosaic virus (TMV), has been used to deliver *SpCas9* and sgRNA to induce gene editing transiently in the infiltrated leaves of *N. benthamiana* [13]. However, TMV with *SpCas9* or sgRNA is unable to move from initial infiltrated leaves into systemic tissue [13]. Another RNA virus, potato virus X (PVX), has been used to deliver *SpCas9* and sgRNA into *N. benthamiana* leaves [14] and one or more sgRNAs into *N. benthamiana* expressing *SpCas9* [15] to induce high-efficiency somatic editing. However, in both cases, tissue culture was used to regenerate edited plants from PVX-infected tissues [14, 15].

We recently used an RNA virus, tobacco rattle virus (TRV), to develop a TRV-based vector to deliver sgRNAs into *N. benthamiana* plants expressing *SpCas9*. This induced efficient somatic editing; however, editing was very low [16]. Another virus related to TRV, pea early browning virus (PEBV), has been used to express sgRNAs in transgenic *N. benthamiana* and *Arabidopsis thaliana* Col-0 plants expressing *SpCas9* [17]. However, the heritable transmission of PEBV-mediated gene editing was not reported in this study.

In collaboration with the Voytas group at University of Minnesota, our group recently showed that adding phloem mobile RNA sequences that are known to move to the shoot apical meristem into the sgRNAs can greatly enhance heritable editing efficiency [18]. TRV vector-based expression of sgRNAs augmented with mobile RNA sequences such as *Arabidopsis Flowering Locus T* (*AtFT*), tRNA methionine ($tRNA^{Met}$), and tRNA isoleucine ($tRNA^{Ile}$) induced high-efficiency somatic editing in the *SpCas9*-expressing *N. benthamiana* plants and 65–100% heritable biallelic editing in the next generation [18]. These findings demonstrated that expression of sgRNAs augmented with mobile RNA sequences using TRV vector; one could obtain biallelic heritable mutations in a single generation with high efficiency.

Recently, we showed that TRV augmented with $tRNA^{Ileu}$ also induces high-efficiency somatic and heritable editing in *Arabidopsis* [19]. Furthermore, the TRV with $tRNA^{Ileu}$ can induce high-efficiency heritable base-edits in *Arabidopsis* [20], and TRV with sgRNA flanked by self-splicing tRNA can induce epigene editing in *Arabidopsis* [21]. Expression of sgRNAs fused to *AtFT* using PVX has also been shown to induce heritable editing with 22% efficiency in *N. benthamiana* [15]. In *N. attenuata*, TRV-based delivery of sgRNA induced heritable editing only in plants that expressed Cas9 under *Ribosomal protein S5 A* (*RPS5A*) promoter that is known to be active in meristematic cells but not in plants that expressed *SpCas9* under cauliflower mosaic virus 35S promoter [22].

Among DNA viruses, geminiviral non-infectious DNA replicons have been used to enhance gene editing frequency when expressed in the transgenic *Nicotiana sp*, wheat, potato, and tomato plants [23–29]. In addition, a modified cabbage leaf curl geminivirus (CabLCV) vector has been used to deliver sgRNA into transgenic *N. benthamiana* plants expressing *SpCas9* [30]. Although this vector can induce editing of target genes in the newly emerging leaves, the heritable transmission of mutations is not possible with this system [30]. The geminivirus *cotton leaf crumple virus* (CLCrV) expressing a sgRNA fused to N terminus 102 bp of *AtFT* has been shown to induce low-efficiency heritable editing in *Arabidopsis*, but only in the second generation (M_2) [31].

Currently, few engineered viruses have been shown to deliver sgRNAs into Cas9-expressing monocot plants. Among these, only *the barley stripe mosaic virus* (BSMV) has been shown to induce heritable editing in wheat [32–34]. BSMV and *foxtail mosaic virus* (FoMV) have been shown to induce somatic editing in *SpCas9*-expressing maize, but no heritable editing was observed [35–37].

The RNA and DNA viruses discussed above can deliver sgRNAs but not Cas nucleases because of the limited cargo capacity of these viruses. Recently, a plant-negative strand virus, *sonchus yellow net nucleorhabdovirus* (SYNV)-based vector, has been engineered to deliver both *SpCas9* nuclease and sgRNAs into *N. benthamiana* plants [38]. Although the SYNV-based approach can induce efficient somatic editing, the heritable edits could be recovered only through the regeneration of plants from SYNV-infected somatic tissue [38].

Here, we describe a detailed protocol for heritable gene editing using a TRV-based vector in *SpCas9* expressing *N. benthamiana* (hereafter referred to as *N. benthamiana::SpCas9*) and *Arabidopsis thaliana* (hereafter referred to as *Arabidopsis::SpCas9*) plants based on our recent reports [18, 19]. The method we describe here can easily be adapted to study gene function in these plants.

2 Materials

2.1 Editing in N. benthamiana::SpCas9 and Arabidopsis::SpCas9 Plants

1. Soil (Sunshine® Mix #1, SunGro Horticulture, or similar).
2. 11″ W × 22″ L × 2.5″ D planting trays.
3. 5.25″ × 5.25″ pots.
4. 2.5″ × 2.5″ plastic pots.

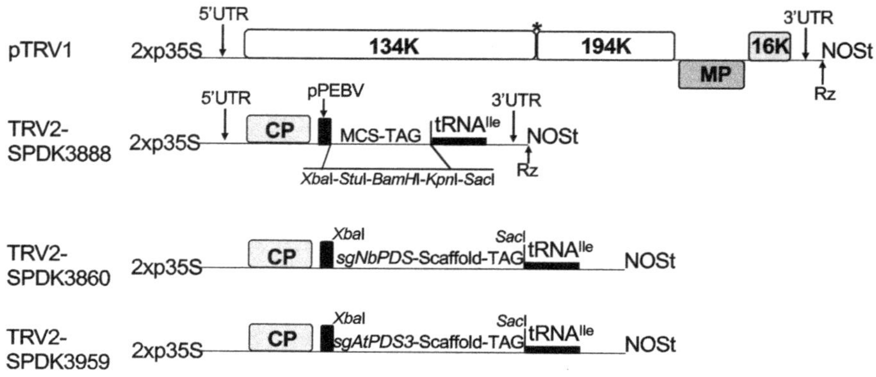

Fig. 1 Schematic representation of TRV1 and TRV2 derivative vectors. The cDNA of RNA1 and RNA2 of TRV is under the control of a duplicated CaMV 35S promoter (2xp35S) and nopaline synthase terminator (NOSt). A self-cleaving ribozyme (Rz) sequence is included before the NOSt to generate a precise 3′ end of TRV in the cells. 5′ UTR, 5′ untranslated region; 3′ UTR, 3′ untranslated region; *, indicates translational readthrough generating 194 K protein; MP, cell-to-cell movement protein, 16 K, 16-kDa cysteine-rich protein; CP, coat protein; pPEBV, *Pea early browning virus* subgenomic promoter; MCS-TAG, multiple cloning sites followed by a stop codon, $tRNA^{Ile}$, *Arabidopsis thaliana* tRNA isoleucine; sg*NbPDS*, single-guide RNA targeted to *N. benthamiana Phytoene Desaturase* gene; sg*AtPDS3*, single-guide RNA targeted to *Arabidopsis thaliana Phytoene Desaturase* 3 gene; scaffold, 76-bp sequence that binds to *Sp*Cas9 nuclease

5. pTRV1 (pYL192, Addgene) (Fig. 1): This is a TRV RNA1 vector that should be mixed with pTRV2 derivatives for infiltration into plants.
6. pTRV2::pPEBV::MCS::$tRNA^{Ileu}$ (SPDK3888; Addgene) (Fig. 1): This is TRV RNA2 vector that is used for expressing guides fused to $tRNA^{Ileu}$ under the *pea early browning virus* subgenomic promoter (pPEBV). Guide and scaffold sequences can be cloned into multiple cloning sites (MCS). For this, primers and *Xba*I and *Sac*I restriction enzymes are required.
7. Luria Bertani (LB) liquid media: 10 g/L tryptone, 5 g/L yeast extract, and 10 g/L NaCl dissolve in ddH_2O. Adjust pH to 7.0 and autoclave to sterilize.
8. LB agar plates with appropriate antibiotics: LB liquid media plus 15 g DIFCO Agar per 1 L and autoclave. Cool the media to 65 °C and add appropriate antibiotics and pour plates.
9. Filter—0.22 μm.
10. 50 mg/mL Kanamycin: Dissolve in ddH_2O, filter-sterilize (0.22 μm filter), and store at −20 °C.
11. 50 mg/mL Rifampicin: Dissolve in DMSO, filter-sterilize (0.22 μm filter), and store at −20 °C.
12. Glycerol.
13. Sterile Petri Dish, 100 × 15 mm.
14. 15-mL and 50-mL polypropylene centrifuge tubes.

15. 1 M magnesium chloride ($MgCl_2$): Prepare in double-deionized water (ddH_2O), autoclave to sterilize and store at room temperature.
16. 1 M MES [2-(N-Morpholino) ethane sulfonic acid]: Prepare in ddH_2O, adjust pH to 5.6 with NaOH, filter-sterilize (0.22 μm filter), and store at −20 °C.
17. 250 mM acetosyringone (AS, 3′, 5′ dimethoxy 4′-hydroxy acetophenone): Prepare in dimethyl sulfoxide (DMSO) and store at −20 °C.
18. Infiltration media (IM): 10 mM $MgCl_2$, 10 mM MES, and 250 μM AS prepared in ddH_2O.
19. Phusion polymerase.
20. Phire Plant Direct PCR Kit.

2.2 Editing in N. benthamiana::SpCas9 Plants

1. Seeds of *SpCas9*-expressing *N. benthamiana*.
2. 3.5″ × 3.5″ pots.
3. *Agrobacterium tumefaciens* strain GV2260.
4. 100 mg/mL Carbenicillin: Dissolve in ddH_2O, filter-sterilize (0.22-μm filter), and store at −20 °C.
5. 50 mg/mL Streptomycin: Dissolve in ddH_2O, filter-sterilize (0.22-μm filter), and store at −20 °C.
6. pTRV2::pPEBV::sg*NbPDS*::tRNA^{Ileu} (SPDK3860; Addgene #199245) (Fig. 1): Positive control; this is a TRV RNA2 vector that expresses guide targeted to *N. benthamiana Phytoene Desaturase* (*NbPDS*) gene fused to tRNA^{Ileu}.
7. 1-mL disposable Covidien Monoject™ Tuberculin Syringe.
8. 25-G (0.5 mm × 25 mm) BD safety glide needle.

2.3 Editing in Arabidopsis::SpCas9 Plants

1. Seeds of *SpCas9*-expressing *Arabidopsis thaliana* Columbia-0.
2. Murashige and Skoog (MS) plates: 2.2 g/L MS basal media with Gamborg's vitamins (Caisson Labs) and 10 g/L sucrose dissolve in ddH_2O. Adjust pH to 5.8 using 1 M potassium hydroxide (KOH). Add 0.25% Phytoblend agar and autoclave to sterilize. Cool the medium to 65 °C and pour 100 × 25 mm deep plates.
3. Phytoblend agar (Caisson Labs).
4. Sucrose.
5. 1 M potassium hydroxide (KOH): Prepare in ddH_2O and autoclave to sterilize.
6. 100% and 70% ethanol.
7. Triton X-100.
8. 3 M Micropore tape.

9. *Agrobacterium tumefaciens* strain GV3101.
10. 50 mg/mL Gentamicin: Dissolve in ddH_2O, filter-sterilize (0.22-μm filter), and store at −20 °C.
11. pTRV2::pPEBV::sg*AtPDS3*::tRNA^{Ileu} (SPDK3959; Addgene #199246) (Fig. 1): positive control; this is a TRV RNA2 vector that expresses guide targeted to *A. thaliana* PDS3 (*AtPDS3*) gene fused to tRNA^{Ileu}.

3 Methods

3.1 Growing Plants

3.1.1 Growing N. benthamiana::SpCas9 Plants

Sow 30–40 *SpCas9* expressing *N. benthamiana* seeds into moistened soil in a 5.25″ × 5.25″ pot, place it in a planting tray, and cover it with a transparent humidity dome. Keep the tray in a growth chamber with a photoperiod of 12-h light/12-h dark (100 μmol/m^2/s) at 23–25 °C and 40% humidity.

3.1.2 Growing Arabidopsis::SpCas9 Plants

1. Inside a laminar hood, transfer about 80 to 100 seeds in 1.5-mL Eppendorf tube, and add 1 mL of 70% ethanol and 0.01% Triton X-100 solution. Place them on an Eppendorf Mixer 5432 or a similar mixer for 4–5 min (*see* **Note 1**).
2. Let the seeds sink to the bottom of the tube and carefully discard the solution by vacuum aspiration (*see* **Note 2**).
3. Add 1 mL 100% ethanol and keep them on a stirrer for 3 min (*see* **Note 3**).
4. Carefully discard the ethanol by vacuum aspiration.
5. Repeat **steps 3** and **4** once.
6. Open the lid of each tube and place them in the laminar hood overnight or until they are completely dried.
7. Spread the sterilized seeds on MS plates and seal them with micropore tape.
8. Cover the plates in aluminum foil and keep them in the cold room or refrigerator for 3–4 days (*see* **Note 4**).
9. Keep plates in a growth chamber set at 24 °C with a 12-h light and 12-h dark condition with a light intensity of 100 μE m^{-2} s^{-1} for 7 to 10 days.

3.2 Construction of pTRV2 Containing sgRNA Target Plasmids for Editing

1. Select 3–4 sgRNA targets for your target gene of interest using CRISPOR (http://crispor.tefor.net/) or CRISPR-P (http://crispr.hzau.edu.cn/CRISPR2/) (*see* **Notes 5** and **6**).
2. Design forward primer with *Xba*I site and reverse primer with *Sac*I site and use SPDK3860 as a template to PCR guide plus scaffold using Phusion polymerase and clone into *Xba*I-*Sac*I cut SPDK3888 (Fig. 2a). Alternatively, an oligo containing

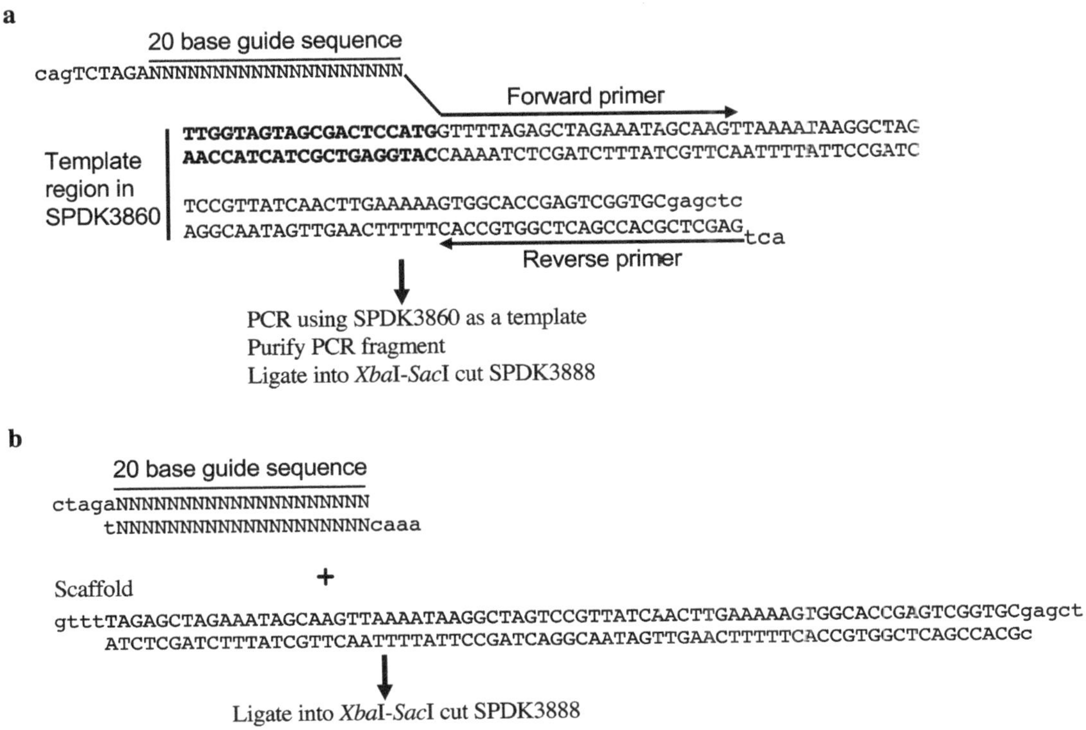

Fig. 2 Strategies for cloning sgRNA targets into pTRV2 vectors. (**a**) PCR guide plus the scaffold using SPDK3860 as a template with forward and reverse primers, as shown. Purify PCR fragment and clone into *Xba*I-*Sac*I cut SPDK3888. (**b**) Design oligos that include target guide sequence (top) and oligos that correspond to scaffold (bottom) with overhangs and anneal. Ligate the oligos into *Xba*I-*Sac*I cut SPDK3888

guide sequence and an oligo containing scaffold sequence could be annealed and ligated directly into *Xba*I-*Sac*I cut SPDK3888 (Fig. 2b) (*see* **Note 7**).

3. Confirm the sequence of all constructs by Sanger sequencing using primers SP9338 (5′- CATAATTATACT GATTTGTCTCTCG) and SP8626 (5′-CAAAAGACTTACC GATCAATCAAG-3′).

3.3 Introducing TRV Constructs into Agrobacterium

1. *For Editing in N. benthamiana::SpCas9 Plants:* Transform pTRV1, SPDK3860 (positive control), SPDK3888 (negative control), and SPDK3888 + sgRNA target generated in Subheading 3.2 separately into *A. tumefaciens* strain GV2260 using the freeze–thaw method of transformation. Briefly, add 500 ng of plasmid DNA into 50 μL GV2260 competent cells and incubate on ice for 30 min. Put the tubes in a floater into liquid nitrogen. After 1 min, move the tubes to 37 °C water bath for 5 min. Add 250 μL LB liquid media and incubate in a 28 °C shaker for 2–2.5 h. Plate 100–150 μL onto a fresh LB agar plate with KSRC antibiotics (50 mg/L kanamycin,

50 mg/L streptomycin, 25 mg/L rifampicin, and 50 mg/L carbenicillin) and grow at 28 °C for 2–3 days. Pick and streak two colonies onto a fresh KSRC containing LB agar plate and grow at 28 °C overnight. Inoculate 2 mL LB liquid culture with KSRC, grow overnight at 28 °C, make glycerol stocks, and store at −80 °C (*see* **Note 8**).

2. *For Editing in Arabidopsis::SpCas9 Plants:* Transform pTRV1, SPDK3959 (positive control), SPDK3888 (negative control), and SPDK3888 + sgRNA target generated in Subheading 3.2 separately into 50 μL of *A. tumefaciens* GV3101 competent cells using heat-shock method similar to the one used for transformation of *E. coli*. Add 250 μL LB liquid media and incubate in a shaker at 28 °C for 1 h. Plate 100–150 μL cells onto LB agar plate with KRG antibiotics (50 mg/L kanamycin, 25 mg/L rifampicin, and 50 mg/L gentamicin) and grow at 28 °C for 2 days. Pick and streak two colonies onto a fresh LB plate with KRG antibiotics and incubate overnight at 28 °C. Make glycerol stock and store at −80 °C.

3.4 Editing Target Gene of Interest in N. benthamiana::SpCas9 Plants

The steps involved in generating heritable editing in the target gene of interest using *N. benthamiana::SpCas9* plants and TRV system are summarized in Fig. 3.

3.4.1 Infiltration of Agrobacterium Harboring TRV Constructs into N. benthamiana::SpCas9 Plants

1. Streak out Agrobacterium GV2260 containing pTRV1, SPDK3860, and SPDK3888 with sgRNA targets from Subheading 3.3 **step 1** onto a fresh LB plate with KSRC antibiotics. Incubate overnight at 28 °C.
2. Inoculate 6 mL LB liquid media with KSRC antibiotics and incubate in a shaker at 28 °C for 14–16 h (*see* **Note 9**).
3. Collect Agrobacterium cells by centrifugation at 3500×*g* for 20 min. Resuspend the pellet in agroinfiltration media to $OD_{600} = 1.0$.
4. Incubate the Agrobacterium cultures on a flat shaker at 50–60 rpm for a minimum of 3 h at room temperature (*see* **Note 10**).
5. Mix pTRV1 with TRV2 derivatives (SPDK3860 or SPDK3888 or SPDK3888 + sgRNA target) in a 1:1 ratio.
6. Using a 25-G needle, prick the abaxial side of the three leaves to be infiltrated on each plant. Use plants 2 weeks old after transplanting for infiltration (*see* **Notes 11** and **12**).
7. Use a 1-mL needleless syringe to infiltrate mixed Agrobacterium solution from **step 5** into three leaves with a needle prick from **step 6** (Fig. 3a). Repeat infiltration until the entire leaf is

a

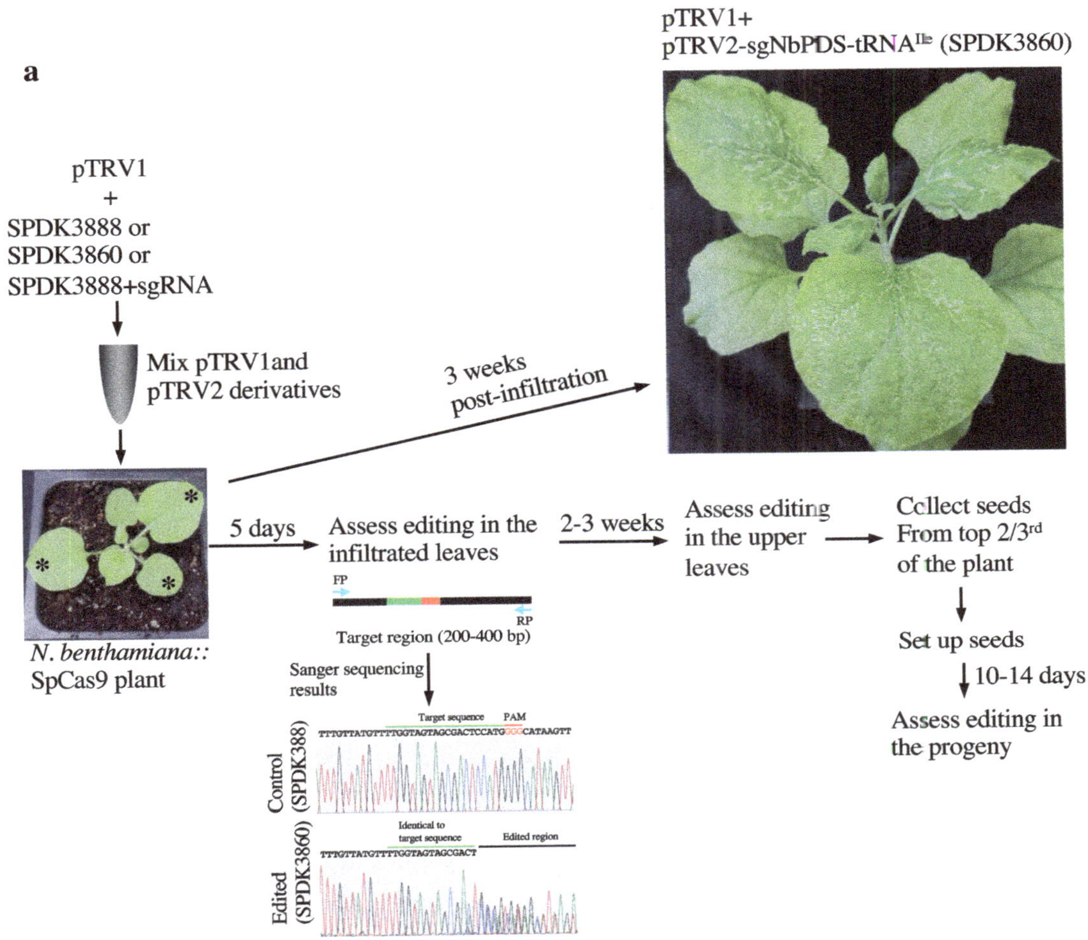

b

ID	Sequence	Group
White 1	AAGGTACA	1
White 2	ACAGCAGA	1
White 3	ACGCTCGA	1
White 4	ACTATGCA	1
White 5	AGATCGCA	1
Green 1	AGTCACTA	1
Green 2	ATTGAGGA	1
Green 3	CAAGACTA	1
Green 4	CACTTCGA	1
Col-0	CATACCAA	1

Fig. 3 Heritable genome editing using *N. benthamiana::SpCas9* plants with TRV system. (**a**) Key steps to generate heritable editing in a target gene of interest are depicted. Black asterisks indicate leaves that should be infiltrated. In the target region schematic, green and red represent target guide sequence (N_{20}) and PAM (NGG), respectively; *FP* forward primer, *RP*, reverse primer. An example of a Sanger sequence trace file includes samples from negative control SPDK3888 with no editing (top) and positive control SPDK3860 with editing in *NbPDS* (bottom); the sequence below the green and red line represents target and PAM sequences. The plant image on the top right shows the typical editing phenotype of *NbPDS*; white photobleached spots or

infiltrated with the Agrobacterium solution. Infiltrate three plants per sgRNA target (*see* **Note 13**).

8. Keep the infiltrated plants in the lab overnight. Transfer plants to the growth chamber set at 24 °C with a 14-h light and 10-h dark condition with a light intensity of 100 μE m^{-2} s^{-1} or light cart with a day/night cycle timer at room temperature.

3.4.2 Assessment of Somatic Editing in the TRV Infiltrated and Systemic Leaves of N. benthamiana::SpCas9 *Plants*

1. Five days post-agrobacterium infiltration with TRV vectors, take a 0.5-mm-diameter leaf sample from the infiltrated leaves and use Phire Plant Direct PCR Kit to PCR the DNA fragment flanking the sgRNA target site. Aim for 200–400 bp fragments (Fig. 3a) (*see* **Note 14**).
2. Purify the PCR fragment using a Zymo PCR Purification Kit and sequence it with Sanger sequencing, using the forward and reverse primers used for amplifying the PCR fragment. Typical trace files from Sanger sequencing showing editing in *NbPDS* is shown in Fig. 3a.
3. Analyze the Sanger sequence data using ICE analysis (https://ice.synthego.com/#/) (REF) to determine the efficiency of editing in the infiltrated leaves (*see* **Note 15**).
4. About 2–3 weeks post-infiltration of TRV vectors, assess the somatic editing in the systemic leaves by following **steps 1–3** described in Subheading 3.4.2 (*see* **Note 16**).

3.4.3 Assessment of Heritable Editing in Progeny from Seeds Collected from TRV-Infiltrated N. benthamiana::SpCas9 *Plants*

1. Collect seeds from the top 1/3rd of TRV-infiltrated plants that show high somatic editing efficiency, as described in [18].
2. Sow individual seeds into moistened soil in a 2.5″ × 2.5″ pot, place it in an 11″ W × 22″ L × 2.5″ D planting tray, and cover the tray with a transparent humidity dome. Set up 32 pots that fit into a single tray. Keep the tray in a growth chamber with a photoperiod of 14 h light/10 h dark (100 μmol/m^2/s) at 24 ° C and 40–60% humidity (*see* **Note 17**).
3. Ten to fourteen days post-sowing the seed, assess the editing in individual progeny plants by following **steps 1–3** described in Subheading 3.4.2 (*see* **Note 18**).
4. To determine the type of mutation and percentage of mutation type in each progeny plant, sequence the PCR product flanking the target site using Amplicon-EZ next-generation sequencing (NGS) from Azenta Life Sciences (Formerly Genewiz) or from another similar amplicon NGS sequencing servicing company.

Fig. 3 (continued) regions correspond to tetra-allelic editing of *NbPDS* mediated by TRV. (**b**) Representative barcode file used for demultiplexing by fastq-multx

3.4.4 Analysis of NGS Sequencing and Determination of Editing Efficiency and Type of Editing

1. For demultiplexing, we use fastq-multx (https://github.com/brwnj/fastq-multx).
2. Create separate .txt files that correspond to your forward and reverse barcodes, respectively. They should contain three columns: Id, Sequence, and group (Fig. 3b).
   (a) Id = Unique sample identifier.
   (b) Sequence = Sequence that is unique to the sample, e.g., barcode.
   (c) Group = Leave as 1.
3. Use the command fastq-multx -l fwd_barcode.txt ForwardReads.fq ReverseReads.fq -o %R1%R2.
4. Each Id in the barcode file will be given two separate files, R1 and R2, containing all reads which contain the barcode sequence.
5. Use the command fastq-multx -l rev_barcode.txt Sample1_R1.fq Sample1_R2.fq -o %R1%R2 (*see* **Note 19**).
6. Upload both the R1 and R2 files for each sample to (http://www.rgenome.net/cas-analyzer/#!).
7. Add your reference sequence to the template (i.e., the sequenced amplicon).
8. Add the target sequence (i.e., gRNA target).
9. Adjust the comparison range to an appropriate length; for our analyses, we use a 40 bp comparison range (*see* **Note 20**).
10. Submit the program and record the editing efficiency and type of editing.

3.5 Editing Target Gene of Interest in Arabidopsis::SpCas9 Plants

The steps involved in generating heritable editing in a target gene of interest using *A. thaliana::SpCas9* plants and the TRV system are summarized in Fig. 4.

3.5.1 Flooding of Agrobacterium with TRV Constructs into Arabidopsis::SpCas9 Plants

1. Streak out Agrobacterium GV3101 containing pTRV1, SPDK3959, SPDK3888, and SPDK3888+sgRNA target generated in Subheading 3.2 onto a fresh LB plate with KRG antibiotics. Incubate overnight at 28 °C.
2. Inoculate Agrobacterium GV3101 containing various constructs into 5 mL LB liquid media with KRG antibiotics and incubate in a shaker at 28 °C for 14–16 h. For each TRV2 derivative, inoculate 5 mL pTRV1.
3. Collect Agrobacterium cells by centrifugation at 3500×*g* for 20 min. Resuspend the pellet in 5 mL sterile water by pipetting up and down (no vortexing).

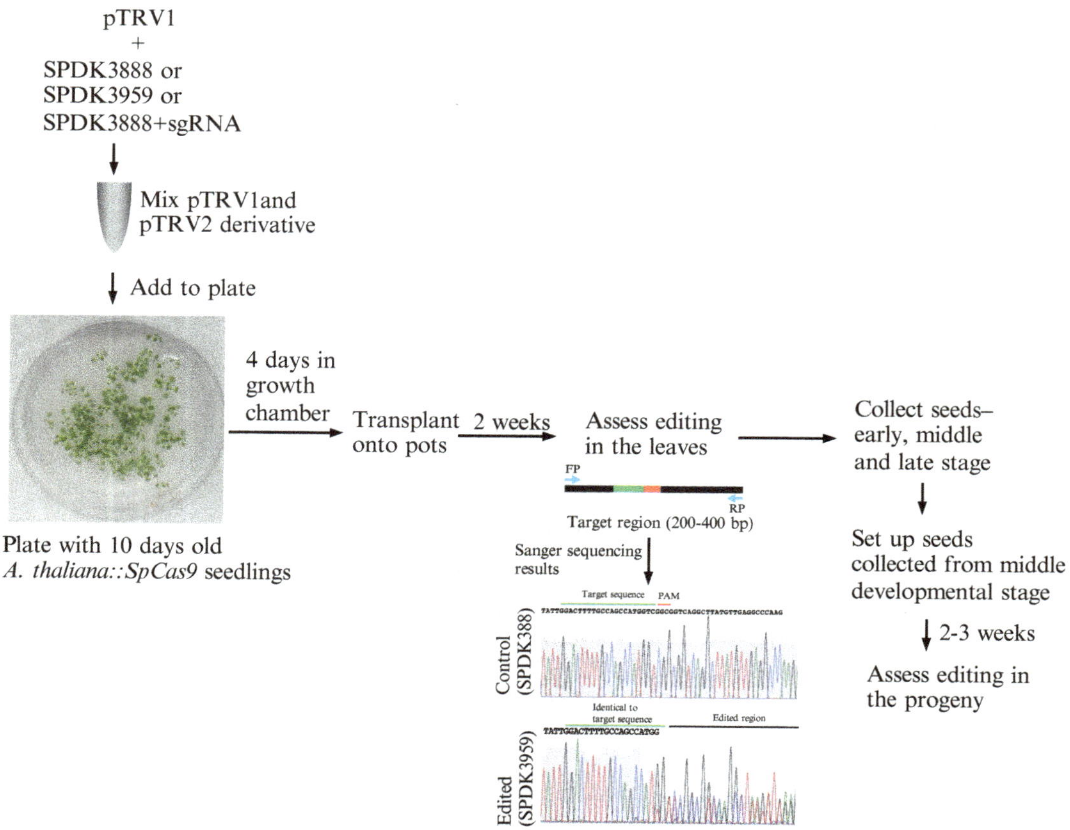

Fig. 4 Heritable genome editing using *A. thaliana::SpCas9* plants with TRV system. Key steps to generate heritable editing in a target gene of interest are depicted. In the target region schematic, green and red represent target guide sequence (N_{20}) and PAM (NGG), respectively; *FP* forward primer, *RP*, reverse primer. A Sanger sequence trace file example includes samples from negative control SPDK3888 with no editing (top) and positive control SPDK3959 with editing in *AtPDS3* (bottom); sequences below the green and red line represent target and PAM sequences

4. Centrifuge at 3500×*g* for 10 min. Resuspend the pellet in agroinfiltration media to $OD_{600} = 1.5$. Prepare 5 mL culture for each construct.
5. Incubate the cultures on a flat shaker at 50–60 rpm for 3 h at room temperature.
6. Mix pTRV1 with TRV2 derivatives (SPDK3959 or SPDK3888 or SPDK3888 + sgRNA target) in a 1:1 ratio.
7. Use a 1-mL pipette to dispense the mixed TRV culture onto an MS plate containing 10-day-old *Arabidopsis::SpCas9* seedlings (*see* Subheading 3.1.2) (Fig. 4). Add enough culture so that all the seedlings are completely submerged in the TRV culture.
8. Seal the plates with micropore tape and incubate them in a growth chamber set at 24 °C with a 12-h light/12-h dark condition with a light intensity of 100 μE m^{-2} s^{-1} for 4 days.

9. Transplant at least 20–30 seedlings from the plate onto 2.5″ × 2.5″ pots containing moistened soil and keep the pots in 11″ W × 22″ L × 2.5″ D trays. Cover the tray with a transparent humidity dome. Keep the tray in a growth chamber with a photoperiod of 14 h light/10 h dark (100 μmol/m^2/s) at 24 °C and 40–60% humidity. Remove the humidity dome after 3–4 days (*see* **Note 21**).

3.5.2 Assessment of Somatic Editing in the TRV-Flooded Arabidopsis::SpCas9 Plants

Two weeks post-transplanting of TRV Agro-flooded plants, assess somatic editing by following **steps 1–3** described in Subheading 3.4.2. Typical trace files from Sanger sequencing showing editing in *AtPDS3* are shown in Fig. 4 (*see* **Note 22**).

3.5.3 Assessment of Heritable Editing in Progeny from Seeds Collected from TRV-Flooded Arabidopsis::SpCas9 Plants

1. Collect seeds in three batches—early, middle, and late-stage developed seeds as described in [19].
2. Soak 40–50 seeds from the middle batch in 1.5-mL Eppendorf tube containing 1 mL water. Keep it in the cold room or refrigerator for 3 days.
3. Place individual seeds in 2.5″ × 2.5″ pot with moistened soil and place the pots in an 11″ W × 22″ L × 2.5″ D tray, and cover the tray with a transparent humidity dome. Set up 32 pots that fit into a single tray. Keep the tray in a growth chamber with a photoperiod of 14 h light/10 h dark (100 μmol/m^2/s) at 24 ° C and 40–60% humidity. Take out the humidity dome once the seeds germinate and reach 2–3 leaf stage.
4. About 2–3 weeks post-sowing, assess the editing in individual progeny plants by following **steps 1–3** in Subheading 3.4.2, **step 4** in Subheading 3.4.3, and **steps 1–10** in Subheading 3.4.4.

4 Notes

1. Use ethanol-resistant marker to label the tubes. Mixing can be performed using any rotating shaker.
2. A 200- or 1000-μL pipet could also be used to remove the liquid.
3. Longer incubation will bleach the seeds.
4. Cold treatment, also called stratification, will help to break dormancy and synchronize germination.
5. It is better to design 3–4 guides and test the efficiency of editing in the infiltrated leaves. If possible, target the beginning of the coding region of the gene for editing. Follow up only with those guides that show the high efficiency of somatic editing in the infiltrated leaves.

6. To target two or three genes, generate sgRNA targets for each gene with a 23-bp spacer in between the targets. See the design example as described in [19].
7. Other cloning methods, such as in-fusion or Golden Gate, can also be used to clone, guide, and scaffold into SPDK3888.
8. *A. tumefaciens* strain GV3101 can also be used for *N. benthamiana*. However, we sometimes see necrosis of infiltrated leaves with GV3101.
9. For each TRV2 derivative construct to be infiltrated, inoculate 6 mL of pTRV1.
10. A minimum of 3-h incubation is required for optimal induction of *vir* genes.
11. A 200-ul pipette tip or razor blade could also be used to make a small prick or slit.
12. Younger plants are better at obtaining better infection and editing efficiency. Infectivity is lower when older plants are used.
13. To prevent cross-contamination, for each sgRNA target in TRV use a new needless syringe and fresh pair of gloves.
14. CTAB or any other alternative methods could be used to get the template DNA for PCR.
15. Use Sequences from plants infiltrated with negative control SPDK3888 to assess the editing efficiency. The efficiency of editing varies between 60% to 95%. It is better to follow up with plants that show more than 75% editing efficiency.
16. Two weeks after infiltration of pTRV1 plus SPDK3860, photobleaching due to editing of *NbPDS* will be visible in the systemic leaves.
17. When screening for *PDS* phenotype, cover the top of the pot with dark soil before sowing the seeds so that the white seedlings phenotype will be visible better.
18. If the editing efficiency is more than 90%, it generally means the plant contains biallelic homozygous mutation in the target gene of interest.
19. Completing this second demultiplexing step is not required for the analysis. However, it does ensure that all analyzed reads contain the proper barcode on both ends of the sequence.
20. The comparison range can be set to any value; however, a smaller comparison range will minimize the quantification of low-abundance SNPs not caused by Cas9-mediated editing. It is likely that these erroneous SNPs are caused by PCR or sequencing errors.

21. Editing efficiency in *Arabidopsis* is lower than in *N. benthamiana*. Therefore, at least 25–30 seedlings after treatment with TRV should be genotyped to find some plants that will have high somatic editing.
22. After 2 weeks of flooding Agrobacterium harboring pTRV1 plus SPDK3959, photobleaching due to editing of *AtPDS3* will be visible in the leaves.

Acknowledgments

We thank Neha Dinesh-Kumar for editing the manuscript. This work was supported by grants from Agricultural Innovation through Gene Editing program grant no. 2020-67013-31544/ project accession no. 1022332, USDA National Institute of Food and Agriculture (to S.P.D.-K. and U.N.), USDA-NIFA predoctoral fellowship grant 2021-67034-35187 (to NM), and National Science Foundation (NSF) grant IOS-2139987 (to S.P.D-K).

References

1. Wang JY, Doudna JA (2023) CRISPR technology: a decade of genome editing is only the beginning. Science 379:eadd8643
2. Zhu H, Li C, Gao C (2020) Applications of CRISPR-Cas in agriculture and plant biotechnology. Nat Rev Mol Cell Biol 21:661–677
3. Wang M, Gao SL, Zeng WZ, Yang YQ, Ma JF, Wang Y (2020) Plant virology delivers diverse toolsets for biotechnology. Viruses 12:1338
4. Cody WB, Scholthof HB (2019) Plant virus vectors 3.0: transitioning into synthetic genomics. Ann Rev Phytopathol 57:211–230
5. Ibrahim A, Odon V, Kormelink R (2019) Plant viruses in plant molecular pharming: toward the use of enveloped viruses. Front Plant Sci 10:803
6. Abrahamian P, Hammond RW, Hammond J (2020) Plant virus-derived vectors: applications in agricultural and medical biotechnology. Annu Rev Virol 7:513–535
7. Khakhar A, Voytas DF (2021) RNA viral vectors for accelerating plant synthetic biology. Front Plant Sci 12:668580
8. Torti S, Schlesier R, Thummler A, Bartels D, Romer P, Koch B, Werner S, Panwar V, Kanyuka K, Wiren NV, Jones JDG, Hause G, Giritch A, Gleba Y (2021) Transient reprogramming of crop plants for agronomic performance. Nat Plants 7:159–171
9. Tengzhi X, Nagalakshmi U, Dinesh-Kumar SP (2021) Virus-induced gene silencing (VIGS). In: Bamford DH, Zuckerman M (eds) Encyclopedia of virology, vol 3, 4th edn. Academic Press, Oxford, pp 123–131
10. Rossner C, Lotz D, Becker A (2022) VIGS goes viral: how VIGS transforms our understanding of plant science. Annu Rev Plant Biol 73:703–728
11. Zaidi SS, Mansoor S (2017) Viral vectors for plant genome engineering. Front Plant Sci 8: 539
12. Zhang C, Liu S, Li X, Zhang R, Li J (2022) Virus-induced gene editing and its applications in plants. Internat J Mol Sci 23:10202
13. Cody WB, Scholthof HB, Mirkov TE (2017) Multiplexed gene editing and protein overexpression using a tobacco mosaic virus viral vector. Plant Physiol 175:23–35
14. Ariga H, Toki S, Ishibashi K (2020) Potato virus X vector-mediated DNA-free genome editing in plants. Plant Cell Physiol 61:1946–1953
15. Uranga M, Aragones V, Selma S, Vazquez-Vilar M, Orzaez D, Daros JA (2021) Efficient Cas9 multiplex editing using unspaced sgRNA arrays engineering in a potato virus X vector. Plant J 106:555–565
16. Ali Z, Abul-faraj A, Li L, Ghosh N, Piatek M, Mahjoub A, Aouida M, Piatek A, Baltes NJ, Voytas DF, Dinesh-Kumar S, Mahfouz MM (2015) Efficient virus-mediated genome

editing in plants using the CRISPR/Cas9 system. Mol Plant 8:1288–1291

17. Ali Z, Eid A, Ali S, Mahfouz MM (2018) Pea early-browning virus-mediated genome editing via the CRISPR/Cas9 system in *Nicotiana benthamiana* and *Arabidopsis*. Virus Res 244: 333–337
18. Ellison EE, Nagalakshmi U, Gamo ME, Huang PJ, Dinesh-Kumar S, Voytas DF (2020) Multiplexed heritable gene editing using RNA viruses and mobile single guide RNAs. Nat Plants 6:620–624
19. Nagalakshmi U, Meier N, Liu JY, Voytas DF, Dinesh-Kumar SP (2022) High-efficiency multiplex biallelic heritable editing in Arabidopsis using an RNA virus. Plant Physiol 189:1241–1245
20. Liu D, Xuan S, Prichard LE, Donahue LI, Pan C, Nagalakshmi U, Ellison EE, Starker CG, Dinesh-Kumar SP, Qi Y, Voytas DF (2022) Heritable base-editing in Arabidopsis using RNA viral vectors. Plant Physiol 189: 1920–1924
21. Ghoshal B, Vong B, Picard CL, Feng S, Tam JM, Jacobsen SE (2020) A viral guide RNA delivery system for CRISPR-based transcriptional activation and heritable targeted DNA demethylation in *Arabidopsis thaliana*. PLoS Genet 16:e1008983
22. Oh Y, Kim SG (2021) RPS5A promoter-driven Cas9 produces heritable virus-induced genome editing in *Nicotiana attenuata*. Mol Cells 44: 911–919
23. Baltes NJ, Gil-Humanes J, Cermak T, Atkins PA, Voytas DF (2014) DNA replicons for plant genome engineering. Plant Cell 26:151–163
24. Butler NM, Atkins PA, Voytas DF, Douches DS (2015) Generation and inheritance of targeted mutations in potato (*Solanum tuberosum* L.) using the CRISPR/Cas system. PLoS One 10:e0144591
25. Cermak T, Baltes NJ, Cegan R, Zhang Y, Voytas DF (2015) High-frequency, precise modification of the tomato genome. Genome Biol 16:232
26. Cermak T, Curtin SJ, Gil-Humanes J, Cegan R, Kono TJY, Konecna E, Belanto JJ, Starker CG, Mathre JW, Greenstein RL, Voytas DF (2017) A multipurpose toolkit to enable advanced genome engineering in plants. Plant Cell 29:1196–1217
27. Wang M, Lu Y, Botella JR, Mao Y, Hua K, Zhu JK (2017) Gene targeting by homology-directed repair in rice using a geminivirus-based CRISPR/Cas9 system. Mol Plant 10: 1007–1010
28. Gil-Humanes J, Wang Y, Liang Z, Shan Q, Ozuna CV, Sanchez-Leon S, Baltes NJ, Starker C, Barro F, Gao C, Voytas DF (2017) High-efficiency gene targeting in hexaploid wheat using DNA replicons and CRISPR/Cas9. Plant J 89:1251–1262
29. Dahan-Meir T, Filler-Hayut S, Melamed-Bessudo C, Bocobza S, Czosnek H, Aharoni A, Levy AA (2018) Efficient *in planta* gene targeting in tomato using geminiviral replicons and the CRISPR/Cas9 system. Plant J 95:5–16
30. Yin K, Han T, Liu G, Chen T, Wang Y, Yu AY, Liu Y (2015) A geminivirus-based guide RNA delivery system for CRISPR/Cas9 mediated plant genome editing. Sci Rep 5:14926
31. Lei J, Dai P, Li Y, Zhang W, Zhou G, Liu C, Liu X (2021) Heritable gene editing using FT mobile guide RNAs and DNA viruses. Plant Methods 17:20
32. Li T, Hu J, Sun Y, Li B, Zhang D, Li W, Liu J, Li D, Gao C, Zhang Y, Wang Y (2021) Highly efficient heritable genome editing in wheat using an RNA virus and bypassing tissue culture. Mol Plant 14:1787–1798
33. Chen H, Su Z, Tian B, Liu Y, Pang Y, Kavetskyi V, Trick HN, Bai G (2022) Development and optimization of a *Barley stripe mosaic virus*-mediated gene editing system to improve Fusarium head blight resistance in wheat. Plant Biotechnol J 20:1018–1020
34. Wang W, Yu Z, He F, Bai G, Trick HN, Akhunova A, Akhunov E (2022) Multiplexed promoter and gene editing in wheat using a virus-based guide RNA delivery system. Plant Biotechnol J 20(12):2332–2341
35. Hu J, Li S, Li Z, Li H, Song W, Zhao H, Lai J, Xia L, Li D, Zhang Y (2019) A barley stripe mosaic virus-based guide RNA delivery system for targeted mutagenesis in wheat and maize. Mol Plant Pathol 20:1463–1474
36. Mei Y, Beernink BM, Ellison EE, Konecna E, Neelakandan AK, Voytas DF, Whitham SA (2019) Protein expression and gene editing in monocots using foxtail mosaic virus vectors. Plant Direct 3:e00181
37. Beernink BM, Lappe RR, Bredow M, Whitham SA (2022) Impacts of RNA mobility signals on virus-induced somatic and germline gene editing. Front Genome Ed 4:925088
38. Ma X, Zhang X, Liu H, Li Z (2020) Highly efficient DNA-free plant genome editing using virally delivered CRISPR-Cas9. Nat Plants 6: 773–779

Chapter 21

Immunoprecipitation Followed by Mass Spectrometry: An Approach for Identifying Host–Viral Protein–Protein Interactions

Laura Medina-Puche and Rosa Lozano-Durán

Abstract

As obligate intracellular parasites, viruses rely on the efficient manipulation of the cell they invade in order to multiply and spread. Protein–protein interactions between viral proteins (or their complexes) and cellular proteins are at the interface between virus and host and hence crucial for the outcome of the infection. Multiple techniques can be used to study protein–protein interactions in vivo in the context of the infected cell; among them, immunoprecipitation followed by mass spectrometry (IP-MS) has proven an efficient approach for the unbiased identification of protein complexes containing a viral protein of interest. In this chapter, we discuss how to employ IP-MS to define the interactome of plant virus proteins by using transient expression in the experimental host *Nicotiana benthamiana*, using the geminivirus tomato yellow leaf curl virus (TYLCV) as an example.

Key words Immunoprecipitation, Mass spectrometry, IP-MS, Protein complexes, Protein–protein interactions, *Nicotiana benthamiana*

1 Introduction

Since viruses are strict intracellular parasites, they have evolved strategies to efficiently manipulate the cells they infect, on which they fully rely. Such strategies commonly depend on the physical association between viral and cellular proteins and ultimately determine the outcome of the virus–host interaction. It clearly transpires, therefore, that the study of protein–protein interactions is crucial for the understanding of the viral infection as well as the host anti-viral responses. These interactions, whenever possible, must be studied in the context of the infection, which changes the cellular environment, potentially enabling associations that are simply not possible in uninfected cells due, for example, to the absence of a partner or a given post-translational modification.

Elizabeth P. B. Fontes and Kristiina Mäkinen (eds.), *Plant-Virus Interactions*, Methods in Molecular Biology, vol. 2724,
https://doi.org/10.1007/978-1-0716-3485-1_21,

In the absence of widely used, well-established plant cell lines amenable to easy transformation protocols to study viral infection, the plant virology field has massively profited from the use of the experimental species *Nicotiana benthamiana*, which allows *Agrobacterium tumefaciens*-mediated transient transformation of virtually all cells in a leaf patch, therefore enabling a concomitant viral infection and expression of a tagged protein of interest, if required, in a local, temporally controlled manner.

Multiple techniques have been developed to detect protein–protein interactions, both in vivo, in vitro, and in surrogate systems. Immunoprecipitation followed by mass spectrometry (IP-MS) approaches has recently experienced a surge in virology studies due to technological advances in mass spectrometry-based proteomics and experimental pipelines for immunopurification of protein complexes, as well as the intrinsic advantages that this technique offers [1]. Among these, the *in vivo* detection of the interactions, the possibility to infer the interactomes of viral proteins in isolation, in combination, or in the context of the infection, as well as the chance to detect post-translational modifications of the proteins of interest and both direct and indirect interactions, including higher-order protein complexes, make it particularly suitable for unbiased screens for interacting partners in plant cells.

A conventional IP-MS workflow starts with the expression of the viral protein of interest in *N. benthamiana*, be it from the infectious clone directly or from a binary vector if the protein needs to be fused to a tag to enable immunopurification. Following the accumulation of the transgenically expressed protein, the cells are disrupted, total protein is isolated, and immunopurification (with antibodies against the protein of interest or against the fused tag) ensues. The purified protein complexes are then digested and can be separated by liquid chromatography (LC) and analyzed by tandem MS (MS/MS) (Fig. 1). Bioinformatic analysis and data processing will provide a list of putative interactors, which must be experimentally confirmed (Fig. 2).

In this chapter, we describe the use of IP-MS to define the interactome of plant virus proteins by using transient expression in *N. benthamiana*, with the geminivirus tomato yellow leaf curl virus (TYLCV) as a model case; experimental details, as well as a simplified overview of the general workflow (Figs. 1 and 2), are provided.

2 Materials

2.1 Equipment

1. Standard lab plasticware and adjustable micropipettes.
2. Thermocycler.
3. Electrophoresis equipment.
4. Protein transfer equipment.

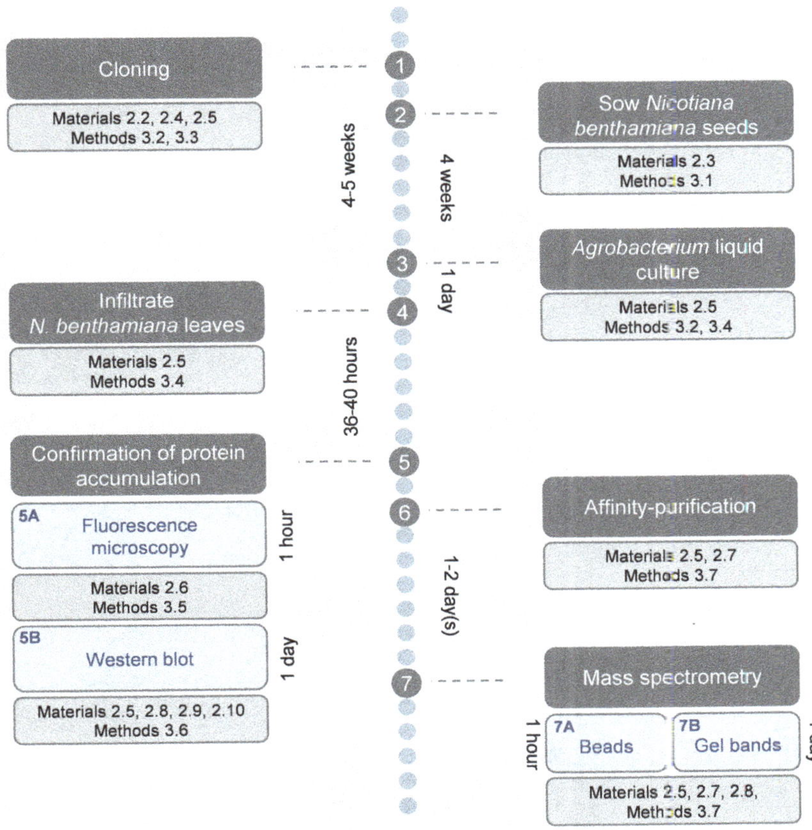

Fig. 1 Schematic protocol overview. In brief, the *Agrobacterium tumefaciens* strain(s) carrying the plasmid (s) of interest is/are liquid-cultured in LB medium supplemented with the appropriate antibiotics and incubated at 28 °C overnight. After centrifugation, the bacterial pellet is resuspended in agroinfiltration buffer, OD_{600} adjusted accordingly, and infiltrated in the abaxial side of a leaf of a 3- to 4-week-old *N. benthamiana* plant. 36–40 h later, 4- or 8-mm leaf disks are collected to confirm protein accumulation under the fluorescence microscope or by Western blot. The rest of the leaf tissue is collected for further protein extraction, followed by immunoprecipitation–mass spectrometry (IP-MS) analysis. The estimated time required for the major steps is indicated. Reference to relevant Materials and Methods sections is indicated in boxes

5. Temperature-adjustable ovens set at 28 and 37 °C.
6. Space for working in sterile conditions (e.g., a lab bench with Bunsen or alcohol burner, or a laminar flow hood).
7. Fume hood.
8. Temperature-adjustable shakers for liquid cultures set at 28 and 37 °C with speed at 200 rpm.
9. Centrifuge and microfuge.
10. Spectrophotometer.
11. Plastic spectrophotometer cuvettes (1 mL).

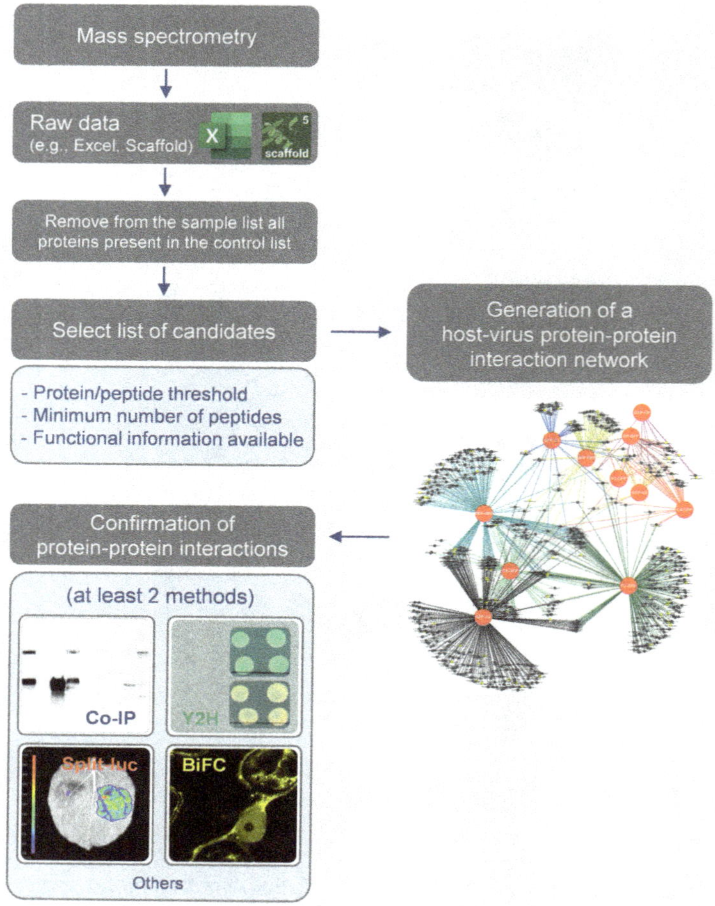

Fig. 2 Schematic overview of a recommended pipeline for the data analysis. The main steps for filtering the data, selecting a list of candidate interactors, and the subsequent confirmation of the protein–protein interactions are indicated. The network representation of TYLCV-*N. benthamiana* protein–protein interactions is reprinted from [4]. Viral proteins (fused to GFP) are depicted as red circles; *N. benthamiana* proteins are depicted as gray or yellow circles. Best reciprocal hits from *Arabidopsis thaliana* proteins identified as hubs in the Arabidopsis interactome [7] are depicted as yellow nodes; non-hub proteins are depicted as black nodes. Lines (edges) indicate interactions between viral and host proteins

12. 1-mL syringe (needleless).
13. Corkborer (4 and 8 mm diameter).
14. Balance.
15. Revolver lab rotator.
16. Glass slides (76 × 26 mm).
17. Coverslips (24 × 50 mm).

18. Imaging system to detect fluorescence (e.g., a fluorescence microscope or a confocal laser scanning microscope (CLSM)).
19. Enhanced chemiluminescent (ECL) substrates, or X-ray films and X-ray developer and fixer solutions.
20. Imaging system to capture and analyze protein sample gels and membranes in high resolution or a darkroom for X-ray film processing.
21. Water purification system.
22. Autoclave.

2.2 Cloning

If a specific antibody against the viral protein of interest is available, transient expression of the protein in *N. benthamiana* can be carried out directly from the infectious viral clone. If this is not the case, the viral gene of interest must be fused to a tag for immunopurification and expressed from the viral clone (if possible) or from a binary vector.

1. Primers to amplify the coding sequence of interest (CDS) (*see* **Note 1**). To amplify the corresponding CDS, please note that the forward primers include the initial CACC sequence for TOPO cloning. Remove or replace this sequence when following other cloning strategies. To generate C-terminal tag-fused versions, remove the CDS stop codon.
2. Proofreading Taq polymerase.
3. Standard gel extraction kit.
4. Gateway cloning kit (pENTR/D-TOPO® entry vector; Thermo Scientific) (*see* **Note 1**).
5. *Escherichia coli* competent cells.
6. Standard plasmid mini-prep kit. Alternatively, plasmid DNA isolation can be carried out following homemade protocols and solutions (e.g., the alkaline lysis protocol described in [2]).
7. LR Clonase II Single Site Gateway® (Thermo Scientific) (*see* **Note 1**).
8. Gateway® compatible destination binary vectors for *in planta* expression (*see* **Note 1**).
9. *Agrobacterium tumefaciens* competent cells.

2.3 Plant Growth

1. Plastic growth pots.
2. Plastic trays.
3. Transparent plastic lids or cling film.
4. Soil mixture: three parts of potting substrate and one part of vermiculite.
5. *Nicotiana benthamiana* seeds (wild-type or transgenic expressing a protein of interest) (*see* **Note 2**).

2.4 Bacterial Strains

1. *E. coli* DH5α. This strain is used for the maintenance and amplification of general plasmids and vectors in general molecular cloning procedures.
2. *E. coli* DB3.1. This strain is used for the maintenance and amplification of vectors from the Gateway® cloning system containing the *ccdB* gene (*see* **Note 1**).
3. *A. tumefaciens* GV3101 (containing pMP90, Rif^R, $Gent^R$). This bacterial strain is used for efficient transient expression in *N. benthamiana* leaves. Other *A. tumefaciens* strains, such as AGL1, MOG101, or C58C1, are also frequently used for agroinfiltration.

2.5 Media, Buffers, and Other Solutions

1. Luria Broth (LB), liquid and solid (agar plates): dissolve 5 g NaCl, 5 g tryptone, and 2.5 g yeast extract in 500 mL distilled water; 7.5 g agar for plates only. Autoclave at 121 °C for 15 min and cool to 55 °C before adding the appropriate antibiotics. Pour plates directly after.
2. Antibiotics. Kanamycin: 50 mg/mL in water; spectinomycin: 50 mg/mL in water; rifampicin: 50 mg/mL in dimethylsulfoxide (DMSO); gentamycin: 50 mg/mL in water. All stock antibiotic solutions are aliquoted and stored at −20 °C for 4–-6 months. Before use, the antibiotic aliquot should be thawed on ice (except for rifampicin) and then added to the molten cooled-down medium. Avoid thawing/freezing cycles of the stock solutions. Rifampicin and kanamycin are light sensitive: keep the aliquots protected from light (e.g., in aluminum foil-wrapped or opaque containers).
3. Acetosyringone stock solution (AS, 150 mM): dissolve 0.3924 g of acetosyringone powder in 12 mL 95% ethanol, 8 mL of sterile distilled water. Filter-sterilize and aliquot to store at −20 °C for up to 1 month.
4. Infiltration buffer (magnesium chloride ($MgCl_2$, 10 mM), 2-[N-morpholino] ethanesulfonic acid [MES, 10 mM (pH 5.6)], and acetosyringone (AS, 150 μM)). MgCl2 (1 M) and MES (1 M, pH 5.6): prepare stock solutions in distilled water and autoclave at 121 °C for 15 min. Dissolve 9.521 g $MgCl_2$ in 100 mL of distilled water, and store at room temperature (23–25 °C) for a maximum 1 year. Dissolve 1.952 g MES in 10 mL of distilled water and store at 4 °C for a maximum 1 month. Infiltration buffer is freshly prepared at room temperature (23–25 °C) and used immediately.
5. Ammonium persulfate (APS, 10% w/v): dissolve 1 g APS in 10 mL of distilled water and store at 4 °C for up to 3 weeks. APS undergoes decaying in solution, so replace the stock solution every 2–3 weeks.

6. Bovine serum albumin (BSA, 1% w/v): dissolve 100 mg BSA in 10 mL of distilled water and store in 1 mL aliquots at 4 or −20 °C. If stored at 4 or −20 °C, BSA powder and BSA solution are stable for 2 years.
7. Dithiothreitol (DTT, 1 M): dissolve 1.54 g DTT in 10 mL of distilled water. Prepare 1 mL aliquots and store them at −20 °C for 4–6 months. Wear gloves and prepare in a fume hood.
8. Ethylenediaminetetraacetic acid [EDTA, pH 8 (0.5 M)]: dissolve 14.61 g EDTA in 100 mL of distilled water. Adjust pH to 8 with 1 M NaOH solution. Sterilize by autoclaving at 121 °C for 20 min and store at 4 °C for 4–6 months. Initially, the solution will be turbid and turn clear only when the pH reaches 8.
9. Sodium chloride (NaCl, 5 M): dissolve 29.22 g NaCl in 100 mL distilled water. Sterilize by autoclaving at 121 °C for 20 min and store at room temperature (23–25 °C) for 4–6 months.
10. Sodium dodecyl sulfate (SDS, 10% w/v): dissolve 10 g SDS in 100 mL distilled water. Store at room temperature (23–25 °C) for 4–6 months. SDS is highly hazardous. Wear gloves and masks and work in a fume hood while handling the powder to avoid inhalation.
11. N,N,N′,N′-tetramethylethylenediamine (TEMED): use fresh from the bottle.
12. Tris–HCl [pH 8.8 (1.5 M)]: dissolve 181.65 g Tris Base in 400 mL distilled water and then add concentrated HCl to bring pH to 8.8. Make up the volume to 1 L, autoclave for sterilization and store at room temperature (23–25 °C) for 4–6 months.
13. Tris–HCl [pH 6.8 (1 M)]: dissolve 121.1 g Tris Base in 400 mL distilled water and then add concentrated HCl to bring pH to 6.8. Make up the volume to 1 L, autoclave for sterilization, and store at room temperature (23–25 °C) for 4–6 months.
14. Sodium fluoride (NaF, 10 mM): to prepare a 1 M stock solution and dissolve 0.4196 g NaF in 10 mL distilled water. Store at room temperature (23–25 °C) for 4–6 months.
15. Sodium orthovanadate (Na_3VO_4, 10 mM): to prepare a 1 M stock solution and dissolve 1839 g Na_3VO_4 in 10 mL distilled water. Store at room temperature (23–25 °C) for 4–6 months.
16. Sodium molybdate (Na_2MoO_4, 2 mM): to prepare a 200 mM stock solution and dissolve 0,4118 g Na_2MoO_4 in 10 mL distilled water. Store at room temperature (23–25 °C) for 4–6 months.

17. Glycerol (50% v/v): take 50 mL glycerol and add distilled water to make up the volume to 100 mL. Sterilize by autoclaving at 121 °C for 20 min and store at room temperature (23–25 °C) for 4–6 months.
18. Triton X-100 (0.5% v/v): use fresh from the bottle. Add 0.5–100 mL buffer.
19. Tween-20: use fresh from the bottle. Add 1 mL to 1 L TBST (1×) buffer.
20. Protease inhibitor cocktail: use 1 tablet for 100 mL buffer or the corresponding volume.
21. Protein extraction buffer (1×): add 1 mL Tris–HCl (pH 8, 1 M), 1.5 mL NaCl (1 M), 1 mL glycerol, 0.05 mL EDTA (pH 8, 0.5 M), 0.05 mL DTT (1 M), 0.05 mL protease inhibitor cocktail, 0.1 mL PMSF (0.1 M), 0–2% detergent (Igepal or Triton X-100) depending on the purpose, and distilled water to make up the volume to 10 mL. For phosphorylation work, add 100 μL Na_2MoO_4 (200 mM), 100 μL NaF (1 M), and 100 μL Na_3VO_4 (1 M). Prepare freshly at a cold temperature (ice or 4 °C room). Use immediately.
22. Wash buffer (1×): add 1 mL Tris–HCl (pH 8, 1 M), 1.5 mL NaCl (1 M), 1 mL glycerol, 0.01 mL DTT (1 M), 0.05 mL protease inhibitor cocktail, 0–2% detergent (Igepal or Triton X-100) depending on the purpose, and distilled water to make up the volume to 10 mL. For phosphorylation work, add 100 μL Na_2MoO_4 (200 mM), 100 μL NaF (1 M), and 100 μL Na_3VO_4 (1 M). Prepare freshly at cold temperature (ice or 4 °C room). Use immediately.
23. Laemmli buffer sample (4×) [3]. Add 2 mL Tris–HCl [pH 6.8 (1 M)], 4 mL glycerol (100% v/v), 0.8 g SDS (powder), 0.4 mL DTT (1 M), 0.2 mL bromophenol blue (1%), and distilled water to make up the volume to 100 mL. Prepare 1 mL aliquots and store at −20 °C for maximum 6 months.
24. SDS–PAGE resolving gel (10% concentration as a reference; choose the appropriate concentration based on the predicted size of the protein of interest). Add 3.3 mL acrylamide mix (30%), 2.5 mL Tris–HCl (pH 8.8, 1.5 M), 0.1 mL SDS (10%), 0.1 mL APS (10%), 0.004 mL TEMED, and distilled water to make up the volume to 10 mL. Prepare freshly at room temperature (23–25 °C). Use immediately.
25. SDS–PAGE stacking gel (4.5% concentration). Add 0.83 mL acrylamide mix (30%), 0.63 mL Tris–HCl (pH 6.8 m 1 M), 0.05 mL SDS (10%), 0.05 mL APS (10%), 0.005 mL TEMED, and distilled water to make up the volume to 5 mL. Prepare freshly at room temperature (23–25 °C). Use immediately.

26. Running buffer (1×). Add 25 mL Tris–HCl (pH 8.3, 1 M), 10 mL SDS (10% v/v), 144 g glycine, and distilled water to make up the volume to 1 L. Prepare freshly at room temperature (23–25 °C). Use immediately.
27. Transfer buffer (1×). Add 25 mL Tris–HCl (pH 8.3, 1 M), 144 g glycine, 200 mL methanol, and distilled water to make up the volume to 1 L. Freshly prepared at room temperature (23–25 °C). Use immediately.
28. Coomassie brilliant blue (CBB) staining solution. Add 1 g Coomassie brilliant blue R-250, 500 mL methanol, 100 mL acetic acid, and distilled water to make up the volume to 1 L. Store at room temperature (23–25 °C) for a maximum 1 year.
29. Distaining solution. Add 400 mL methanol, 100 mL glacial acetic acid, and distilled water to make up the volume to 1 L. Store at room temperature (23–25 °C) for a maximum 1 year.
30. TBS (10×). Add 24.28 g Tris–Base, 80 g NaCl, and 700 mL distilled water. Adjust pH to 7.6 and add distilled water to make up the volume to 1 L.

    TBS (1×). Take 100 mL TBS 10× and add 900 mL distilled water. Store at room temperature (23–25 °C) for a maximum 1 year.
31. TBST (1×). Add 100 mL from TBS 10×, 1 mL Tween-20, and distilled water to make up the volume to 1 L. Store at room temperature (23–25 °C) for a maximum 1 year.

2.6 Imaging

This is an optional step to couple the analysis of the subcellular localization and accumulation of the protein of interest to the inference of its interactome.

1. Imaging system to detect fluorescence (e.g., a fluorescent microscope or a confocal laser scanning microscope (CLSM)).
2. 20× and 40× objectives.
3. Fluorescent proteins/dyes. Different fluorescent proteins and dyes can be used in combination, depending on the number of elements to be observed in the experiment. Please note that for the rest of the protocol, GFP will be mentioned as the default option, but alternative/additional fluorescent proteins can be used. Please, *see* **Note 3** for considerations on how to choose fluorescent proteins/dyes to be combined for simultaneous imaging.

2.7 Immunoprecipitation

1. Agarose beads or magnetic agarose beads (*see* **Note 4**).
2. Miracloth roll or disposable chromatography columns (e.g., Bio-Rad Bio-Spin Disposable Chromatography Columns).

2.8 Electrophoresis

1. Protein electrophoresis equipment.

2.9 Protein Transfer System

Western blot analysis is a recommended optional step to confirm protein accumulation before continuing with the IP-MS analysis. For more details, *see* **Note 5**.

Protein transfer is a crucial step in Western blot analysis. In our lab, we preferentially use semi-dry transfer; please note that the solutions indicated above in Subheading 2.5 of this chapter are those required to carry out the semi-dry transfer, although alternatives are possible. For more details, *see* **Note 6**.

1. Bio-Rad Semi-Dry Transfer Cell, or a similar device.
2. PVDF or nitrocellulose (NC) membranes.
3. Extra thick blot filter paper.

2.10 Western Blotting Analysis

Protein detection can be achieved by different methods, among which chemiluminescent detection is our preferred option. The subsequent protein visualization can rely on imaging equipment or X-ray films.

1. Enhanced chemiluminescent (ECL) substrates.
2. (Optional) X-ray films and X-ray developer and fixer solutions.
3. Western blot imaging system, or darkroom for X-ray film processing.

3 Methods

3.1 Plant Growth Conditions

N. benthamiana plants are grown in a controlled growth chamber in long-day conditions (16 h light/8 h dark) at 25 °C (*see* **Note 7**).

1. Sow *N. benthamiana* seeds in a well-watered soil mixture [as a reference, three parts of potting substrate and one part of the inorganic substrate (e.g., vermiculite)].
2. Cover the pots sown with a transparent plastic lid or with cling film.
3. Stratify for 2 days at 4 °C in the dark.
4. Transfer the pots to the controlled growth room chamber.
5. Remove the transparent plastic lids or the cling film once the seedlings begin to emerge.
6. Transfer the seedlings individually to plastic pots (*see* **Note 8**) filled with the same well-watered soil mixture and cover them with transparent plastic lids or cling film for 2–3 days.
7. Check the plants regularly and water/fertilize them as necessary (*see* **Note 8**).

3.2 Bacterial Growth Conditions

E. coli strains DH5α and DB3.1 and *A- tumefaciens* strain GV3101 harboring the corresponding vectors are liquid-cultured in LB medium (1% tryptone, 0.5% yeast extract, and 1% NaCl) with the appropriate antibiotics at 37 and 28 °C overnight, respectively.

3.3 Plasmids and Cloning

1. Design specific primers to amplify the CDS of interest according to the cloning system of your preference (*see* Subheading 2.2 and **Note 1**).
2. PCR-amplify the CDS of interest from the viral genome of interest using a Taq proofreading polymerase. If the viral genome is unavailable, synthesize the CDS of interest.
3. Run the PCR product in an agarose gel and purify the band corresponding to the expected size.
4. Clone the purified CDS fragment into a Gateway entry vector (*see* Subheading 2.2 and **Note 1**).
5. Transform the Gateway reaction into *E. coli* DH5α competent cells following the procedure of your choice, plate them in solid LB medium supplemented with the appropriate antibiotics, and leave them to grow overnight in a 37 °C oven.
6. Check colonies by PCR using appropriate primers.
7. Prepare a liquid culture of a confirmed colony harboring the corresponding construct in 5 mL of LB liquid media supplemented with the appropriate antibiotics and leave it to grow overnight at 37 °C and 200 rpm. The following day, extract the plasmid from the liquid-cultured bacteria and confirm the integrity and correctness of the cloned sequence by Sanger sequencing.
8. Next, sub-clone the CDS fragment into a Gateway-compatible binary destination vector through a LR reaction (*see* Subheading 2.2 and **Note 1**).
9. Next, proceed as indicated in **steps 5–7**.
10. Transform the destination vector into *A. tumefaciens* competent cells, plate them in solid LB medium supplemented with the appropriate antibiotics, and allow them to grow in a 28 °C oven for 2 days.
11. Then, proceed as described in **step 6**.

3.4 Agrobacterium-Mediated Transient Gene Expression in N. benthamiana

Transient expression assays are performed as described in [4] with minor modifications. As the viral infection reshapes the cell environment where plant and viral proteins coexist, the host–virus protein–protein interaction network must be identified in the presence of the virus whenever possible.

1. Three days before the experiment, streak out *A. tumefaciens* harboring the corresponding binary vectors from the

corresponding glycerol stocks or freshly transformed colonies on LB agar plates supplemented with appropriate antibiotics and incubate in a 28 °C oven for 2 days.

2. Inoculate a liquid culture by picking a single colony with a sterile toothpick or pipette tip and dropping it in 5 mL LB liquid media supplemented with the appropriate antibiotics. Incubate overnight at 28 °C in a shaker (200 rpm).
3. Centrifuge bacterial cultures at 4000*g* for 10 min.
4. Resuspend the pellet in 5 mL of infiltration buffer (10 mM $MgCl_2$, 10 mM MES pH 5.6, 150 μM acetosyringone) and measure the optical density (OD_{600}) in a spectrophotometer.
5. Adjust the bacterial concentration to a final OD_{600} of 0.2–1 (*see* **Note 9**) with infiltration buffer and incubate at room temperature in the dark for 2–4 h (*see* **Note 10**).
6. Infiltrate the abaxial side of four-week-old *N. benthamiana* plant leaves with a 1-mL needleless syringe (*see* **Note 11**). For experiments that required co-infiltration, mix the *A. tumefaciens* suspensions carrying different constructs at 1:1 ratio before infiltration.
7. Return the infiltrated *N. benthamiana* plants to the growth room to allow transfer and expression of the cassette containing the transgene(s) of interest. Samples will be collected between 36 and 40 h post-infiltration.

3.5 Visualization of Protein Accumulation and Subcellular Localization (Confirmation of Protein Accumulation by Fluorescence Microscopy)

1. Imaging can start between 36 to 40 h post-infiltration (*see* **Note 12**). Cut small leaf disks from the plants expressing the protein of interest with a corkborer (4 mm diameter).
2. Place the disks on a microscope slide, put a drop of water, and cover with a coverslip.
3. For subcellular localization and/or confirmation of protein accumulation, plant tissues expressing fluorescent tag-fused proteins can be imaged with a fluorescence microscope or a confocal laser scanning microscope (CLSM) using the corresponding preset settings. As a reference, in our lab, we use CLSM from Leica (Leica TCS SP8 confocal microscope provided with LAS-X 2.01 software) or Zeiss (Zeiss LSM 880 upright confocal microscope provided with ZEN black edition software).
4. For experiments that require co-expression of two (or more) fluorescent tag-fused proteins, the preset sequential scan settings are recommended.

3.6 Quick Protein Extraction and Western Blot (Confirmation of Protein Accumulation by Western Blot)

1. Protein extraction can start 36 to 40 h post-infiltration (*see* **Note 12**). Cut three to six leaf disks from the plants expressing the protein of interest with a corkborer (8 mm diameter).
2. Grind the samples by your preferred method keeping them frozen during the entire procedure.
3. Add 120 μL protein extraction buffer, mix well by vortex, and let thaw on ice for 10 min.
4. Centrifuge at 4 °C for 10 min at 16,000*g*.
5. Take 100 μL of the clear supernatant, add 30 μL of 4× Laemmli buffer, vortex, and boil for 10 min (*see* **Note 13**).
6. Centrifuge at 4 °C for 1 min at 16,000*g*. Load the supernatant on a 10% 1.5 mm acrylamide gel (*see* **Note 14**) and run at 120 V.
7. Wash the acrylamide gel in 1× semi-dry transfer buffer supplemented with 20% methanol for 15 min.
8. Proceed to protein transfer to a previously activated PVDF membrane via wet or semi-dry transfer methods. For reference, we use Bio-Rad Semi-Dry Transfer Cell and PVDF membranes, but other similar systems can be used.
9. Block the membrane in 5% milk-TBST for at least 1 h. Then, wash it with TBST, add the primary antibody in 5% milk-TBST, and incubate for 1–3 h at room temperature in gentle continuous shaking (*see* **Note 15**).
10. Remove the primary antibody, rinse, and wash the membrane 3 times in TBST for 5 min each. Then, add the secondary antibody in 5% milk-TBST and incubate for 1 h at room temperature in gentle continuous shaking (*see* **Note 15**).
11. Remove the secondary antibody, rinse, and wash the membrane three times in TBST for 5 min each and one time in TBS for 5 min.
12. ECL develop the membrane and visualize the result in a Western blot imager. Alternatively, expose the ECL-developed membrane to an X-ray film and visualize the results following film processing.

3.7 Protein Extraction and Immunoprecipitation–Mass Spectrometry (IP-MS) Analysis

1. Infiltrate fully expanded young leaves of 4-week-old *N. benthamiana* plants with *A. tumefaciens* carrying constructs to express the corresponding viral protein of interest fused to an appropriate tag. To analyze plant host–viral protein–protein interactions in the context of the viral infection, co-infiltrate *A. tumefaciens* carrying the infectious clone of the corresponding virus in the respective experiments.
2. Collect samples 36–40 h post-infiltration; accumulation of tag-fused proteins is confirmed by confocal microscopy (*see* Subheading 3.5, **Note 3**) or by Western blot (*see* Subheading

3.6, **Note 15**). Collect the rest of the infiltrated *N. benthamiana* leaves, freeze them immediately in liquid nitrogen, and keep them at −80 °C until further use (*see* **Notes 12** and **16**).

3. Perform protein extraction and IP-MS analysis as previously described in [4] and as detailed in the following steps.
4. Grind the samples in liquid nitrogen using your preferred method, keeping them frozen during the entire procedure.
5. Extract total proteins in protein extraction buffer (2 mL per 1 g of tissue).
6. Clean the extracts by filtration using Miracloth or chromatography columns.
7. Incubate the cleared extracts with immunoconjugate agarose beads for 1 h at 4 °C under continuous and gentle rotation. For GFP-fused proteins, we recommend using GFP-Trap beads (Chromotek, Germany) (*see* **Note 4**).
8. Wash the beads using wash buffer with detergent (0–2% NP-40) 3 times and wash buffer without detergent twice.
9. Run the immunoprecipitated proteins in an acrylamide gel. Stain the gel with Coomassie blue, excise bands covering each lane with a scalpel, and subject to mass spectrometry analysis. During this process, take extreme precautions to avoid contamination with keratin from inappropriate sample handling (*see* **Note 17**). Alternatively, the beads loaded with the precipitated proteins can be sent directly to MS analysis.

3.8 Bioinformatic Peptide Identification Process

Matching raw MS data to peptide sequences is performed using Mascot software with the annotated proteins from the *N. benthamiana* draft genome sequence v. 0.4.4, which can be obtained from the International Solanaceae Genomics Project (SOL) (https://solgenomics.net/). A decoy database can be constructed with those protein sequences originated from randomized and reversed raw *N. benthamiana* protein sequences. The decoy hits are used to estimate the false-positive rate in the peptide identification process. Because frequently a large number of potential protein–protein interactions are identified from the IP-MS data, the MiST software (Mass spectrometry interaction STatistics) [5] can be used to restrict the putative protein–protein interactions to those which are high confidence; since it has been demonstrated that viral proteins can convergently target host targets, we recommend removing the specificity criterion.

We recommend filtering high-quality MS data based on four criteria: (1) protein threshold—larger than 99%; (2) minimum number of peptides—1; (3) peptide threshold—>95%; and (4) DECOY FDR—lower than 1%; (5) functional information available. Data were analyzed by Scaffold 4.9.

3.9 Confirmation of Protein–Protein Interactions

Usually, a combination of techniques is necessary to confirm and characterize protein–protein interactions identified by IP-MS. To confirm intraviral or plant host–viral protein–protein interactions, proteins of interest must be co-expressed alone or in the presence of the virus. While methods to study protein–protein interactions are numerous, using at least two different approaches, at least one being in vivo, is recommended (*see* **Note 18**).

4 Notes

1. Following the procedures described in [6]. We mostly use the Gateway cloning system; however, any other cloning method would be equally applicable.
2. The Solanaceae species *N. benthamiana* is the most widely used experimental host in plant virology due at least in part to the large number of diverse plant viruses that can successfully infect it. Another aspect to consider is its amenability to genetic manipulation and to be a platform for *Agrobacterium*-mediated transient assays such as virus-induced gene silencing (VIGS) or transient protein expression. For these reasons, its use as a plant model has expanded considerably in recent years, rapidly gaining in popularity, particularly in studies requiring protein localization or interaction, or plant-based systems for protein expression and purification.
3. If the emission spectra of fluorescent proteins/dyes overlap, bleed-through will occur when imaging. We recommend using sequencing scanning to avoid this crosstalk. The Spectra-Viewer (https://www.thermofisher.com/order/spectra-viewer) can be consulted for choosing the fluorescent proteins/dyes to use. Choosing a practical combination of fluorescent proteins/dyes is crucial for the success of the experiment.
4. Agarose beads for very low background and high binding capacity IP, or magnetic agarose beads for magnetic separation and high binding capacity IP.
5. Western blot analysis can be an alternative to the imaging step when microscopy facilities are unavailable, fluorescent tags are not used, or when working with untagged proteins. We strongly recommend confirming the appropriate accumulation of the protein of interest by one of the two methods before proceeding to the immunoprecipitation step.
6. During protein transfer, proteins previously separated in a gel by electrophoresis are transferred to a membrane sheet. There are a variety of methods where electroblotting (wet, semi-dry, and dry electroblotting) has emerged as the most used due to its higher speed and efficiency compared to other methods.

7. Standardized transient protein expression experiments require four-week-old *N. benthamiana* plants. Therefore, to timely perform this protocol, researchers must sow seeds 5 weeks in advance. To ensure a continuous supply of plants, we recommend sowing new batches of seeds every week. Second, to successfully implement this protocol, researchers must grow healthy *N. benthamiana* plants. Since the growth of *N. benthamiana* plants and *Agrobacterium*-mediated T-DNA transformation requires optimal environmental conditions such as light, humidity, and temperature, we highly recommend the use of environmentally controlled cultivation facilities.
8. It is recommended to use pots of a size not smaller than 7 × 7 cm and to supplement the watering solution with fertilizer once every 2 weeks. The use of plants older than 4 weeks is not recommended.
9. We recommend optimizing in advance the optical density required for each case. Use the lowest concentration of *Agrobacterium* that does not compromise transformation efficiency whenever possible.
10. Incubation is required to allow the phenolic compound acetosyringone to induce the expression of the *Agrobacterium vir* genes, required for the plant cell transformation through T-DNA transfer with higher efficiency.
11. We suggest infiltrating two leaves per plant corresponding to the (already fully expanded) third and fourth youngest leaves. In order to reduce plant-to-plant variation, we recommend infiltrating at least six individual leaves from three different plants per sample for each biological replicate. After the infiltration, gently dry the excess bacterial suspension with a paper towel.
12. We recommend to experimentally optimize the post-agroinfiltration sample harvest time in order to ensure the highest protein accumulation.
13. Optimize boiling time and temperature so as not to lead to precipitation of the protein of interest.
14. The percentage of the acrylamide gel will vary depending on the molecular weight of the protein of interest. For more information, please visit the following websites: https://www.bio-rad.com/de-de/applications-technologies/introduction-polyacrylamide-gels?ID=LUSPBRM5B; https://www.thermofisher.com/de/de/home/life-science/protein-biology/protein-biology-learning-center/protein-biology-resource-library/pierce-protein-methods/overview-electrophoresis.html.

15. Antibody concentration and incubation time have to be optimized for each individual case. Overnight incubations should be done at 4 °C. For the detection of GFP-tagged proteins, we usually use the following primary and secondary antibodies at the indicated dilutions: mouse anti-green fluorescent protein (GFP) (M0802-3a, Abiocode, Agoura Hills, CA, USA) (1: 10,000), and goat polyclonal anti-mouse coupled to horseradish peroxidase (A2554, Sigma, St. Louis, MO, USA) (1: 15,000) [6].
16. For low-abundance proteins, scale up the amount of tissue.
17. A common problem in mass spectrometry analyses is keratin contamination from the skin, hair, dust, clothing, and latex gloves. If keratins are present in high concentrations, they can interfere with mass spectrometry analyses, causing a loss in signal and masking proteins/peptides of interest, reducing the total number of proteins/peptides identified. Please check common tips on how to avoid keratin contamination.
18. To confirm specific protein–protein interactions, we recommend a number of techniques: co-immunoprecipitation (co-IP), bimolecular fluorescence complementation assay (BiFC), split-luciferase assay (Split-luc) (*see* Chap. 17 in this volume), and yeast-two hybrid (Y2H) are among our most frequently used ones [6].

Acknowledgments

The authors thank Huang Tan and Man Gao for sharing materials and Liping Wang and Man Gao for critically reading the manuscript.

References

1. Lum KK, Cristea IM (2016) Proteomic approaches to uncovering virus-host protein interactions during the progression of viral infection. Expert Rev Proteomics 13(3):325–340
2. Birnboim HC, Doly J (1979) A rapid alkaline extraction procedure for screening recombinant plasmid DNA. Nucleic Acids Res 7(6): 1513–1523
3. Laemmli UK (1970) Cleavage of structural proteins during the assembly of the head of bacteriophage T4. Nature 227(5259):680–685
4. Wang L et al (2017) Inference of a geminivirus-host protein-protein interaction network through affinity purification and mass spectrometry analysis. Viruses 9(10):275
5. Verschueren E et al (2015) Scoring large-scale affinity purification mass spectrometry datasets with MiST. Curr Protoc Bioinformatics 49: 8.19.1–8.19.16
6. Wang L et al (2022) Combinatorial interactions between viral proteins expand the potential functional landscape of the tomato yellow leaf curl virus proteome. PLoS Pathog 18(10): e1010909
7. Arabidopsis Interactome Mapping Consortium (2011) Evidence for network evolution in an Arabidopsis interactome map. Science 333(6042):601–607

Chapter 22

In Situ Protein Microarray for Identifying the Geminivirus–Arabidopsis Interactome

Junshi Yazaki and Maximiller Dal-Bianco

Abstract

The protein–protein interactions (PPI) by protein array technology complement other PPI assay technologies such as AP-MS and Y2H. The in situ protein array technology (NAPPA) enables low-cost, rapid, and comprehensive protein detection. It allows standardized and simultaneous assay of a wide range of proteins with a broad range of expression in cells. This technology facilitates the detection of protein–protein interactions within species and between heterologous species such as host–microbe. Here, we described the technique that identified a syntaxin-6 protein-mediated begomovirus infection using an array containing 4600 *Arabidopsis* genes. The protein microarray assay also identified several other viral protein–host protein interactions.

Key words Host–virus interaction, Protein–protein interaction, In situ protein synthesis, Cell-free protein expression, HaloTag-NAPPA

1 Introduction

Proteins are the central system for the function of all biological processes and are critical for understanding biological systems. One of the most pressing challenges in understanding biological systems is providing detailed functional assignments to many proteins whose functions remain unknown. Even though protein–protein interactions are of fundamental importance, their complexity makes the development of this field slow [1].

Protein arrays are advantageous for assessing the diversity of interactions. For example, this technology can be used in antigen–antibody screening [2], protein–protein interactions [3], post-translational modification target identification [4], and protein-nucleic acid-binding identification [5].

We previously constructed an in situ protein array (nucleic acid programmable protein array: NAPPA) that enables screening for normalized protein abundances and can effectively detect protein–

Elizabeth P. B. Fontes and Kristiina Mäkinen (eds.), *Plant-Virus Interactions*, Methods in Molecular Biology, vol. 2724, https://doi.org/10.1007/978-1-0716-3485-1_22,

protein interactions, even in small numbers of proteins in the cell, such as transcription factors [6, 7]. Since this technique uses on-demand synthetic proteins from ORF clones, the cost is low, and more than several thousand proteins can be screened quickly with little effort. In *Arabidopsis*, ORF clones are well established, and NAPPA can be created immediately with HaloTag ORF clones (https://www.arabidopsis.org/abrc/halo_tagged_orf_clones.jsp).

NAPPA can identify molecular interactions within the same species and analyze interactions between species, e.g., host–virus interactions [8]. Host–pathogenic virus interactions are important determinants of health and disease [9]. The NAPPA technology can identify host proteins that interact with viral proteins involved in infection, which will significantly help scientists understand how the virus suppresses the plant antiviral mechanism.

This chapter reports the procedure to detect plant–virus interactions during *Arabidopsis* infection by a begomovirus. We used the movement protein (MP) and the nuclear shuttle protein (NSP) ORF clones from cabbage leaf curl virus (CabLCV) as queries to screen against 4600 ORF-NAPPAs. Among the candidate targets, we found a plant-specific syntaxin-6 protein displaying a proviral function and confirmed previously identified MP-NSP-Host PPI [8].

2 Materials

2.1 *In Vitro* Expression Clones

1. Target ORF clones from *Arabidopsis thaliana*: All pIX-HaloTag expression clones are available from the ABRC stock center (https://www.arabidopsis.org/abrc/halo_tagged_orf_clones.jsp).
2. Query begomovirus ORF clones: 3xHA fused clones of the nuclear shuttle protein (NSP) and the movement protein (MP) were prepared by recombination of the virus ORF pDONR-NSP and pDONR-MP clones to the pIX-3xHA in vitro expression vector (https://www.arabidopsis.org/servlets/TairObject?type=vector&id=1001200524) using Gateway cloning system (Invitrogen).
3. Plasmid DNA from target and query clones: Prepare the DNA plasmid from pIX-HaloTag expression clones, derived from *Arabidopsis thaliana* ORF collection, using QIAprep 96-well format miniprep kit (Qiagen). pIX-3xHA expression clones are prepared with QIAprep Spin Miniprep Kit (Qiagen). P1 buffer can be stored at 4 °C. Add 700 μL RNase inhibitor and 700 μL LyseBlue.
4. Terrific broth (TB) medium. Add 55 g TB for each liter of distilled water. Add 8 mL of 50% glycerol. Autoclave 15 min and add 1 mL ampicillin. Store at 4 °C for a month.

5. Isopropanol for DNA precipitation.
6. 70%, 95% ethanol.
7. Sodium acetate.
8. QIAprep 96-well format miniprep kit with TurboFilter, QIA-prep, and Qiagen 96-deep well block plate (Qiagen).
9. 48-pin stamp.

2.2 HaloTag-NAPPA Microarray Fabrication

1. Prepare 66 mg/mL bovine serum albumin (BSA), 100 mM Halo amine ligand, and 50 mg/mL BS3 cross-linker. Mix 3.3 mL of 66 mg/mL BSA, 1.2 mL of 100 mM Halo amine ligand, 1.5 mL of 50 mg/mL BS3 cross-linker, and 6 mL of 100% DMSO to end with 12 mL total volume of master mix for 6000 clones.
2. Spotting solution: Mix plasmid DNA from pIX-HaloTag expression clones and the master mix in a one-to-one ratio.
3. Glass slide coated with aminosilane.
4. Microarray spotter.

2.3 In Vitro Protein Synthesis of HaloTag-NAPPA

1. Divider dish (BD Biosciences).
2. Super block (Thermo Fisher Scientific).
3. Transcription and translation (TNT) T7 coupled wheat germ extract kit (Promega).
4. RNase inhibitor (Promega).
5. Rocking shaker (Thermo Fisher Scientific).
6. Incubator for protein expression.
7. PBS with 0.1% Tween-20 (PBST).
8. PBST with 5% non-fat dry milk (Milk-PBST). Weigh 5 g non-fat milk and dissolve in 100 mL PBS with 0.1% (w/v) Tween-20 (PBST). Chill before use. Store at 4 °C for 3 days.

2.4 Query Virus Protein Expression and Detection of PPI

1. QIAprep Spin Miniprep.
2. Transcription and translation (TNT) T7 coupled wheat germ extract kit (Promega).
3. RNase inhibitor.
4. Incubator for protein expression.
5. Anti-HA mouse monoclonal antibody (Covance, HA.11 clo-ne16B12) (*see* **Note 1**).
6. Horseradish peroxidase (HRP)-conjugated anti-mouse IgG (GE Healthcare) (*see* **Note 2**).
7. Cy3-conjugated tyramide signal amplification (TSA) substrate (Molecular Probes) (*see* **Note 3**).
8. PBST.

9. Milk-PBST.
10. Scanner (Array-Pro Analyzer and Array-Pro software; Tecan).

3 Methods

3.1 Development of In Situ Protein Arrays

1. Prepare the Qiagen 96-deep well block plate with 1200 μL TB media in each well. Sterilize the 48-pin stamp in 95% ethanol and expose it to flames at a Bunsen burner. After cooling, stamp the *E. coli* sample into a 96-deep well block plate. Incubate overnight at 37 °C in a shaker. Harvest the *E. coli* cells by centrifugation for 20 min at 275 g at RT. Discard the supernatant and dry the plate.
2. Prepare plasmid DNA with QIAprep 96 Turbo Miniprep Kit. Resuspend pelleted *E. coli* cells in 300 μL P1 resuspend buffer. Vortex to dissolve the pellet and add 300 μL P2 lysate buffer and seal. Vortex gently and incubate at RT for 5 min. Add 400 μL neutralization buffer N3 and vortex.
3. Remove cell lysate with the Turbo Filter plate and bind the plasmid DNA to the QIAprep plate. After washing 2 times with 1 mL washing buffer PE, dry the QIAprep plate, elute the DNA by adding 200 μL elution buffer PE, and incubate for 10 min. Centrifuge 10 min at 2750×*g* at RT to yield the plasmid DNA.
4. Combine 20–30 μg of plasmid DNA from one to three preparation plates in collection tubes from the kit. Add 10% total volume of sodium acetate to tubes (i.e., 45 μL sodium acetate in a total volume of 450 μL). Add 100% total volume of isopropanol to tubes to precipitate plasmid DNA (i.e., 450 μL isopropanol in a total volume of 450 μL). Centrifuge at 2750×*g* at 4 °C for 1 h. Change plate direction and centrifuge at 2750×*g* at 4 °C for 1 h. Discard the isopropanol. Add 500 μL of 75% ethanol and centrifuge at 2750×*g* at 4 °C for 20 min. Decant the 75% ethanol and dry. Resuspend the DNA pellet in 20 μL RNase-free water.
5. Spot the plasmid DNA with the master mix to an aminosilane-coated glass slide using microarray spotter or similar device [10, 11]. Plasmid DNA should be combined with the master mix at a 1 to 1 volume mixing ratio.

3.2 In Situ Protein Expression on NAPPA Array

1. Block the glass slides with 10 mL Superblock in a divider dish for at least 1 h at RT or overnight in a cold room. Rinse the glass slides with MilliQ water three times to keep the slides wet and later dry the glass slides using airflow such as a ball valve hose.

2. Prepare the in vitro expression reagents in the TNT T7 coupled wheat germ extract kit as below.

   225 μL wheat germ extract (final concentration: 15%).

   30 μL 10 × buffer.

   7.5 μL minus Met amino acid mix.

   7.5 μL minus Leu/Cys amino acid mix.

   15 μL RNase inhibitor.

   15 μL of T7 polymerase.

   DEPC water to final volume of 1500 μL per slide.

3. Submerge the glass slide in the dish with 1.5 mL of in vitro expression mix. Rock for 2 h at 30 °C. Then, wash the NAPPA array three times with Milk-PBST and block overnight at 4 °C.

3.3 In Vitro Expression of Virus Query Proteins and Detection of Host–Virus PPI

1. Prepare query plasmid DNA from *E. coli* cells with QIAprep Spin Miniprep. Express the 3xHA tagged virus query proteins individually using the TNT T7 coupled wheat germ extract kit for 2 h at 30 °C (*see* Fig. 1). Expression mix for query protein (TNT kit) is below.

   50 μL wheat germ (−80 °C).

   4 μL TNT buffer.

   6 μL minus Met amino acid mix.

   6 μL minus Leu/Cys amino acid mix.

   2 μL RNase inhibitor.

   2 μL T7 polymerase.

   2 μL 3 × HA-query DNA (1 μg/μL).

   Diethylpyrocarbonate (DEPC) water to final volume of 100 μL.

   Incubate at 30 °C for 2 h. Prepare immediately before use.

2. Combine two tubes of query proteins, and then, add 1.5 mL milk-PBST for each query protein. Transfer the glass slides from Subheading 3.2, **step 3** to new divider dishes and add 1.5 mL of query proteins with milk-PBST. For the interaction reaction, rock for 2 h at 4 °C. Retrieve the slides, immerse in milk-PBST in new dishes, and wash 3 times with milk-PBST. Block in milk-PBST and rock overnight at 4 °C.

3. Transfer the glass slides to fresh divider dishes and add 1.5 mL of 1:500 dilution primary antibody (anti-HA mouse monoclonal antibody), rock for 1 h at RT (*see* **Note 1**). Transfer the glass slides to fresh dishes, block in milk-PBST again for 5 min, and repeat 3 times. Transfer the glass slides to fresh divider dishes and add 1.5 mL of 1:200 dilution of anti-mouse HRP-coupled secondary antibody and rock for 1 h at RT (*see* **Note 2**).

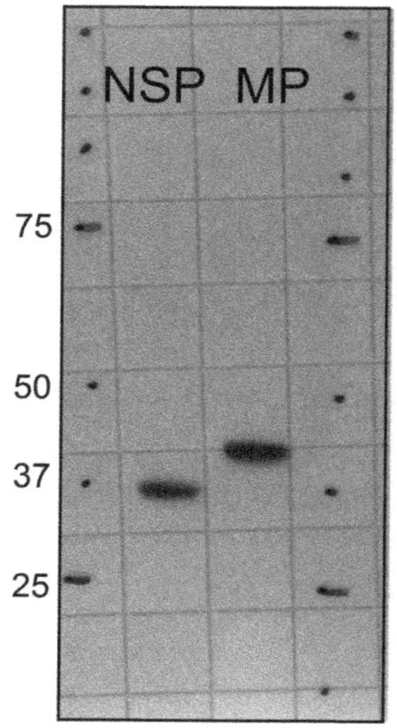

Fig. 1 Query protein expression. In vitro synthesized 3xHA-NSP (nuclear shuttle protein) and 3xHA-MP (virus movement protein) were separated by SDS–PAGE and then screened by Western blot analysis using an anti-HA antibody to validate the query protein expression. (Reproduced from Gouveia-Mageste et al. [8] with the permission of Oxford University Press, license number 5503290696299)

Transfer the glass slides to fresh divider dishes, block with PBST for 2 min, and repeat at 5 min, 10 min, 15 min, and 15 min. Rinse the glass slide three times in MilliQ water.

4. Transfer the glass slides to fresh dishes and add 1.5 mL TSA mix to each glass slide (*see* **Note 3**). Rock at RT for 10 min. Transfer the glass slides to fresh dishes and rinse the glass slides with MilliQ water three times to keep the slides wet, followed by drying the glass slides using a ball valve hose. Scan using Tecan scanner and Array-Pro software (*see* Fig. 2). To extract the data for spot identification, use an array-pro analyzer scanner and Tecan image analysis software or similar. Determine the position and boundaries of each spot on the NAPPA array and measure the signal intensity.

4 Notes

1. Primary antibody: Add 24 μL anti-HA antibody into 12 mL milk-PBST mixture (for one glass slide: 3 μL anti-HA antibody/1.5 mL milk-PBST) in a 15-mL conical tube.
2. Secondary antibody: 1:200 dilution of anti-mouse HRP-coupled secondary antibody.

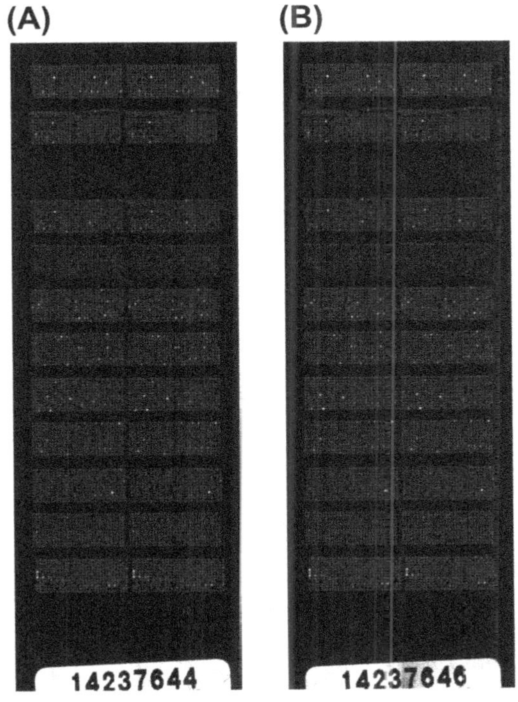

Fig. 2 Protein interaction detection with high-density HaloTag-NAPPA arrays. (**a**) The array showing protein interactors of 3xHA-tagged MP probe. (**b**) The array showing protein interactors of NSP probe. Each protein array contains ~4600 unique ORFs spotted in duplicate (9200 spots), and protein synthesis, capture, and interaction detection were conducted (*see* Subheading 3.2 and 3.3). Candidate interactors were detected using anti-HA antibodies and can be found as white spots in the arrays. The signal intensity from each candidate interactor is calculated by the image analysis software. (Reproduced from Gouveia-Mageste et al. [8] with the permission of Oxford University Press, license number 5503290696299)

3. Substrate application: 1:500 dilution of Cy3-conjugated TSA substrate. For eight glass slides, add 24 μL TSA into 9 ml PBS plus 3 mL diluent provided in the kit (25% total volume of the diluent) in a 15-mL conical tube (for one glass slide: 3 μL TSA/1.5 mL buffer). Mix well.

Acknowledgments

This protocol has been derived and adapted from Yazaki et al. [6, 7] and Gouveia-Mageste et al. [8]. This work was supported partially by the Ministry of Education, Science, Sports and Culture, Grant-in-Aid for Scientific Research (C), 2019-2021 [19 K0574 to J.Y.]. The work was supported by the Conselho Nacional de

Desenvolvimento Científico e Tecnológico [220164/2012-6 and 312641/2021-4 to MD), Fundação de Amparo à Pesquisa do Estado de Minas Gerais (APQ-01416-16, PPM-640-18 and RED-00124-22 to MD), and Caramuru Alimentos (Grant to MD). We would like to thank Editage (www.editage.com) for English language editing.

References

1. Elhabashy H, Merino F, Alva V, Kohlbacher O, Lupas AN (2022) Exploring protein-protein interactions at the proteome level. Structure 30(4):462–475. https://doi.org/10.1016/j.str.2022.02.004
2. Ruano-Gallego D, García-Villadangos M, Moreno-Paz M et al (2021) A multiplex antigen microarray for simultaneous IgG and IgM detection against SARS-CoV-2 reveals higher seroprevalence than reported. Microb Biotechnol 14(3):1228–1236. https://doi.org/10.1111/1751-7915.13801
3. Popescu SC, Popescu GV, Bachan S et al (2007) Differential binding of calmodulin-related proteins to their targets revealed through high-density Arabidopsis protein microarrays. Proc Natl Acad Sci U S A 104(11):4730–4735. https://doi.org/10.1073/pnas.0611615104
4. Lin YY, Lu JY, Zhang J et al (2009) Protein acetylation microarray reveals that NuA4 controls key metabolic target regulating gluconeogenesis. Cell 136(6):1073–1084. https://doi.org/10.1016/j.cell.2009.01.033
5. Gong W, He K, Covington M, Dinesh-Kumar SP et al (2008) The development of protein microarrays and their applications in DNA-protein and protein-protein interaction analyses of Arabidopsis transcription factors. Mol Plant 1(1):27–41. https://doi.org/10.1093/mp/ssm009
6. Yazaki J, Galli M, Kim AY et al (2016) Mapping transcription factor interactome networks using HaloTag protein arrays. Proc Natl Acad Sci U S A 113(29):E4238–E4247. https://doi.org/10.1073/pnas.1603229113
7. Yazaki J, Galli M, Kim AY, Ecker JR (2018) Profiling interactome networks with the HaloTag-NAPPA in situ protein array. Curr Protoc Plant Biol 3(3):e20071. https://doi.org/10.1002/cppb.20071
8. Gouveia-Mageste BC, Martins LGC, Dal-Bianco M et al (2021) A plant-specific syntaxin-6 protein contributes to the intracytoplasmic route for the begomovirus CabLCV. Plant Physiol 187(1):158–173. https://doi.org/10.1093/plphys/kiab252
9. Fontes EPB, Teixeira RM, Lozano-Durán R (2021) Plant virus-interactions: unraveling novel defense mechanisms under immune-suppressing pressure. Curr Opin Biotechnol 70:108–114. https://doi.org/10.1016/j.copbio.2021.03.007
10. Yazaki J, Kishimoto N, Nakamura K et al (2000) Embarking on rice functional genomics via cDNA microarray: use of 3′ UTR probes for specific gene expression analysis. DNA Res 7(6):367–370. https://doi.org/10.1093/dnares/7.6.367
11. Yazaki J, Kishimoto N, Nagata Y et al (2003) Genomics approach to abscisic acid- and gibberellin-responsive genes in rice. DNA Res 10(6):249–261. https://doi.org/10.1093/dnares/10.6.249

Index

Elizabeth P. B. Fontes and Kristiina Mäkinen (eds.), *Plant-Virus Interactions*, Methods in Molecular Biology, vol. 2724, https://doi.org/10.1007/978-1-0716-3485-1,

GPSR Compliance
The European Union's (EU) General Product Safety Regulation (GPSR) is a set of rules that requires consumer products to be safe and our obligations to ensure this.

If you have any concerns about our products, you can contact us on

ProductSafety@springernature.com

In case Publisher is established outside the EU, the EU authorized representative is:

Springer Nature Customer Service Center GmbH
Europaplatz 3
69115 Heidelberg, Germany

www.ingramcontent.com/pod-product-compliance
Lightning Source LLC
Chambersburg PA
CBHW081939250425
25626CB00011B/65
* 9 7 8 1 0 7 1 6 3 4 8 4 4 *